PENGUIN BOOKS

RATIONALITY

'Eloquent in his defence of clear thinking . . . reason is a tool that human beings have to learn to use with care, something this book will help any reader to do' Julian Baggini, *Financial Times*

'Masterly . . . illuminating' Raymond Tallis, *TLS*

'In our uncertain age, which can so often feel so dark and disturbing, Steven Pinker has distinguished himself as a voice of positivity' *The New York Times*

'As with all of his books, this one, this one is erudite, lucid, funny, and dense with fascinating material . . . His characteristic brew of Yiddish jokes, brainy comics and incisive argumentation is a pleasure to read, even when the subjects are technical and mathematical. It's no small achievement to make formal logic, game theory, statistics and Bayesian reasoning delightful topics full of charm and relevance' *Washington Post*

'An impassioned and zippy introduction to the tools of rational thought . . . Punchy, funny, and invigorating' *The Sunday Times*

'An engaging analysis of the highest of our faculties and perhaps (ironically) the least understood' *Wall Street Journal*

'Almost every sentence in *Rationality* is crisp and intelligible, which is quite a feat, given that explaining logic to humans is like teaching them Sanskrit' *New Statesman*

'Steven Pinker is among the best science writers in history, and with *Rationality* he applies his talents to one of the most important and misunderstood human abilities . . . If you've ever considered taking drugs to make yourself smarter, read *Rationality* instead' Jonathan Haidt, co-author of *The Coddling of the American Mind*

'*Rationality* is a terrific book, much-needed for our time. In addition to drawing together the tools for overcoming obstacles to rational thinking, Pinker breaks new ground with the evidence he provides linking rationality and moral progress' Peter Singer, author of *The Life You Can Save* and *The Most Good You Can Do*

'Pinker manages to be scrupulously rigorous yet steadily accessible and entertaining whether probing the rationality of Andrew Yang's presidential platform, Dilbert cartoons, or Yiddish proverbs. The result is both a celebration of humans' ability to make things better with careful thinking and a penetrating rebuke to muddleheadedness' *Publishers Weekly*

'A reader-friendly primer in better thinking through the cultivation of that rarest of rarities: a sound argument' *Kirkus Reviews*

ABOUT THE AUTHOR

Steven Pinker is an experimental cognitive scientist. Currently Johnstone Professor of Psychology at Harvard, he has also taught at Stanford and MIT. He has won many prizes for his research, teaching, and his eleven books, including *The Language Instinct*, *How the Mind Works*, *The Blank Slate*, *The Better Angels of Our Nature*, and *Enlightenment Now*. He is a member of the National Academy of Sciences, a two-time Pulitzer Prize finalist, a Humanist of the Year, a recipient of nine honorary doctorates, one of *Foreign Policy*'s 'World's Top 100 Public Intellectuals' and *Time*'s '100 Most Influential People in the World Today'.

STEVEN PINKER

Rationality

What it Is, Why it Seems Scarce, Why it Matters

PENGUIN BOOKS

PENGUIN BOOKS

UK | USA | Canada | Ireland | Australia
India | New Zealand | South Africa

Penguin Books is part of the Penguin Random House group of companies
whose addresses can be found at global.penguinrandomhouse.com.

First published in the United States of America by Viking,
an imprint of Penguin Random House LLC 2021
First published in Great Britain by Allen Lane 2021
First published in Penguin Books 2022

003

Grateful acknowledgment is made for permission to reprint the following:
Excerpt on page 55 from *The Odyssey* by Homer, translated by Emily Wilson.
Copyright © 2018 by Emily Wilson. Used by permission of W. W. Norton & Company, Inc.

Charts rendered by Ilavenil Subbiah

The moral right of the author has been asserted

Printed and bound in Great Britain by Clays Ltd, Elcograf S.p.A.

The authorized representative in the EEA is Penguin Random House Ireland,
Morrison Chambers, 32 Nassau Street, Dublin D02 YH68

A CIP catalogue record for this book is available from the British Library

ISBN: 978-0-141-98986-0

www.greenpenguin.co.uk

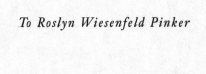

To Roslyn Wiesenfeld Pinker

What is a man,
If his chief good and market of his time
Be but to sleep and feed? A beast, no more.
Sure he that made us with such large discourse,
Looking before and after, gave us not
That capability and godlike reason
To fust in us unus'd.

—HAMLET

CONTENTS

Preface   *xi*

1. How Rational an Animal?   *1*

2. Rationality and Irrationality   *35*

3. Logic and Critical Thinking   *73*

4. Probability and Randomness   *111*

5. Beliefs and Evidence
   (Bayesian Reasoning)   *149*

6. Risk and Reward
   (Rational Choice and Expected Utility)   *173*

7. Hits and False Alarms
   (Signal Detection and Statistical Decision Theory)   *201*

8. Self and Others
   (Game Theory)   *227*

9. Correlation and Causation   *245*

10. What's Wrong with People?   *283*

11. Why Rationality Matters   *319*

Notes   *341*

References   *361*

Index of Biases and Fallacies   *389*

Index   *391*

PREFACE

Rationality ought to be the lodestar for everything we think and do. (If you disagree, are your objections rational?) Yet in an era blessed with unprecedented resources for reasoning, the public sphere is infested with fake news, quack cures, conspiracy theories, and "post-truth" rhetoric.

How can we make sense of making sense—and its opposite? The question is urgent. In the third decade of the third millennium, we face deadly threats to our health, our democracy, and the livability of our planet. Though the problems are daunting, solutions exist, and our species has the intellectual wherewithal to find them. Yet among our fiercest problems today is convincing people to accept the solutions when we do find them.

Commentaries by the thousands have lamented our shortfall of reason, and it's become conventional wisdom that people are simply irrational. In social science and the media, the human being is portrayed as a caveman out of time, poised to react to a lion in the grass with a suite of biases, blind spots, fallacies, and illusions. (The *Wikipedia* entry for cognitive biases lists almost two hundred.)

Yet as a cognitive scientist I cannot accept the cynical view that the human brain is a basket of delusions. Hunter-gatherers—our ancestors and contemporaries—are not nervous rabbits but cerebral problem solvers. A list of the ways in which we're stupid can't explain why we're so smart: smart enough to have discovered the laws of nature, transformed the planet, lengthened and enriched our lives, and, not least, articulated the rules of rationality that we so often flout.

To be sure, I am among the first to insist that we can understand human nature only by considering the mismatch between the environment in which we evolved and the environment we find ourselves in today. But the world to which our minds are adapted is not just the Pleistocene savannah. It's any nonacademic, nontechnocratic milieu—which is to say, most of human experience—in which the modern instruments of rationality like statistical formulas and datasets are unavailable or inapplicable. As we shall see, when people are given problems that are closer to their lived reality and framed in the ways in which they naturally encounter the world, they are not as witless as they appear. Not that this gets us off the hook. Today we do have refined instruments of reason, and we are best off, as individuals and as a society, when we understand and apply them.

This book grew out of a course I taught at Harvard which explored the nature of rationality and the puzzle of why it seems to be so scarce. Like many psychologists, I love to teach the arresting, Nobel Prize–winning discoveries of the infirmities that afflict human reason, and consider them to be among the deepest gifts to knowledge that our science has contributed. And like many, I believe that the benchmarks of rationality that people so often fail to measure up to should be a goal of education and popular science. Just as citizens should grasp the basics of history, science, and the written word, they should command the intellectual tools of sound reasoning. These include logic, critical thinking, probability, correlation and causation, the optimal ways to adjust our beliefs and commit to decisions with uncertain evidence, and the yardsticks for making rational choices alone and with others. These tools of reasoning are indispensable in avoiding folly in our personal lives and public policies. They help us calibrate risky choices, evaluate dubious claims, understand baffling paradoxes, and gain insight into life's vicissitudes and tragedies. But I knew of no book that tried to explain them all.

The other inspiration for this book was my realization that for all its fascination, the cognitive psychology curriculum left me ill equipped to answer the questions I was most frequently asked when I told people I was teaching a course on rationality. Why do people believe that Hillary Clinton ran a child sex ring out of a pizzeria, or that jet contrails are really mind-altering drugs dispersed by a secret government program? My standard lecture bullet points like "the gambler's fallacy" and "base-rate neglect" offered little insight into just the enigmas that are making human irrationality so pressing an issue today. Those enigmas drew me into new territories, including the nature of rumor, folk wisdom, and conspiratorial thinking; the contrast between rationality within an individual and in a community; and the distinction between two modes of believing: the reality mindset and the mythology mindset.

Finally, though it may seem paradoxical to lay out rational arguments for rationality itself, it's a timely assignment. Some people pursue the opposite paradox, citing reasons (presumably rational ones, or why should we listen?) that rationality is overrated, such as that logical personalities are joyless and repressed, analytical thinking must be subordinated to social justice, and a good heart and reliable gut are surer routes to well-being than tough-minded logic and argument. Many act as if rationality is obsolete—as if the point of argumentation is to discredit one's adversaries rather than collectively reason our way to the most defensible beliefs. In an era in which rationality seems both more threatened and more essential than ever, *Rationality* is, above all, an affirmation of rationality.

A MAJOR THEME of this book is that none of us, thinking alone, is rational enough to consistently come to sound conclusions: rationality emerges from a community of reasoners who spot each other's fallacies. In that spirit I thank the reasoners who made this book more rational. Ken Binmore, Rebecca Newberger Goldstein, Gary King, Jason Nemirow, Roslyn Pinker, Keith Stanovich, and Martina Wiese incisively commented on the first draft. Charleen Adams, Robert Aumann, Joshua Hartshorne, Louis Liebenberg, Colin McGinn, Barbara Mellers, Hugo Mercier, Scott Ogawa, Judea Pearl, David Ropeik, Michael Shermer, Susanna Siegel, Barbara Spellman, Lawrence Summers, Philip Tetlock, and Juliani Vidal reviewed chapters in their areas of expertise. Many questions arose as I planned and wrote the book, and they were answered by Daniel Dennett, Emily-Rose Eastop, Baruch Fischhoff, Reid Hastie, Nathan Kuncel, Ellen Langer, Jennifer Lerner, Beau Lotto, Daniel Loxton, Gary Marcus, Philip Maymin, Don Moore, David Myers, Robert Proctor, Fred Shapiro, Mattie Toma, Jeffrey Watumull, Jeremy Wolfe, and Steven Zipperstein. I counted on the expert transcription, fact-checking, and reference hunting by Mila Bertolo, Martina Wiese, and Kai Sandbrink, and on original data analyses by Bertolo, Toma, and Julian De Freitas. Also appreciated were the questions and suggestions from the students and teaching staff of General Education 1066: Rationality, especially Mattie Toma and Jason Nemirow.

Special thanks go to my wise and supportive editor, Wendy Wolf, for working with me on this book, our sixth; to Katya Rice, for copyediting our ninth; and to my literary agent, John Brockman, for his encouragement and advice on our ninth. I appreciate as well the support over many years from Thomas Penn, Pen Vogler, and Stefan McGrath of Penguin UK. Ilavenil Subbiah once again designed the graphics, and I thank her for her work and her encouragement.

Rebecca Newberger Goldstein played a special role in the conception of this book, because it is she who impressed on me that realism and reason are ideals that must be singled out and defended. Love and gratitude go as well to the other members of my family: Yael and Solly; Danielle; Rob, Jack, and David; Susan, Martin, Eva, Carl, and Eric; and my mother, Roslyn, to whom this book is dedicated.

RATIONALITY

1

HOW RATIONAL AN ANIMAL?

Man is a rational animal. So at least we have been told.
Throughout a long life I have searched diligently for evidence in favor of this statement. So far, I have not had the
good fortune to come across it.

—BERTRAND RUSSELL[1]

He that can carp in the most eloquent or acute manner at
the weakness of the human mind is held by his fellows as
almost divine.

—BARUCH SPINOZA[2]

omo sapiens means wise hominin, and in many ways we have
earned the specific epithet of our Linnaean binomial. Our
species has dated the origin of the universe, plumbed the
nature of matter and energy, decoded the secrets of life, unraveled
the circuitry of consciousness, and chronicled our history and diversity. We have applied this knowledge to enhance our own flourishing, blunting the scourges that immiserated our ancestors for most
of our existence. We have postponed our expected date with death

from thirty years of age to more than seventy (eighty in developed countries), reduced extreme poverty from ninety percent of humanity to less than nine, slashed the rates of death from war twentyfold and from famine a hundredfold.[3] Even when the ancient bane of pestilence rose up anew in the twenty-first century, we identified the cause within days, sequenced its genome within weeks, and administered vaccines within a year, keeping its death toll to a fraction of those of historic pandemics.

The cognitive wherewithal to understand the world and bend it to our advantage is not a trophy of Western civilization; it's the patrimony of our species. The San of the Kalahari Desert in southern Africa are one of the world's oldest peoples, and their foraging lifestyle, maintained until recently, offers a glimpse of the ways in which humans spent most of their existence.[4] Hunter-gatherers don't just chuck spears at passing animals or help themselves to fruit and nuts growing around them.[5] The tracking scientist Louis Liebenberg, who has worked with the San for decades, has described how they owe their survival to a scientific mindset.[6] They reason their way from fragmentary data to remote conclusions with an intuitive grasp of logic, critical thinking, statistical reasoning, causal inference, and game theory.

The San engage in persistence hunting, which puts to use our three most conspicuous traits: our two-leggedness, which enables us to run efficiently; our hairlessness, which enables us to dump heat in hot climates; and our big heads, which enable us to be rational. The San deploy this rationality to track the fleeing animals from their hoofprints, effluvia, and other spoor, pursuing them until they keel over from exhaustion and heat stroke.[7] Sometimes the San track an animal along one of its habitual pathways, or, when a trail goes cold, by searching in widening circles around the last known prints. But often they track them by reasoning.

Hunters distinguish dozens of species by the shapes and spacing of their tracks, aided by their grasp of cause and effect. They may infer that a deeply pointed track comes from an agile springbok, which needs a good grip, whereas a flat-footed track comes from a heavy kudu, which has to support its weight. They can sex the animals from the configuration of their tracks and the relative location of their urine to their hind feet and droppings. They use these categories to make syllogistic deductions: steenbok and duiker can be run down in the rainy season because the wet sand forces open their hooves and stiffens their joints; kudu and eland can be run down in the dry season because they tire easily in loose sand. It's the dry season and the animal that left these tracks is a kudu; therefore, this animal can be run down.

The San don't just pigeonhole animals into categories but make finer-grained logical distinctions. They tell individuals apart within a species by reading their hoofprints, looking for telltale nicks and variations. And they distinguish an individual's permanent traits, like its species and sex, from transient conditions like fatigue, which they infer from signs of hoof-dragging and stopping to rest. Defying the canard that premodern peoples have no concept of time, they estimate the age of an animal from the size and crispness of its hoofprints, and can date its spoor by the freshness of tracks, the wetness of saliva or droppings, the angle of the sun relative to a shady resting place, and the palimpsest of superimposed tracks from other animals. Persistence hunting could not succeed without those logical niceties. A hunter can't track just any gemsbok from among the many that have left tracks, but only the one he has been pursuing to exhaustion.

The San also engage in critical thinking. They know not to trust their first impressions, and appreciate the dangers of seeing what they want to see. Nor will they accept arguments from authority: anyone, including a young upstart, may shoot down a conjecture or come up with his own until a consensus emerges from the disputation. Though it's mainly the men who hunt, the women are just as knowledgeable at interpreting spoor, and Liebenberg reports that one young woman, !Nasi, "put the men to shame."[8]

The San adjust their credence in a hypothesis according to how diagnostic the evidence is, a matter of conditional probability. A porcupine foot, for instance, has two proximal pads while a honey badger has one, but a padprint may fail to register on hard ground. This means that though the probability that a track will have one padprint given that it was made by a honey badger is high, the inverse probability, that a track was made by a honey badger given that it has one padprint, is lower (since it could also be an incomplete porcupine track). The San do not confuse these conditional probabilities: they know that since two padprints could only have been left by a porcupine, the probability of a porcupine given two padprints is high.

The San also calibrate their credence in a hypothesis according to its prior plausibility. If tracks are ambiguous, they will assume they come from a commonly occurring species; only if the evidence is definitive will they conclude that they come from a rarer one.[9] As we shall see, this is the essence of Bayesian reasoning.

Another critical faculty exercised by the San is distinguishing causation from correlation. Liebenberg recalls: "One tracker, Boroh//xao, told me that when the [lark] sings, it dries out the soil, making the roots good to eat. Afterwards, !Nate and /Uase told me that Boroh//xao was wrong—it is not the *bird* that dries out the soil, it is the *sun* that dries out the soil. The bird is only *telling* them that the soil will dry out in the coming months and that it is the time of the year when the roots are good to eat."[10]

The San use their knowledge of the causal texture of their environment not just to understand how it is but to imagine how it might be. By playing out scenarios in their mind's eye, they can think several steps ahead of the animals in their world and devise intricate snares to trap them. One end of a springy branch is anchored in the ground and the stick is bent in half; the other is tied to a noose camouflaged with twigs and sand and held in place by a trigger. They place the snares at the openings of barriers they have built around an antelope's resting place, and guide the animal into the deadly spot with a hurdle the antelope must clear. Or they lure an ostrich to a snare by spotting its tracks under a camelthorn tree (whose pods are an ostrich delicacy) and leaving a conspicuous bone that's too big for the ostrich to swallow, which draws its attention to a smaller but still unswallowable bone, which leads to a still smaller bone, the bait in the snare.

Yet for all the deadly effectiveness of the San's technology, they have survived in an unforgiving desert for more than a hundred thousand years without exterminating the animals they depend on. During a drought, they think ahead to what would happen if they killed the last plant or animal of its kind, and they spare the members of the threatened species.[11] They tailor their conservation plans to the different vulnerabilities of plants, which cannot migrate but recover quickly when the rains return, and animals, which can survive a drought but build back their numbers slowly. And they enforce these conservation efforts against the constant temptation of poaching (everyone feeling they should exploit the scarce species, because if they don't, everyone else will) with an extension of the norms of reciprocity and collective well-being that govern all their resources. It is unthinkable for a San hunter not to share meat with an empty-handed bandmate, or to exclude a neighboring band driven from their drought-stricken territory, since they know that memories are long and some day fortunes may reverse.

· · ·

THE SAPIENCE OF THE SAN makes the puzzle of human rationality acute. Despite our ancient capacity for reason, today we are flooded with reminders of the fallacies and follies of our fellows. People gamble and play the lottery, where they are guaranteed to lose, and fail to invest for their retirement, where they are guaranteed to win. Three quarters of Americans believe in at least one phenomenon that defies the laws of science, including psychic healing (55 percent), extrasensory perception (41 percent), haunted houses (37 percent), and ghosts (32 percent)—which also means that some people believe in houses haunted by ghosts without believing in ghosts.[12] In social media, fake news (such as JOE BIDEN CALLS TRUMP SUPPORTERS "DREGS OF SOCIETY" and FLORIDA MAN ARRESTED FOR TRANQUIL-IZING AND RAPING ALLIGATORS IN THE EVERGLADES) is diffused farther and faster than the truth, and humans are more likely to spread it than bots.[13]

It has become commonplace to conclude that humans are simply irrational—more Homer Simpson than Mr. Spock, more Alfred E. Neuman than John von Neumann. And, the cynics continue, what else would you expect from descendants of hunter-gatherers whose minds were selected to avoid becoming lunch for leopards? But evolutionary psychologists, mindful of the ingenuity of foraging peoples, insist that humans evolved to occupy the "cognitive niche": the ability to outsmart nature with language, sociality, and know-how.[14] If contemporary humans seem irrational, don't blame the hunter-gatherers.

How, then, can we understand this thing called rationality which would appear to be our birthright yet is so frequently and flagrantly flouted? The starting point is to appreciate that rationality is not a power that an agent either has or doesn't have, like Superman's X-ray vision. It is a kit of cognitive tools that can attain particular goals in particular worlds. To understand what rationality is, why it seems scarce, and why it matters, we must begin with the ground truths of rationality itself: the ways an intelligent agent *ought* to reason, given its goals and the world in which it lives. These "normative" models come from logic, philosophy, mathematics, and artificial intelligence, and they are our best understanding of the "correct" solution to a problem and how to find it. They serve as an aspiration for those who want to be rational, which should mean everyone. A major goal of this book is to explain the most widely applicable normative tools of reason; they are the subjects of chapters 3 to 9.

Normative models also serve as benchmarks against which we can assess how human schlemiels *do* reason, the subject matter of psychology and the other behavioral sciences. The many ways in which ordinary people fall short of these benchmarks have become famous through the Nobel Prize–winning research of Daniel Kahneman, Amos Tversky, and other psychologists and behavioral economists.[15] When people's judgments deviate from a normative model, as they so often do, we have a puzzle to solve. Sometimes the disparity reveals a genuine irrationality: the human brain cannot cope with the complexity of a problem, or it is saddled with a bug that cussedly drives it to the wrong answer time and again.

But in many cases there is a method to people's madness. A problem may have been presented to them in a deceptive format, and when it is translated into a mind-friendlier guise, they solve it. Or the normative model may itself be correct only in a particular environment, and people accurately sense that they are not in that one, so the model doesn't apply. Or the model may be designed to bring about a certain goal, and, for better or worse, people are after a different one. In the chapters to come, we will see examples of all these extenuating circumstances. The penultimate chapter will lay out how some of today's florid outbursts of irrationality may be understood as the rational pursuit of goals other than an objective understanding of the world.

Though explanations of irrationality may absolve people of the charge of outright stupidity, to understand is not to forgive. Sometimes we can hold people to a higher standard. They can be taught to spot a deep problem across its superficial guises. They can be goaded into applying their best habits of thinking outside their comfort zones. And they can be inspired to set their sights higher than self-defeating or collectively destructive goals. These, too, are aspirations of the book.

Since a recurring insight of the study of judgment and decision making is that humans become more rational when the information they're dealing with is more vivid and relevant, let me turn to examples. Each of these classics—from math, logic, probability, and forecasting—exposes a quirk in our reasoning and will serve as a preview of the normative standards of rationality (and the ways in which people depart from them) in the chapters to come.

Three Simple Math Problems

Everyone remembers being tormented in high school by algebra problems about where the train that left Eastford traveling west at 70 miles per hour will meet the train that left Westford, 260 miles away, traveling east at 60 miles per hour. These three are simpler; you can do them in your head:

- A smartphone and a case cost $110 in total. The phone costs $100 more than the case. How much does the case cost?

- It takes 8 printers 8 minutes to print 8 brochures. How long would it take 24 printers to print 24 brochures?
- On a field there is a patch of weeds. Every day the patch doubles in size. It takes 30 days for the patch to cover the whole field. How long did it take for the patch to cover half the field?

The answer to the first problem is $5. If you're like most people, you guessed $10. But if that were right, the phone would cost $110 ($100 more than the case), and the total for the pair would be $120.

The answer to the second question is 8 minutes. It takes a printer 8 minutes to print a brochure, so as long as there are as many printers as there are brochures and they are working simultaneously, the time it takes to print the brochures is the same.

The answer to the third problem is 29 days. If the weed patch doubles every day, then by working backwards from when the field was completely covered, we may infer that it was half covered the day before.

The economist Shane Frederick gave these questions (with different examples) to thousands of university students. Five out of six got at least one of them wrong; one in three got them *all* wrong.[16] Yet each question has a simple answer that almost everyone understands when it's pointed out. The problem is that people's heads are turned by superficial features of the problem which they mistakenly think are relevant to the answer, such as the round numbers 100 and 10 in the first problem and the fact that the number of printers is the same as the number of minutes in the second.

Frederick calls his low-tech battery the Cognitive Reflection Test, and suggests that it exposes a cleavage between two cognitive systems, later made famous by Kahneman (his sometime coauthor) in the 2011 bestseller *Thinking, Fast and Slow*. System 1 operates rapidly and effortlessly, and it seduces us with the wrong answers; System 2 requires concentration, motivation, and the application of learned rules, and it allows us to grasp the right ones. No one thinks these are literally two anatomical systems in the brain; they are two modes of operation which cut across many brain structures. System 1 means snap judgments; System 2 means thinking twice.

The lesson of the Cognitive Reflection Test is that blunders of reasoning may come from thoughtlessness rather than ineptitude.[17] Even students at the math-proud Massachusetts Institute of Technology averaged only two out of three correct. Performance does correlate with math skill, as you'd expect, but it also correlates with patience. People who describe themselves as not impulsive, and who would rather wait for a larger payment in a month than get a smaller one right away, are less likely to fall into the traps.[18]

The first two items feel like trick questions. That is because they give details which, in the back-and-forth of conversation, would be relevant to what the speaker is asking, but in these examples are designed to lead the hearer astray. (People do better when the smartphone costs, say, $73 more than the case and the combination costs $89.)[19] But of course real life is also baited with garden paths and siren songs that lure us from good decisions, and resisting them is a part of being rational. People who fall for the alluring but wrong answers on the Cognitive Reflection Test appear to be less rational in other ways, such as turning down lucrative offers that require a bit of waiting or a bit of risk.

And the third problem, the one with the weed patch, is not a trick question but taps a real cognitive infirmity. Human intuition doesn't grasp exponential (geometric) growth, namely something that rises at a rising rate, proportional to how large it already is, such as compound interest, economic growth, and the spread of a contagious disease.[20] People mistake it for steady creep or slight acceleration, and their imaginations don't keep up with the relentless doubling. If you deposit $400 a month into a retirement account that earns 10 percent annually, how big will your nest egg be after forty years? Many people guess around $200,000, which is what you get by multiplying 400 by 12 by 110% by 40. Some know that that can't be right and adjust their guess upward, but never enough. Almost no one gets the correct answer: $2.5 *million*. People with a shaky grasp of exponential growth have been found to save less for retirement and to take on more credit-card debt, two roads to penury.[21]

A failure to visualize exponential blastoff can trap experts as well—even experts in cognitive biases. When Covid-19 arrived in the United States and Europe in February 2020, several social scientists (including two heroes of this book, though not Kahneman himself) opined that people were irrationally panicking because they had read about a gruesome case or two and got carried away by the "availability bias" and "probability neglect." The objective risk at the time, they noted, was lower than that of the flu or strep throat, which everyone accepts calmly.[22] The fallacy of the fallacy scolds was to underestimate the accelerating rate at which a disease as contagious as Covid can spread, with each patient not only infecting new ones but turning each of them into an infector. The single confirmed American death on March 1 grew in successive weeks to 2, 6, 40, 264, 901, and 1,729 deaths per day, adding up to more than 100,000 deaths by June 1 and soon making it the most lethal hazard in the country.[23] Of course the authors of these obscure op-eds cannot be blamed for the insouciance which lulled so many leaders and citizens into dangerous complacency, but their comments show how deeply rooted cognitive biases can be.

Why do people misunderestimate exponential growth, as George W. Bush might have put it? In the great tradition of the physician in the Molière play who explained that opium makes people sleepy because of its dormitive power, social scientists attribute the errors to an "exponential growth bias." Less circularly, we might point to the ephemerality of exponential processes in natural environments (prior to historical innovations like economic growth and compound interest). Things that can't go on forever don't, and organisms can multiply only to the point where they deplete, foul, or saturate their environments, bending the exponential curve into an S. This includes pandemics, which peter out once enough susceptible hosts in the herd are killed or develop immunity.

A Simple Logic Problem

If anything lies at the core of rationality, it must surely be logic. The prototype of a rational inference is the syllogism "If P then Q. P. Therefore, Q." Consider a simple example.

Suppose the coinage of a country has a portrait of one of its eminent sovereigns on one side and a specimen of its magnificent fauna on the other. Now consider a simple if-then rule: "If a coin has a king on one side, then it has a bird on the other." Here are four coins, displaying a king, a queen, a moose, and a duck. Which of the coins do you have to turn over to determine whether the rule has been violated?

If you're like most people, you said "the king" or "the king and the duck." The correct answer is "the king and the moose." Why?

Everyone agrees you have to turn over the king, because if you failed to find a bird on the reverse it would violate the rule in so many words. Most people know there's no point in turning over the queen, because the rule says "If king, then bird"; it says nothing about coins with a queen. Many say you should turn over the duck, but when you think about it, that coin is irrelevant. The rule is "If king, then bird," not "If bird, then king": if the duck shared the coin with a queen, nothing would be amiss. But now consider the moose. If you turned that coin over and found a king on the obverse, the rule "If king, then bird" would have been transgressed. The answer, then, is "the king and the moose." On average, only 10 percent of people make those picks.

The Wason selection task (named after its creator, the cognitive psychologist Peter Wason) has been administered with various "If P then Q" rules for sixty-five years. (The original version used cards with a letter on one side and a number on the other and a rule like "If there is a D on one side, there is a 3 on the other.") Time and again people turn over the P, or the P and the Q, and fail to turn over the not-Q.[24] It's not that they're incapable of understanding the right answer. As with the Cognitive Reflection Test, as soon as it is explained to them they slap themselves on the forehead and accept it.[25] But their unreflective intuition, left to its own devices, fails to do the logic.

What does this tell us about human rationality? A common explanation is that it reveals our *confirmation bias*: the bad habit of seeking evidence that ratifies a belief and being incurious about evidence that might falsify it.[26] People think that dreams are omens because they recall the time when they dreamt a relative had a mishap and she did, but they forget about all the times when a relative was fine after they dreamt she had a mishap. Or they think immigrants commit a lot of crime because they read in the news about an immigrant who robbed a store, but don't think about the larger number of stores robbed by native-born citizens.

Confirmation bias is a common diagnosis for human folly and a target for enhancing rationality. Francis Bacon (1561–1626), often credited with developing the scientific method, wrote of a man who was taken to a church and shown a painting of sailors who had escaped a shipwreck thanks to their holy vows. "Aye," he remarked, "but where are they painted that were drowned after their vows?"[27] He observed, "Such is the way of all superstitions, whether in astrology, dreams, omens, divine judgments, or the like; wherein men, having a delight in such vanities, mark the events where they are fulfilled, but where they fail, although this happened much oftener, neglect and pass them by."[28] Echoing a famous argument by the philosopher Karl Popper, most scientists today insist that the dividing line between science and pseudoscience is whether advocates of a hypothesis deliberately search for evidence that could falsify it and accept the hypothesis only if it survives.[29]

How can humans make it through the day with an inability to apply the most elementary rule of logic? Part of the answer is that the selection task is a peculiar challenge.[30] It doesn't ask people to apply the syllogism to make a useful deduction ("Here's a coin with a king; what's on the other side?") or to test the rule in general ("Is the rule true of the country's coinage?"). It asks whether the rule applies specifically to each of a handful of items before them on the table. The other part of the answer is that people do apply logic when the rule involves the shoulds and shouldn'ts of human life rather than arbitrary symbols and tokens.

Suppose the Post Office sells fifty-cent stamps for third-class mail but requires ten-dollar stamps for Express Mail. That is, properly addressed mail must follow the rule "If a letter is labeled Express Mail, it must have a ten-dollar stamp." Suppose the label and the stamp don't fit on the same side of the envelope, so a postal worker has to turn envelopes over to check to see if the sender has followed the rule. Here are four envelopes. Imagine that you are a postal worker. Which ones do you have to turn over?

Express	3rd class	50¢	$10

The correct answer once again is P and not-Q, namely the Express envelope and the one with the fifty-cent stamp. Though the problem is logically equivalent to the four-coin problem, this time almost everyone gets it right. The content of a logical problem matters.[31] When an if-then rule implements a contract involving permissions and duties—"If you enjoy a benefit, you must pay a cost"—then a violation of the rule (take the benefit, don't pay the cost) is equivalent to cheating, and people intuitively know what it takes to catch a cheater. They don't check up on people who aren't enjoying the benefit or people who have paid a cost, neither of whom could be trying to get away with something.

Cognitive psychologists debate exactly what kinds of content temporarily turn people into logicians. They can't be just any concrete scenarios, but must embody the kinds of logical challenges that we became attuned to as we developed into adults and perhaps when we evolved into humans. Monitoring a privilege or duty is one of these logic-unlocking themes; monitoring danger is another. People know that to verify compliance with the precaution "If you ride a bicycle, then you must wear a helmet," they have to check that a child on a bicycle is wearing a helmet and that a child without a helmet does not get onto a bicycle.

Now, a mind that can falsify a conditional rule when the violations are tantamount to cheating or danger is not exactly a logical mind. Logic, by definition, is about the form of statements, not their content: how Ps and Qs are connected by IF, THEN, AND, OR, NOT, SOME, and ALL, regardless of what the Ps and Qs stand for. Logic is a crowning achievement of human knowledge. It organizes our reasoning with unfamiliar or abstract subject matter, such as the laws of government and science, and when implemented in silicon it turns inert matter into thinking machines. But what the untutored human mind commands is not a general-purpose, content-free tool, with formulas like "[IF P THEN Q] is equivalent to NOT-[P AND NOT Q]," into which any P and Q can be plugged. It commands a set of more specialized tools that bake together the content relevant to the problem with the rules of logic (without those rules, the tools wouldn't work). It isn't easy for people to extricate the rules and wield them in novel, abstract, or apparently meaningless problems. That's what education and other rationality-enhancing institutions are for. They augment the *ecological rationality* we are born and grow up with—our horse sense, our street smarts—with the broader-spectrum and more potent tools of reasoning perfected by our best thinkers over the millennia.[32]

A Simple Probability Problem

One of the most famous television game shows from the heyday of the genre from the 1950s to the 1980s was *Let's Make a Deal*. Its host, Monty Hall, achieved a second kind of fame when a dilemma in probability theory, loosely based on the show, was named after him.[33] A contestant is faced with three doors. Behind one of them is a sleek new car. Behind the other two are goats. The contestant picks a door, say Door 1. To build suspense, Monty opens one of the other two

doors, say Door 3, revealing a goat. To build the suspense still further, he gives the contestant an opportunity either to stick with their
original choice or to switch to the unopened door. You are the contestant. What should you do?

Almost everyone stays.[34] They figure that since the car was placed
behind one of the three doors at random, and Door 3 has been eliminated, there is now a fifty-fifty chance each that the car will be behind Door 1 or Door 2. Though there's no harm in switching, they
think, there's no benefit either. So they stick with their first choice
out of inertia, pride, or anticipation that their regret after an unlucky
switch would be more intense than their delight after a lucky one.

The Monty Hall dilemma became famous in 1990 when it was
presented in the "Ask Marilyn" column in *Parade*, a magazine inserted in the Sunday edition of hundreds of American newspapers.[35]
The columnist was Marilyn vos Savant, known at the time as "the
world's smartest woman" because of her entry in the *Guinness Book of
World Records* for the highest score on an intelligence test. Vos Savant
wrote that you should switch: the odds of the car being behind Door 2
are two in three, compared with one in three for Door 1. The column drew ten thousand letters, a thousand of them from PhDs,
mainly in mathematics and statistics, most of whom said she was
wrong. Here are some examples:

> You blew it, and you blew it big! Since you seem to have dif-
> ficulty grasping the basic principle at work here, I'll explain.
> After the host reveals a goat, you now have a one-in-two
> chance of being correct. Whether you change your selection
> or not, the odds are the same. There is enough mathematical
> illiteracy in this country, and we don't need the world's high-
> est IQ propagating more. Shame!
>
> —SCOTT SMITH, PHD, UNIVERSITY OF FLORIDA

I am sure you will receive many letters on this topic from
high school and college students. Perhaps you should keep a
few addresses for help with future columns.

    —W. Robert Smith, PhD, Georgia State University

Maybe women look at math problems differently than men.

    —Don Edwards, Sunriver, Oregon[36]

Among the objectors was Paul Erdös (1913–1996), the renowned
mathematician who was so prolific that many academics boast of
their "Erdös number," the length of the shortest chain of coauthor-
ships linking them to the great theoretician.[37]

But the mansplaining mathematicians were wrong and the world's
smartest woman was right. You should switch. It's not that hard to
see why. There are three possibilities for where the car could have
been placed. Let's consider each door and count up the number of
times out of the three that you would win with each strategy. You
picked Door 1, but of course that's just a label; as long as Monty follows the rule "Open an unselected door with a goat; if both have
goats, pick one at random," the odds come out the same whichever
door you picked.

Suppose your strategy is "Stay" (left column in the figure). If the
car is behind Door 1 (top left), you win. (It doesn't matter which of
the other doors Monty opened, because you're not switching to either.) If the car is behind Door 2 (middle left), you lose. If the car is
behind Door 3 (bottom left), you lose. So the odds of winning with
the "Stay" strategy are one in three.

Now suppose your strategy is "Switch" (right column). If the car
is behind Door 1, you lose. If the car is behind Door 2, Monty would
have opened Door 3, so you would switch to Door 2 and win. If the
car is behind Door 3, he would have opened Door 2, so you would

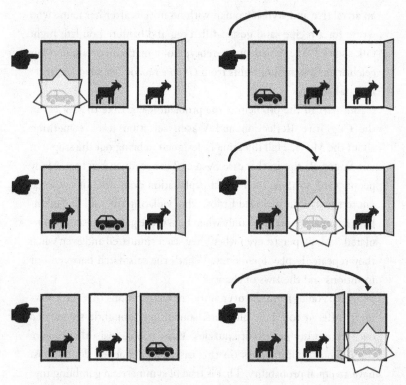

switch to Door 3 and win. The odds of winning with the "Switch" strategy are two in three, double the odds of staying.

It's not rocket surgery.[38] Even if you don't work through the logical possibilities, you could play a few rounds yourself with cutouts and toys and tot up the outcomes, as Hall himself did to convince a skeptical journalist. (Nowadays, you can play it online.)[39] Or you could pursue the intuition "Monty knows the answer and gave me a clue; it would be foolish not to act on it." Why did the mathematicians, professors, and other bigshots get it so wrong?

Certainly there were failures of critical thinking coming from sexism, ad hominem biases, and professional jealousy. Vos Savant is an attractive and stylish woman with no initials after her name who wrote for a recipe- and gossip-filled rag and bantered on late-night talk shows.[40] She defied the stereotype of a mathematician, and her celebrity and bragging rights from *Guinness* made her a big fat target for a takedown.

But part of the problem is the problem itself. Like the teasers in the Cognitive Reflection and Wason selection tests, something about the Monty Hall dilemma is designed to bring out the stupid in our System 1. But in this case System 2 is not much brighter. Many people can't swallow the correct explanation even when it's pointed out to them. This included Erdös, who, violating the soul of a mathematician, was convinced only when he saw the game repeatedly simulated.[41] Many persist even when they see it simulated and even when they repeatedly play for money. What's the mismatch between our intuitions and the laws of chance?

A clue comes from the overconfident justifications that the know-it-alls offered for their blunders, sometimes thoughtlessly carried over from other probability puzzles. Many people insist that each of the unknown alternatives (in this case, the unopened doors) must have an equal probability. That is true of symmetrical gambling toys like the faces of a coin or sides of a die, and it is a reasonable starting point when you know absolutely nothing about the alternatives. But it is not a law of nature.

Many visualize the causal chain. The car and goats were placed prior to the reveal, and opening a door can't move them around after the fact. Pointing out the independence of causal mechanisms is a common way to debunk other illusions such as the gambler's fallacy, in which people misguidedly think that after a run of reds the next spin of the roulette wheel will turn up black, when in fact the wheel has no memory, so every spin is independent. As one of vos Savant's correspondents mansplained, "Picture a race with three horses, each having an equal chance of winning. If horse #3 drops dead 50 feet into the race, the chances for each of the remaining two horses are no longer one in three but rather are now one in two." Clearly, he concluded, it would not make sense to switch one's bet from horse #1 to horse #2. But this is not how the problem works. Imagine that after you place your bet on #1, God announces, "It's not going to be horse #3; no comment on horse #1." He could have warned against horse #2 but didn't. Switching your bet doesn't sound so crazy.[42] In *Let's Make a Deal*, Monty Hall is God.

The godlike host reminds us how exotic the Monty Hall problem is. It requires an omniscient being who defies the usual goal of a conversation—to share what the hearer needs to know (in this case, which door hides the car)—and instead pursues the goal of enhancing suspense among third parties.[43] And unlike the world, whose clues are indifferent to our sleuthing, Monty Almighty knows the truth and knows our choice and picks his revelation accordingly.

People's insensitivity to this lucrative but esoteric information pinpoints the cognitive weakness at the heart of the puzzle: we confuse *probability* with *propensity*. A propensity is the disposition of an object to act in certain ways. Intuitions about propensities are a major part of our mental models of the world. People sense that bent branches tend to spring back, that kudu may tire easily, that porcupines usually leave tracks with two padprints. A propensity cannot be perceived directly (either the branch sprang back or it didn't), but it can be inferred by scrutinizing the physical makeup of an object and working through the laws of cause and effect. A drier branch may snap, a kudu has more stamina in the rainy season, a porcupine has two proximal pads which leave padprints when the ground is soft but not necessarily when it is hard.

But probability is different; it is a conceptual tool invented in the seventeenth century.[44] The word has several meanings, but the one that matters in making risky decisions is the strength of one's belief in an unknown state of affairs. Any scrap of evidence that alters our confidence in an outcome will change its probability and the rational way to act upon it. The dependence of probability on ethereal knowledge rather than just physical makeup helps explain why people fail at the dilemma. They intuit the propensities for the car to have ended up behind the different doors, and they know that opening a door could not have changed those propensities. But probabilities are not about the world; they're about our *ignorance* of the world. New information reduces our ignorance and changes the probability. If that sounds mystical or paradoxical, think about the probability that a coin I just flipped landed heads. For you, it's .5. For me, it's 1 (I peeked). Same event, different knowledge, different probability. In the Monty Hall dilemma, new information is provided by the all-seeing host.

One implication is that when the reduction of ignorance granted by the host is more transparently connected to the physical circumstances, the solution to the problem becomes intuitive. Vos Savant invited her readers to imagine a variation of the game show with, say, a thousand doors.[45] You pick one. Monty reveals a goat behind 998 of the others. Would you switch to the door he left closed? This time it seems clear that Monty's choice conveys actionable information. One can visualize him scanning the doors for the car as he decides which one not to open, and the closed door is a sign of his having spotted the car and hence a spoor of the car itself.

A Simple Forecasting Problem

Once we get into the habit of assigning numbers to unknown events, we can quantify our intuitions about the future. Forecasting events is big business. It informs policy, investment, risk management, and ordinary curiosity about what lies in store for the world. Consider each of the following events, and write down your estimate of the likelihood that it will take place in the coming decade. Many of them are pretty unlikely, so let's make finer distinctions at the lower end of the scale and pick one of the following probabilities for each: less than .01 percent, .1 percent, .5 percent, 1 percent, 2 percent, 5 percent, 10 percent, 25 percent, and 50 percent or more.

1. Saudi Arabia develops a nuclear weapon.
2. Nicolás Maduro resigns as president of Venezuela.
3. Russia has a female president.
4. The world suffers a new and even more lethal pandemic than Covid-19.
5. Vladimir Putin is constitutionally prevented from running for another term as president of Russia and his wife takes his place on the ballot, allowing him to run the country from the sidelines.
6. Massive strikes and riots force Nicolás Maduro to resign as president of Venezuela.
7. A respiratory virus jumps from bats to humans in China and starts a new and even more lethal pandemic than Covid-19.
8. After Iran develops a nuclear weapon and tests it in an underground explosion, Saudi Arabia develops its own nuclear weapon in response.

I presented items like these to several hundred respondents in a survey. On average, people thought it was likelier that Putin's wife would be president of Russia than that a woman would be president. They thought it was likelier that strikes would force Maduro to resign than that he would resign. They thought Saudi Arabia was more

likely to develop a nuclear weapon in response to an Iranian bomb
than it was to develop a nuclear weapon. And they thought it was
likelier that Chinese bats would start a pandemic than that there
would be a pandemic.[46]

You probably agree with at least one of these comparisons; 86
percent of the participants who rated all the items did. If so, you violated an elementary law of probability, the conjunction rule: the probability of a conjunction of events (A and B) must be less than or equal
to the probability of either of the events (A, or B). The probability of
picking an even-numbered spade out of a deck of cards, for example
(even and spade), has to be less than the probability of picking a
spade, because some spades are not even numbers.

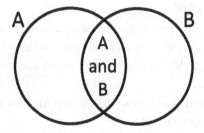

In each pair of world events, the second scenario is a conjunction
of events, one of which is the event in the first scenario. For example,
"Iran tests a nuclear weapon and Saudi Arabia develops a nuclear
weapon" is a conjunction that embraces "Saudi Arabia develops a nuclear weapon" and must have a smaller chance of happening, since
there are other scenarios in which the Saudis might go nuclear (to
counter Israel, to flaunt hegemony over the Persian Gulf, and so on).
By the same logic, Maduro resigning the presidency has to be more
likely than Maduro resigning the presidency after a series of strikes.

What are people thinking? A class of events described by a single

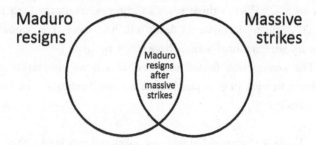

statement can be generic and abstract, with nothing for the mind to hold on to. A class of events described by a conjunction of statements can be more vivid, especially when they spell out a story line we can watch in the theater of our imagination. Intuitive probability is driven by imaginability: the easier something is to visualize, the likelier it seems. This entraps us into what Tversky and Kahneman call the conjunction fallacy, in which a conjunction is more intuitively probable than either of its elements.

The forecasts of pundits are often driven by vivid narratives, probability be damned.[47] A famous 1994 cover story in *The Atlantic* by the journalist Robert Kaplan predicted "The Coming Anarchy."[48] Kaplan forecasted that in the first decades of the twenty-first century, wars would be fought over scarce resources such as water; Nigeria would conquer Niger, Benin, and Cameroon; world wars would be fought over Africa; the United States, Canada, India, China, and Nigeria would break apart, whereupon American regions with many Latinos would erase the border with Mexico while Alberta would merge with Montana; crime would climb in American cities; AIDS would get worse and worse; together with a dozen other calamities, crises, and crackups. Yet as the article was becoming a sensation (including with President Bill Clinton, who passed it around the White House), the number of civil wars, the proportion of people without access to clean water, and the rate of American crime were sinking like stones.[49] Within three years an effective treatment for AIDS would begin to decimate its death toll. And more than a quarter century later, national borders have barely budged.

The conjunction fallacy was first illustrated by Tversky and Kahneman with an example that has become famous as "the Linda problem":[50]

> Linda is 31 years old, single, outspoken and very bright. She majored in philosophy. As a student, she was deeply concerned with issues of discrimination and social justice, and also participated in anti-nuclear demonstrations.
> Please indicate the probability of each of these statements:
>> Linda is a teacher in elementary school.
>> Linda is active in the feminist movement.
>> Linda is a psychiatric social worker.
>> Linda is a bank teller.
>> Linda is an insurance salesperson.
>> Linda is a bank teller and is active in the feminist movement.

Respondents judged that it was likelier that Linda was a feminist bank teller than that she was a bank teller: once again, the probability of A and B was judged to be higher than the probability of A alone. The dated vignette, with its baby-boomer "Linda," backhanded compliment "bright," passé protests, and declining occupation, betrays its early-1980s vintage. But as any psychology instructor knows, the effect is easily replicable, and today, highly intelligent Amanda who marches for Black Lives Matter is still deemed likelier to be a feminist registered nurse than a registered nurse.

The Linda problem engages our intuitions in a particularly compelling way. Unlike the selection task, where people make errors when the problem is abstract ("If P then Q") and get it right when it is couched in certain real-life scenarios, here everyone agrees with the abstract law "prob(A and B) ≤ prob(A)" but are upended when it is made concrete. The biologist and popular science writer Stephen Jay Gould spoke for many when he commented, "I know that the [conjunctive] statement is least probable, yet a little homunculus in my head continues to jump up and down, shouting at me—'but she can't just be a bank teller; read the description.'"[51]

That little homunculus can be exploited by skilled persuaders. A prosecutor with little to work with but a corpse washed up on a beach may spin a yarn on how her husband might, hypothetically, have smothered her and dumped the body so he could marry his mistress and start a business with the insurance money. The defense attorney could tell a competing shaggy-dog story in which she could, in theory, have been the victim of a late-night purse-snatching that went horribly awry. Each conjectural detail should make the scenario less likely, according to the laws of probability, yet each can make it more compelling. As Pooh-Bah says in *The Mikado*, it's all "merely corroborative detail, intended to give artistic verisimilitude to an otherwise bald and unconvincing narrative."[52]

The conjunction rule is a basic law of mathematical probability, and you don't need to think in numbers to understand it. This made Tversky and Kahneman pessimistic about people's intuitive sense of probability, which they argued is driven by representative stereotypes and available memories rather than on a systematic reckoning of possibilities. They rejected the idea that "inside every incoherent person there is a coherent one trying to get out."[53]

Other psychologists are more charitable. As we saw with the Monty Hall dilemma, "probability" has several meanings, including physical propensity, justified strength of belief, and frequency in the long run. Still another sense is provided by the *Oxford English*

Dictionary: "the appearance of truth, or likelihood of being realized, which any statement or event bears in the light of present evidence."[54] People faced with the Linda problem know that "frequency in the long run" is irrelevant: there's only one Linda, and either she is a feminist bank teller or she isn't. In any coherent conversation the speaker would supply biographical details for a reason, namely to lead the listener to a plausible conclusion. According to the psychologists Ralph Hertwig and Gerd Gigerenzer, people may have rationally inferred that the relevant meaning of "probability" in this task is not one of the mathematical senses in which the conjunction rule applies, but the nonmathematical sense of "degree of warrant in light of the present evidence," and they sensibly followed where the evidence pointed.[55]

In support of the charitable reading, many studies, beginning with ones by Tversky and Kahneman themselves, show that when people are *encouraged* to reason about probability in the sense of relative frequency, rather than being left to struggle with the enigmatic concept of the probability of a single case, they are likelier to obey the conjunction rule. Imagine a thousand women like Linda. How many of them do you think are bank tellers? How many of them do you think are bank tellers who are active in the feminist movement? Now the homunculus is quiet; a coherent person tries to get out. The rate of conjunction errors plummets.[56]

So is the conjunction fallacy, the quintessential demonstration of human probability blindness, an artifact of ambiguous wording and leading questions? Tversky and Kahneman insisted that it isn't. They noted that people commit the fallacy even when they are invited to *bet* on the possibilities (yes, a majority prefer to bet that Linda is a feminist bank teller than that she is a bank teller). And even when the question is couched in frequencies, where people could avoid a conjunction error just by counting the bank tellers in their mind's eye, a substantial minority commit it. This rises to a majority when people consider each alternative in isolation rather than seeing one next to the other, and so their noses are not rubbed in the absurdity of a subset outnumbering a superset.[57]

Kahneman has observed that humans are never so irrational as when protecting their pet ideas. So he advocated a new method for resolving scientific controversies to replace the time-honored custom of the rivals taking turns moving the goalposts and talking trash in volleys of rejoinders and replies. In an "adversarial collaboration," the disputants agree in advance on an empirical test that would settle the matter, and invite an arbiter to join them in carrying it out.[58] Fittingly, Kahneman collaborated with Hertwig to see who was right about the Linda problem, recruiting the psychologist Barbara Mellers to act as arbiter. The team of rivals agreed to run three studies that couched the problem in frequencies ("Of 100 people like Linda, how many are . . . ?") rather than asking about lone Linda. In their write-up of the complex results, the trio reported, "We did not think the experiments would resolve all the issues, nor did this miracle occur." But both sides agreed that people are prone to committing the conjunction fallacy, even when dealing with frequencies. And they agreed that under the right circumstances—the alternatives are available for comparison side by side, and the wording of the alternatives leaves nothing to the imagination—people can think their way out of the fallacy.

The Moral from Cognitive Illusions

How do we reconcile the rationality that allows our species to live by its wits in ancient and modern environments with the bloopers and gaffes that these brainteasers reveal—the confirmation bias, the overconfidence, the distractibility by concrete details and conversational habits? The classic errors in reasoning are often called "cognitive illusions," and the parallels with the visual illusions familiar from cereal boxes and science museums are instructive. They run deeper than the obvious fact that our eyes and minds can trick us. They explain how our species can be so smart and yet so easily deluded.

Here are two classic illusions, brought to life by the neuroscientist Beau Lotto.[59] The first is a shading illusion. Believe it or not, the dark stripes on the top of the box and the white stripes on the front are identical shades of gray.

Used by permission of Beau Lotto

The second is a shape illusion: the angles of the four elbows are identical, 90 degrees.

The first takeaway is that we can't always believe our eyes, or, more accurately, the visual System 1 in our brains. The second is that we can recognize our errors using System 2—say, by punching two holes in an index card and laying it over the first figure, and by aligning the corner of the card with the elbows in the second.

But the wrong takeaway is that the human visual system is a buggy

Used by permission of Beau Lotto

contraption that constantly fools us with figments and mirages. The human visual system is one of the wonders of the world. It is a precision instrument that can detect a single photon, recognize thousands of shapes, and negotiate rocky trails and high-speed autobahns. It outperforms our best artificial vision systems, which is why at the time of this writing autonomous vehicles have not been loosed on city streets despite decades of R&D. The vision modules of the robocars are apt to mistake a tractor trailer for a billboard, or a traffic sign plastered with stickers for a refrigerator filled with food.[60]

The shape and shading illusions are not bugs but features. The goal of the visual system is to provide the rest of the brain with an accurate description of the 3-D shapes and material composition of the objects in front of us.[61] This is a hard problem because the information coming into the brain from the retina doesn't reflect reality directly. The brightness of a patch on the retinal image depends not just on the pigmentation of the surface in the world but on the intensity of the illumination falling on it: a gray patch could have arisen from a black surface illuminated by a bright light or from a white surface illuminated by a dim one. (That's the basis for the illusion called #thedress, which took the internet by storm in 2015.)[62]

A shape on the retina depends not just on the 3-D geometry of the object but on its orientation from a vantage point: an acute angle on the retina could be a sharp corner viewed straight on or a right-angled corner foreshortened. The visual system undoes the effects of these distortions, dividing out the intensity of the illumination and inverting the trigonometry of perspective to feed the rest of the brain with a representation that matches real shapes and materials in the world. The intermediate scratch pad in these calculations—the 2-D array of pixels coming in from our retina—is hidden from the reasoning and planning systems of the brain because they would just be distractions.

Thanks to this design, our brains are not very good light meters or protractors, but then they don't have to be (unless we are realist painters). The illusions emerge when people are asked to be just those instruments. The viewer is asked to notice how bright the stripe is, how sharp the angle, *in the picture*. The pictures have been confected so that simple properties—equal brightnesses, right angles—are buried in the scratch pads that the conscious mind ordinarily ignores. If the questions were about the things in the *world* captured in the pictures, our impressions would be correct. The gray stripe really is darker than the white stripe on both the lit and shaded faces of the box; the elbows poised at different tilts really are bent at different angles.

In the same way, cognitive illusions like the ones in this chapter may arise from our setting aside the literal statement of a question as it comes into our brains and thinking through to what a speaker in the social world would reasonably ask. Doing arithmetic on deceptively conspicuous numbers, verifying a proposition about a handful of tokens, choosing from clues offered by a sly and omniscient master, and following a vivid character sketch to a literal but implausible conclusion are a bit like judging the angles and shades of gray in the printed page. They lead to incorrect answers, yes, but they are often correct answers to different and more useful questions. A mind capable of interpreting the intent of a questioner in context is far from unsophisticated. That's why we furiously hit "0" and scream "Operator!" into the phone when the bot on a help line reiterates a list of useless options and only a human can be made to understand why we called.

The explicability of our irrational reactions is not an excuse to fall back on them, any more than we should always trust our eyes. Science and technology have thrillingly extended the powers of the visual system beyond what nature gave us. We have microscopes for the small, telescopes for the distant, photography for the past, lighting for the dark, remote sensing for the invisible. And as we sally into realms outside the envelope in which we evolved, like the very fast and the very high, believing our senses can be fatal. The judgments of depth and orientation that allow our brains to undo the effects of projective geometry in everyday life depend on converging lines, receding texture, and flowing contours arrayed along the ground as we move and look about. When a pilot is suspended thousands of feet in the air with nothing but empty space between him and the earth, and the horizon is obscured by clouds, haze, or mountains, his visual sense gets out of sync with reality. As he flies by the seat of his pants, which cannot distinguish acceleration from gravity, every correction makes things worse, and can send the plane into a "graveyard spiral" within minutes—the fate of an inexperienced and overconfident John F. Kennedy Jr. in 1999. As excellent as our visual systems are, rational pilots know when to discount them and turn their perception over to instruments.[63]

And as excellent as our cognitive systems are, in the modern world we must know when to discount them and turn our reasoning over to instruments—the tools of logic, probability, and critical thinking that extend our powers of reason beyond what nature gave us. Because in the twenty-first century, when we think by the seat of our pants, every correction can make things worse, and can send our democracy into a graveyard spiral.

2

RATIONALITY
AND IRRATIONALITY

> May I say that I have not thoroughly enjoyed serving with
> humans? I find their illogic and foolish emotions a con-
> stant irritant.
>
> —MR. SPOCK

Rationality is uncool. To describe someone with a slang word for the cerebral, like *nerd*, *wonk*, *geek*, or *brainiac*, is to imply they are terminally challenged in hipness. For decades, Hollywood screenplays and rock song lyrics have equated joy and freedom with an escape from reason. "A man needs a little madness or else he never dares cut the rope and be free," said Zorba the Greek. "Stop making sense," advised Talking Heads; "Let's go crazy," adjured the Artist Formerly Known as Prince. Influential academic movements like postmodernism and critical theory (not to be confused with critical thinking) hold that reason, truth, and objectivity are social constructions that justify the privilege of dominant groups. These movements have an air of sophistication about them, implying that Western philosophy and science are provincial, old-fashioned, naïve to the diversity of ways of knowing found across periods and cultures. To be sure, not far from where I live in downtown Boston there is a splendid turquoise and gold mosaic that proclaims, "Follow reason." But it is affixed to the Grand Lodge of the Masons, the fez- and apron-sporting fraternal organization that is the answer to the question "What's the opposite of hip?"

My own position on rationality is "I'm for it." Though I cannot argue that reason is dope, phat, chill, fly, sick, or da bomb, and strictly speaking I cannot even justify or rationalize reason, I will defend the message on the mosaic: we ought to *follow* reason.

Reasons for Reason

To begin at the beginning: what *is* rationality? As with most words in common usage, no definition can stipulate its meaning exactly, and the dictionary just leads us in a circle: most define *rational* as "having reason," but *reason* itself comes from the Latin *ration-*, often defined as "reason."

A definition that is more or less faithful to the way the word is used is "the ability to use knowledge to attain goals." *Knowledge* in turn is standardly defined as "justified true belief."[1] We would not credit someone with being rational if they acted on beliefs that were known to be false, such as looking for their keys in a place they knew the keys could not be, or if those beliefs could not be justified—if they came, say, from a drug-induced vision or a hallucinated voice rather than observation of the world or inference from some other true belief.

The beliefs, moreover, must be held in service of a goal. No one gets rationality credit for merely thinking true thoughts, like calculating the digits of π or cranking out the logical implications of a proposition ("Either $1 + 1 = 2$ or the moon is made of cheese," "If $1 + 1 = 3$, then pigs can fly"). A rational agent must have a *goal*, whether it is to ascertain the truth of a noteworthy idea, called theoretical reason, or to bring about a noteworthy outcome in the world, called practical reason ("what is true" and "what to do"). Even the humdrum rationality of seeing rather than hallucinating is in the service of the ever-present goal built into our visual systems of knowing our surroundings.

A rational agent, moreover, must attain that goal not by doing something that just happens to work there and then, but by using whatever knowledge is applicable to the circumstances. Here is how William James distinguished a rational entity from a nonrational one that would at first appear to be doing the same thing:

> Romeo wants Juliet as the filings want the magnet; and if no obstacles intervene he moves toward her by as straight a line as they. But Romeo and Juliet, if a wall be built between them, do not remain idiotically pressing their faces against its opposite sides like the magnet and the filings with the card. Romeo soon finds a circuitous way, by scaling the wall or otherwise, of touching Juliet's lips directly. With the filings the path is fixed; whether it reaches the end depends on accidents. With the lover it is the end which is fixed; the path may be modified indefinitely.[2]

With this definition the case for rationality seems all too obvious: do you want things or don't you? If you do, rationality is what allows you to get them.

Now, this case for rationality is open to an objection. It advises us to ground our beliefs in the truth, to ensure that our inference from one belief to another is justified, and to make plans that are likely to bring about a given end. But that only raises further questions. What *is* "truth"? What makes an inference "justified"? How do we know that means can be found that really do bring about a given end? But the quest to provide the ultimate, absolute, final reason for reason is a fool's errand. Just as an inquisitive three-year-old will reply to every answer to a "why" question with another "Why?," the quest to find the ultimate reason for reason can always be stymied by a demand to provide a reason for the reason for the reason. Just because I believe P implies Q, and I believe P, why should I believe Q? Is it because I also believe [(P implies Q) and P] implies Q? But why should I believe *that*? Is it because I have still another belief, {[(P implies Q) and P] implies Q} implies Q?

This regress was the basis for Lewis Carroll's 1895 story "What the Tortoise Said to Achilles," which imagined the conversation that would unfold when the fleet-footed warrior caught up to (but could never overtake) the tortoise with the head start in Zeno's second paradox. (In the time it took for Achilles to close the gap, the tortoise moved on, opening up a new gap for Achilles to close, ad infinitum.) Carroll was a logician as well as a children's author, and in this article, published in the philosophy journal *Mind*, he imagines the warrior seated on the tortoise's back and responding to the tortoise's escalating demands to justify his arguments by filling up a notebook with thousands of rules for rules for rules.[3] The moral is that reasoning with logical rules at some point must simply be *executed* by a mechanism that is hardwired into the machine or brain and runs because that's how the circuitry works, not because it consults a rule telling it what to do. We program apps into a computer, but its CPU is not itself an app; it's a piece of silicon in which elementary operations like comparing symbols and adding numbers have been burned. Those operations are designed (by an engineer, or in the case of the brain by natural selection) to implement laws of logic and mathematics that are inherent to the abstract realm of ideas.[4]

Now, Mr. Spock notwithstanding, logic is not the same thing as reasoning, and in the next chapter we'll explore the differences. But they are closely related, and the reasons the rules of logic can't be executed by still more rules of logic (ad infinitum) also apply to the justification of reason by still more reason. In each case the ultimate rule has to be "Just do it." At the end of the day the discussants have no choice but to commit to reason, because that's what they committed themselves to at the beginning of the day, when they opened up a discussion of why we should follow reason. As long as people are arguing and persuading and then evaluating and accepting or rejecting the arguments—as opposed to, say, bribing or threatening each other into mouthing some words—it's too late to ask about the value of reason. They're already reasoning, and have tacitly accepted its value.

When it comes to arguing against reason, as soon as you show up, you lose. Let's say you argue that rationality is unnecessary. Is *that* statement rational? If you concede it isn't, then there's no reason for me to believe it—you just said so yourself. But if you insist I must believe it because the statement is rationally compelling, you've conceded that rationality is the measure by which we should accept beliefs, in which case that particular one must be false. In a similar way, if you were to claim that everything is subjective, I could ask, "Is *that* statement subjective?" If it is, then you are free to believe it, but I don't have to. Or suppose you claim that everything is relative. Is *that* statement relative? If it is, then it may be true for you right here and now but not for anyone else or after you've stopped talking. This is also why the recent cliché that we're living in a "post-truth era" cannot be true. If it were true, then it would not be true, because it would be asserting something true about the era in which we are living.

This argument, laid out by the philosopher Thomas Nagel in *The Last Word*, is admittedly unconventional, as any argument about argument itself would have to be.[5] Nagel compared it to Descartes's argument that our own existence is the one thing we cannot doubt, because the very fact of wondering whether we exist presupposes the existence of a wonderer. The very fact of interrogating the concept of reason using reason presupposes the validity of reason. Because of this unconventionality, it's not quite right to say that we should "believe in" reason or "have faith in" reason. As Nagel points out, that's "one thought too many." The masons (and the Masons) got it right: we should *follow* reason.

Now, arguments for truth, objectivity, and reason may stick in the craw, because they seem dangerously arrogant: "Who the hell are *you* to claim to have the absolute truth?" But that's not what the case for rationality is about. The psychologist David Myers has said that the essence of monotheistic belief is: (1) There is a God and (2) it's not me (and it's also not you).[6] The secular equivalent is: (1) There is objective truth and (2) I don't know it (and neither do you). The same epistemic humility applies to the rationality that leads to truth. Perfect rationality and objective truth are aspirations that no mortal can ever claim to have attained. But the conviction that they are out there licenses us to develop rules we can all abide by that allow us to approach the truth collectively in ways that are impossible for any of us individually.

The rules are designed to sideline the biases that get in the way of rationality: the cognitive illusions built into human nature, and the bigotries, prejudices, phobias, and -isms that infect the members of a race, class, gender, sexuality, or civilization. These rules include the principles of critical thinking and the normative systems of logic, probability, and empirical reasoning that will be explained in the chapters to come. They are implemented among flesh-and-blood people by social institutions that prevent people from imposing their egos or biases or delusions on everyone else. "Ambition must be made to counteract ambition," wrote James Madison about the checks and balances in a democratic government, and that is how other institutions steer communities of biased and ambition-addled people toward disinterested truth. Examples include the adversarial system in law, peer review in science, editing and fact-checking in journalism, academic freedom in universities, and freedom of speech in the public sphere. Disagreement is necessary in deliberations among mortals. As the saying goes, the more we disagree, the more chance there is that at least one of us is right.

THOUGH WE CAN NEVER *PROVE* that reasoning is sound or the truth can be known (since we would need to assume the soundness of reason to do it), we can stoke our confidence that they are. When we apply reason to reason itself, we find that it is not just an inarticulate gut impulse, a mysterious oracle that whispers truths into our ear. We can expose the rules of reason and distill and purify them into normative models of logic and probability. We can even implement them in machines that duplicate and exceed our own rational powers. Computers are literally mechanized logic, their smallest circuits called logic gates.

Another reassurance that reason is valid is that it *works*. Life is not a dream, in which we pop up in disconnected locations and bewildering things happen without rhyme or reason. By scaling the wall, Romeo really does get to touch Juliet's lips. And by deploying reason in other ways, we reach the moon, invent smartphones, and extinguish smallpox. The cooperativeness of the world when we apply reason to it is a strong indication that rationality really does get at objective truths.

And ultimately even relativists who deny the possibility of objective truth and insist that all claims are merely the narratives of a culture lack the courage of their convictions. The cultural anthropologists or literary scholars who avow that the truths of science are merely the narratives of one culture will still have their child's infection treated with antibiotics prescribed by a physician rather than a healing song performed by a shaman. And though relativism is often adorned with a moral halo, the moral convictions of relativists depend on a commitment to objective truth. Was slavery a myth? Was the Holocaust just one of many possible narratives? Is climate change a social construction? Or are the suffering and danger that define these events really real—claims that we know are true because of logic and evidence and objective scholarship? Now relativists stop being so relative.

For the same reason there can be no tradeoff between rationality and social justice or any other moral or political cause. The quest for social justice begins with the belief that certain groups are oppressed and others privileged. These are factual claims and may be mistaken (as advocates of social justice themselves insist in response to the claim that it's straight white men who are oppressed). We affirm these beliefs because reason and evidence suggest they are true. And the quest in turn is guided by the belief that certain measures are necessary to rectify those injustices. Is leveling the playing field enough? Or have past injustices left some groups at a disadvantage that can only be set right by compensatory policies? Would particular measures merely be feel-good signaling that leaves the oppressed groups no better off? Would they make matters worse? Advocates of social justice need to know the answers to these questions, and reason is the only way we can know anything about anything.

Admittedly, the peculiar nature of the argument for reason always leaves open a loophole. In introducing the case for reason, I wrote, "As long as people are arguing and persuading . . . ," but that's a big "as long as." Rationality rejecters can refuse to play the game. They can say, "I don't have to justify my beliefs to you. Your demands for arguments and evidence show that you are part of the problem." Instead of feeling any need to persuade, people who are certain they are correct can impose their beliefs by force. In theocracies and autocracies, authorities censor, imprison, exile, or burn those with the wrong opinions. In democracies the force is less brutish, but people still find means to impose a belief rather than argue for it. Modern universities—oddly enough, given that their mission is to evaluate ideas—have been at the forefront of finding ways to suppress opinions, including disinviting and drowning out speakers, removing controversial teachers from the classroom, revoking offers of jobs and support, expunging contentious articles from archives, and classifying differences of opinion as punishable harassment and discrimination.[7] They respond as Ring Lardner recalled his father doing when the writer was a boy: "'Shut up,' he explained."

If you know you are right, why *should* you try to persuade others through reason? Why not just strengthen solidarity within your coalition and mobilize it to fight for justice? One reason is that you would be inviting questions such as: Are you infallible? Are you *certain* that you're right about *everything*? If so, what makes you different from your opponents, who also are certain they're right? And from authorities throughout history who insisted they were right but who we now know were wrong? If you have to silence people who disagree with you, does that mean you have no good arguments for why they're mistaken? The incriminating lack of answers to such questions could alienate those who have not taken sides, including the generations whose beliefs are not set in stone.

And another reason not to blow off persuasion is that you will have left those who disagree with you no choice but to join the game you are playing and counter *you* with force rather than argument. They may be stronger than you, if not now then at some time in the future. At that point, when you are the one who is canceled, it will be too late to claim that your views should be taken seriously because of their merits.

Stop Making Sense?

Must we *always* follow reason? Do I need a rational argument for why I should fall in love, cherish my children, enjoy the pleasures of life? Isn't it sometimes OK to go crazy, to be silly, to stop making sense? If rationality is so great, why do we associate it with a dour joylessness? Was the philosophy professor in Tom Stoppard's play *Jumpers* right in his response to the claim that "the Church is a monument to irrationality"?

> The National Gallery is a monument to irrationality! Every concert hall is a monument to irrationality! And so is a nicely kept garden, or a lover's favour, or a home for stray dogs! . . . If rationality were the criterion for things being allowed to exist, the world would be one gigantic field of soya beans![8]

The rest of this chapter takes up the professor's challenge. We will see that while beauty and love and kindness are not literally rational, they're not exactly irrational, either. We can apply reason to our emotions and to our morals, and there is even a higher-order rationality that tells us when it can be rational to be irrational.

Stoppard's professor may have been misled by David Hume's famous argument that "reason is, and ought only to be the slave of the passions, and can never pretend to any other office than to serve and obey them."[9] Hume, one of the hardest-headed philosophers in the history of Western thought, was not advising his readers to shoot from the hip, live for the moment, or fall head over heels for Mr. Wrong.[10] He was making the logical point that reason is the means to an end, and cannot tell you what the end should be, or even that you must pursue it. By "passions" he was referring to the source of those ends: the likes, wants, drives, emotions, and feelings wired into us, without which reason would have no goals to figure out how to attain. It's the distinction between thinking and wanting, between believing something you hold to be true and desiring something you wish to bring about. His point was closer to "There's no disputing tastes" than "If it feels good, do it."[11] It is neither rational nor irrational to prefer chocolate ripple to maple walnut. And it is in no way irrational to keep a garden, fall in love, care for stray dogs, party like it's 1999, or dance beneath the diamond sky with one hand waving free.[12]

Still, the impression that reason can oppose the emotions must come from somewhere—surely it is not just a logical error. We keep our distance from hotheads, implore people to be reasonable, and regret various flings, outbursts, and acts of thoughtlessness. If Hume was right, how can the opposite of what he wrote also be true: that the *passions* must often be slaves to *reason*?

In fact, it's not hard to reconcile them. One of our goals can be incompatible with the others. Our goal at one time can be incompatible with our goals at other times. And one person's goals can be incompatible with others'. With those conflicts, it won't do to say that we should serve and obey our passions. Something has to give, and

that is when rationality must adjudicate. We call the first two applications of reason "wisdom" and the third one "morality." Let's look
at each.

Conflicts among Goals

People don't want just one thing. They want comfort and pleasure,
but they also want health, the flourishing of their children, the es-
teem of their fellows, and a satisfying narrative on how they have
lived their lives. Since these goals may be incompatible—cheesecake
is fattening, unattended kids get into trouble, and cutthroat ambition earns contempt—you can't always get what you want. Some
goals are more important than others: the satisfaction deeper, the
pleasure longer lasting, the narrative more compelling. We use our
heads to prioritize our goals and pursue some at the expense of
others.

Indeed, some of our apparent goals are not even really *our* goals—
they are the metaphorical goals of our genes. The evolutionary process selects for genes that lead organisms to have as many surviving
offspring as possible in the kinds of environments in which their
ancestors lived. They do so by giving us motives like hunger, love,
fear, comfort, sex, power, and status. Evolutionary psychologists call
these motives "proximate," meaning that they enter into our conscious experience and we deliberately try to carry them out. They
can be contrasted with the "ultimate" motives of survival and repro-
duction, which are the figurative goals of our genes—what they
would say they wanted if they could talk.[13]

Conflicts between proximate and ultimate goals play out in our
lives as conflicts between different proximate goals. Lust for an attractive sexual partner is a proximate motive, whose ultimate motive

is conceiving a child. We inherited it because our more lustful ancestors, on average, had more offspring. However, conceiving a child may not be among our proximate goals, and so we may deploy our reason to foil that ultimate goal by using contraception. Having a trusted romantic partner we don't betray and maintaining the respect of our peers are other proximate goals, which our rational faculties may pursue by advising our not-so-rational faculties to avoid dangerous liaisons. In a similar way we pursue the proximate goal of a slim, healthy body by overriding another proximate goal, a delicious dessert, which itself arose from the ultimate goal of hoarding calories in an energy-stingy environment.

When we say someone's acting emotionally or irrationally, we're often alluding to bad choices in these tradeoffs. It often feels great in the heat of the moment to blow your stack when someone has crossed you. But our cooler head may realize that it's better to put a lid on it, to achieve things that make us feel even greater in the long run, like a good reputation and a trusting relationship.

Conflicts among Time Frames

Since not everything happens at once, conflicts between goals often involve goals that are realized at different times. And these in turn often feel like conflicts between different selves, a present self and a future self.[14]

The psychologist Walter Mischel captured the conflict in an agonizing choice he gave four-year-olds in a famous 1972 experiment: one marshmallow now or two marshmallows in fifteen minutes.[15] Life is a never-ending gantlet of marshmallow tests, dilemmas that force us to choose between a sooner small reward and a later large reward. Watch a movie now or pass a course later; buy a bauble now or pay the rent later; enjoy five minutes of fellatio now or an unblemished record in the history books later.

The marshmallow dilemma goes by several names, including self-control, delay of gratification, time preference, and discounting the future.[16] It figures into any analysis of rationality because it helps explain the misconception that too much rationality makes for a cramped and dreary life. Economists have studied the normative grounds for self-control—when we *ought* to indulge now or hold off for later—since it is the basis for interest rates, which compensate people for giving up money now in exchange for money later. They have reminded us that often the rational choice is to indulge now: it all depends on when and how much. In fact this conclusion is already a part of our folk wisdom, captured in aphorisms and jokes.

First, a bird in the hand is worth two in the bush. How do you know that the experimenter will keep his promise and reward you for your patience with two marshmallows when the time comes? How do you know that the pension fund will still be solvent when you retire and the money you have put away for retirement will be available when you need it? It's not just the imperfect integrity of trustees that can punish delay of gratification; it's the imperfect knowledge of experts. "Everything they said was bad for you is good for you," we joke, and with today's better nutrition science we know that a lot of pleasure from eggs, shrimp, and nuts was forgone in past decades for no good reason.

Second, in the long run we're all dead. You could be struck by lightning tomorrow, in which case all the pleasure you deferred to next week, next year, or next decade will have gone to waste. As the bumper sticker advises, "Life is short. Eat dessert first."

Third, you're only young once. It may cost more overall to take out a mortgage in your thirties than to save up and pay cash for a house in your eighties, but with the mortgage you get to live in it all those years. And the years are not just more numerous but different. As my doctor once said to me after a hearing test, "The great tragedy in life is that when you're old enough to afford really good audio equipment, you can't hear the difference." This cartoon makes a similar point:

"See, the problem with doing things to prolong your life is that all the extra years come at the end, when you're old."

These arguments are combined in a story. A man is sentenced to be hanged for offending the sultan, and offers a deal to the court: if they give him a year, he will teach the sultan's horse to sing, earning his freedom. When he returns to the dock, a fellow prisoner says, "Are you crazy? You're only postponing the inevitable. In a year there will be hell to pay." The man replies, "I figure over a year, a lot can happen. Maybe the sultan will die, and the new sultan will pardon me. Maybe I'll die; in that case I would have lost nothing. Maybe the horse will die; then I'll be off the hook. And who knows? Maybe I'll teach the horse to sing!"

Does this mean it's rational to eat the marshmallow now after all?

Not quite—it depends on how long you have to wait and how many marshmallows you get for waiting. Let's put aside aging and other changes and assume for simplicity's sake that every moment is the same. Suppose that every year there's a 1 percent chance that you'll be struck by a bolt of lightning. That means there's a .99 chance that you'll be alive in a year. What are the chances that you'll be alive in two years? For that to be true, you will have had to escape the lightning bolt for a second year, with an overall probability of .99 × .99, that is, $.99^2$ or .98 (we'll revisit the math in chapter 4). Three years, .99 × .99 × .99, or $.99^3$ (.97); ten years, $.99^{10}$ (.90); twenty years, $.99^{20}$ (.82), and so on—an exponential drop. So, taking into account the possibility that you will never get to enjoy it, a marshmallow in the hand is worth one and one ninth marshmallows in the decade-hence bush. Additional hazards—a faithless experimenter, the possibility you will lose your taste for marshmallows—change the numbers but not the logic. It's rational to discount the future *exponentially*. That is why the experimenter has to promise to reward your patience with more marshmallows the longer you wait—to pay interest. And the interest compounds exponentially, compensating for the exponential decay in what the future is worth to you now.

This in turn means that there are two ways in which living for the present can be irrational. One is that we can discount a future reward too steeply—put too low a price on it given how likely we are to live to see it and how much enjoyment it will bring. The impatience can be quantified. Shane Frederick, inventor of the Cognitive Reflection Test from the previous chapter, presented his respondents with hypothetical marshmallow tests using adult rewards, and found that a majority (especially those who fell for the seductive wrong answers on the brainteasers) preferred $3,400 then and there to $3,800 a month later, the equivalent of forgoing an investment with a 280 percent annual return.[17] In real life, about half of Americans nearing retirement age have saved *nothing* for retirement: they've planned their lives as if they would be dead by then (as most of our ancestors in fact were).[18] As Homer Simpson said to Marge when she warned him that he would regret his conduct, "That's a problem for future Homer. Man, I don't envy that guy."

The optimal rate at which to discount the future is a problem that we face not just as individuals but as societies, as we decide how much public wealth we should spend to benefit our older selves and future generations. Discount it we must. It's not only that a current sacrifice would be in vain if an asteroid sends us the way of the dinosaurs. It's also that our ignorance of what the future will bring, including advances in technology, grows exponentially the farther out we plan. (Who knows? Maybe we'll teach the horse to sing.) It would have made little sense for our ancestors a century ago to have scrimped for our benefit—say, diverting money from schools and roads to a stockpile of iron lungs to prepare for a polio epidemic—given that we're six times richer and have solved some of their problems while facing new ones they could not have dreamed of. At the same time we can curse some of their shortsighted choices whose consequences we are living with, like despoiled environments, extinct species, and car-centered urban planning.

The public choices we face today, like how high a tax we should pay on carbon to mitigate climate change, depend on the rate at which we discount the future, sometimes called the social discounting rate.[19] A rate of 0.1 percent, which reflects only the chance we'll go extinct, means that we value future generations almost as much as ourselves and calls for investing the lion's share of our current income to boost the well-being of our descendants. A rate of 3 percent, which assumes growing knowledge and prosperity, calls for deferring most of the sacrifice to generations that can better afford it. There is no "correct" rate, since it also depends on the moral choice of how we weight the welfare of living people against unborn ones.[20] But our awareness that politicians respond to election cycles rather than the long term, and our sad experience of finding ourselves un-prepared for foreseeable disasters like hurricanes and pandemics, suggest that our social discounting rate is irrationally high.[21] We leave problems to future Homer, and don't envy that guy.

There's a second way in which we irrationally cheat our future selves, called myopic discounting.[22] Often we're perfectly capable of delaying gratification from a future self to an even more future self. When a conference organizer sends out a menu for the keynote din-ner in advance, it's easy to tick the boxes for the steamed vegetables and fruit rather than the lasagna and cheesecake. The small pleasure of a rich dinner in 100 days versus the large pleasure of a slim body in 101 days? No contest! But if the waiter were to tempt us with the same choice then and there—the small pleasure of a rich dinner in fifteen minutes versus the large pleasure of a slim body tomorrow—we flip our preference and succumb to the lasagna.

The preference reversal is called myopic, or nearsighted, because we see an attractive temptation that is near to us in time all too clearly, while the faraway choices are emotionally blurred and (a bit contrary to the ophthalmological metaphor) we judge them more objectively. The rational process of exponential discounting, even if the discounting rate is unreasonably steep, cannot explain the flip, because if a small imminent reward is more enticing than a large later one, it will still be more enticing when both rewards are pushed into the future. (If lasagna is more enticing than steamed vegetables now, the prospect of lasagna several months from now should be more enticing than the prospect of vegetables several months from now.) Social scientists say that a preference reversal shows that the discounting is *hyperbolic*—not in the sense of being exaggerated, but of falling along a curve called a hyperbola, which is more L-shaped than an exponential drop: it begins with a steep plunge and then levels off. Two exponential curves at different heights never cross (more tempting now, more tempting always); two hyperbolic curves can. The graphs on the next page show the difference. (Note that they plot absolute time as it is marked on a clock or calendar, not time relative to now, so the self who is experiencing things right now is gliding along the horizontal axis, and the discounting is shown in the curves from right to left.)

Admittedly, explaining weakness of the will as a reward gets closer by hyperbolic discounting is like explaining the effect of Ambien by its dormitive power. But the elbow shape of a hyperbola suggests that it may really be a composite of two curves, one plotting the irresistible pull of a treat that you can't get out of your head (the bakery smell, the come-hither look, the glitter in the showroom), the other plotting a cooler assessment of costs and benefits in a hypothetical future. Studies that tempt volunteers in a scanner with adult versions of marshmallow tests confirm that different brain patterns are activated by thoughts of imminent and distant goodies.[23]

Though hyperbolic discounting is not rational in the way that calibrated exponential discounting can be (since it does not capture the ever-compounding uncertainty of the future), it does provide an opening for the rational self to outsmart the impetuous self. The opening may be seen in the leftmost segment of the hyperbolas, the time when both rewards lie far off in the future, during which the large reward is subjectively more appealing than the small one (as rationally it should be). Our calmer selves, well aware of what will happen as the clock ticks down, can chop off the right half of the graph, never allowing the switchover to temptation to arrive. The trick was explained by Circe to Odysseus:[24]

Exponential Discounting

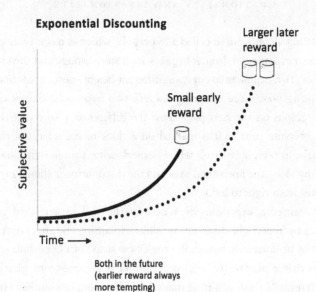

Larger later reward

Small early reward

Subjective value

Time ⟶

Both in the future
(earlier reward always
more tempting)

Hyperbolic Discounting

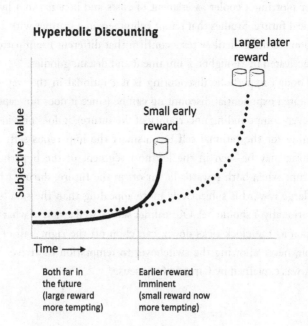

Larger later reward

Small early reward

Subjective value

Time ⟶

Both far in
the future
(large reward
more tempting)

Earlier reward
imminent
(small reward now
more tempting)

First you will reach the Sirens, who bewitch
all passersby. If anyone goes near them
in ignorance, and listens to their voices,
that man will never travel to his home,
and never make his wife and children happy
to have him back with them again. The Sirens
who sit there in their meadow will seduce him
with piercing songs. Around about them lie
great heaps of men, flesh rotting from their bones,
their skin all shriveled up. Use wax to plug
your sailors' ears as you row past, so they
are deaf to them. But if you wish to hear them,
your men must fasten you to your ship's mast
by hand and foot, straight upright, with tight ropes.
So bound, you can enjoy the Sirens' song.

The technique is called Odyssean self-control, and it is more effective than the strenuous exertion of willpower, which is easily overmatched in the moment by temptation.[25] During the precious interlude before the Sirens' song comes into earshot, our rational faculties preempt any possibility that our appetites will lure us to our doom by tying us to the mast with tight ropes, cutting off the option to succumb. We shop when we are sated and pass by the chips and cakes that would be irresistible when we are hungry. We instruct our employers to tithe our paycheck and set aside a portion for retirement so there's no surplus at the end of the month to blow on a vacation.

In fact, Odyssean self-control can step up a level and cut off the option to have the option, or at least make it harder to exercise. Suppose the thought of a full paycheck is so tempting that we can't bring ourselves to fill out the form that authorizes the monthly deduction.

Before being faced with *that* temptation, we might allow our employers to make the choice for us (and other choices that benefit us in the long run) by enrolling us in mandatory savings by default: we would have to take steps to opt out of the plan rather than to opt in. This is the basis for the philosophy of governance whimsically called libertarian paternalism by the legal scholar Cass Sunstein and the behavioral economist Richard Thaler in their book *Nudge*. They argue that it is rational for us to empower governments and businesses to fasten us to the mast, albeit with loose ropes rather than tight ones. Informed by research on human judgment, experts would engineer the "choice architecture" of our environments to make it difficult for us to do tempting harmful things, like consumption, waste, and theft. Our institutions would paternalistically act as if they know what's best for us, while leaving us the liberty to untie the ropes when we are willing to make the effort (which in fact few people exercise).

Libertarian paternalism, together with other "behavioral insights" drawn from cognitive science, has become increasingly popular among policy analysts, because it promises more effective outcomes at little cost and without impinging on democratic principles. It may be the most important practical application of research on cognitive biases and fallacies so far (though the approach has been criticized by other cognitive scientists who argue that humans are more rational than that research suggests).[26]

Rational Ignorance

While Odysseus had himself tied to the mast and rationally relinquished his option to *act*, his sailors plugged their ears with wax and rationally relinquished his option to *know*. At first this seems puzzling. One might think that knowledge is power, and you can never know too much. Just as it's better to be rich than poor, because if you're rich you can always give away your money and be poor, you might think it's always better to know something, because you can always choose not to act on it. But in one of the paradoxes of rationality, that turns out not to be true. Sometimes it really is rational to plug your ears with wax.[27] Ignorance can be bliss, and sometimes what you don't know can't hurt you.

An obvious example is the spoiler alert. We take pleasure in watching a plot unfold, including the suspense, climax, and denouement, and may choose not to spoil it by knowing the ending in advance. Sports fans who cannot see a match in real time and plan to watch a recorded version later will sequester themselves from all media and even from fellow fans who might leak the outcome in a subtle tell. Many parents choose not to learn the sex of their unborn child to enhance the joy of the moment of birth. In these cases we rationally choose ignorance because we know how our own involuntary positive emotions work, and we arrange events to enhance the pleasure they give us.

By the same logic, we can understand our negative emotions and starve ourselves of information that we anticipate would give us pain. Many consumers of genetic testing know they would be better off remaining ignorant of whether the man who calls himself their father is biologically related to them. Many choose not to learn whether they have inherited a dominant gene for an incurable disease that killed a parent, like the musician Arlo Guthrie, whose father, Woody, died of Huntington's. There's nothing they can do about it, and knowledge of an early and awful death would put a pall over the rest of their lives. For that matter most of us would plug our ears if an oracle promised to tell us the day we will die.

We also preempt knowledge that would bias our cognitive faculties. Juries are forbidden to see inadmissible evidence from hearsay, forced confessions, or warrantless searches—"the tainted fruit of the poisoned tree"—because human minds are incapable of ignoring it. Good scientists think the worst of their own objectivity and conduct their studies double blind, choosing not to know which patients got the drug and which the placebo. They submit their papers to anonymous peer review, removing any temptation to retaliate after a bad one, and, with some journals, redact their names, so the reviewers can't indulge the temptation to repay favors or settle scores.

In these examples, rational agents choose to be ignorant to game their own less-than-rational biases. But sometimes we choose to be ignorant to prevent our rational faculties from being exploited by rational adversaries—to make sure they cannot make us an offer we can't refuse. You can arrange not to be home when the mafia wiseguy calls with a threat or the deputy tries to serve you with a subpoena. The driver of a Brink's truck is happy to have his ignorance proclaimed on the sticker "Driver does not know combination to safe," because a robber cannot credibly threaten him to divulge it. A hostage is better off if he does not see the faces of his captors, because that leaves them an incentive to release him. Even misbehaving young children know they're better off not meeting their parents' glares.

Rational Incapacity and Rational Irrationality

Rational ignorance is an example of the mind-bending paradoxes of reason explained by the political scientist Thomas Schelling in his 1960 classic *The Strategy of Conflict*.[28] In some circumstances it can be rational to be not just ignorant but powerless, and, most perversely of all, irrational.

In the game of Chicken, made famous in the James Dean classic *Rebel Without a Cause*, two teenage drivers approach each other at high speed on a narrow road and whoever swerves first loses face (he is the "chicken").[29] Since each one knows that the other does not want to die in a head-on crash, each may stay the course, knowing the other has to swerve first. Of course when both are "rational" in this way, it's a recipe for disaster (a paradox of game theory we will return to in chapter 8). So is there a strategy that wins at Chicken? Yes—relinquish your ability to swerve by conspicuously locking the steering wheel, or by putting a brick on the gas pedal and climbing into the back seat, leaving the other guy no choice but to swerve. The player who lacks control wins. More precisely, the *first* player to lack control wins: if both lock their wheels simultaneously . . .

Though the game of Chicken may seem like the epitome of teenage foolishness, it's a common dilemma in bargaining, both in the marketplace and in everyday life. Say you're willing to pay up to $30,000 for a car and know that it cost the dealer $20,000. Any price between $20,000 and $30,000 works to both of your advantages, but of course you want it to be as close as possible to the lower end of the range and the sales rep to the upper end. You could lowball him, knowing he's better off consummating the deal than walking away, but he could highball you, knowing the same thing. So he agrees that your offer is reasonable but needs the OK from his manager, but when he comes back he says regretfully that the manager is a hard-ass who nixed the deal. Alternatively, you agree that the price is reasonable but you need the OK from your bank, and the loan officer refuses to lend you that much. The winner is the one whose hands are tied. The same can happen in friendships and marriages in which both partners would rather do something together than stay home, but differ in what they most enjoy. The partner with the superstition or hang-up or maddeningly stubborn personality that categorically rules out the other's choice will get his or her own.

Threats are another arena in which a lack of control can afford a paradoxical advantage. The problem with threatening to attack, strike, or punish is that the threat may be costly to carry out, rendering it a bluff that the target of the threat could call. To make it credible, the threatener must be committed to carrying it out, forfeiting the control that would give his target the leverage to threaten him right back by refusing to comply. A hijacker who wears an explosive belt that goes off with the slightest jostle, or protesters who chain themselves to the tracks in front of a train carrying fuel to a nuclear plant, cannot be scared away from their mission.

The commitment to carry out a threat can be not just physical but emotional.[30] The narcissist, borderline, hothead, high-maintenance romantic partner, or "man of honor" who considers it an intolerable affront to be disrespected and lashes out regardless of the consequences is someone you don't want to mess with.

A lack of control can blend into a lack of rationality. Suicide terrorists who believe they will be rewarded in paradise cannot be deterred by the prospect of death on earth. According to the Madman Theory in international relations, a leader who is seen as impetuous, even unhinged, can coerce an adversary into concessions.[31] In 1969 Richard Nixon reportedly ordered nuclear-armed bombers to fly recklessly close to the USSR to scare them into pressuring their North Vietnamese ally to negotiate an end to the Vietnam War. Donald Trump's bluster in 2017 about using his bigger nuclear button to rain fire and fury on North Korea could charitably be interpreted as a revival of the theory.

The problem with the madman strategy, of course, is that both sides can play it, setting up a catastrophic game of Chicken. Or the threatened side may feel it has no choice but to take out the madman by force rather than continue a fruitless negotiation. In everyday life, the saner party has an incentive to bail out of a relationship with a madman or madwoman and deal with someone more reasonable. These are reasons why we are not all madpeople all the time (though some of us get away with it some of the time).

Promises, like threats, have a credibility problem that can call for a surrender of control and of rational self-interest. How can a contractor convince a client that he will pay for any damage, or a borrower convince a lender that she will repay a loan, when they have every incentive to renege when the time comes? The solution is to post a bond that they would forfeit, or sign a note that empowers the creditor to repossess the house or car. By signing away their options, they become trustworthy partners. In our personal lives, how do we convince an object of desire that we will forsake all others till death do us part, when someone even more desirable may come along at any time? We can advertise that we are incapable of rationally choosing someone better because we never rationally chose that person in the first place—our love was involuntary, irrational, and elicited by the person's unique, idiosyncratic, irreplaceable qualities.[32] I can't help falling in love with you. I'm crazy for you. I like the way you walk, I like the way you talk.

The paradoxical rationality of irrational emotion is endlessly thought-provoking and has inspired the plots of tragedies, Westerns, war movies, mafia flicks, spy thrillers, and the Cold War classics *Fail Safe* and *Dr. Strangelove*. But nowhere was the logic of illogic more pithily stated than in the 1941 film noir *The Maltese Falcon*, when detective Sam Spade dares Kasper Gutman's henchmen to kill him, knowing they need him to find the jewel-encrusted falcon. Gutman replies:

That's an attitude, sir, that calls for the most delicate judgment on both sides, because as you know, sir, in the heat of
action men are likely to forget where their best interests lie,
and let their emotions carry them away.[33]

Taboo

Can certain thoughts be not just strategically compromising but evil
to think? This is the phenomenon called *taboo*, from a Polynesian
word for "forbidden." The psychologist Philip Tetlock has shown
that taboos are not just customs of South Sea islanders but active in
all of us.[34]

Tetlock's first kind of taboo, the "forbidden base rate," arises from
the fact that no two groups of people—men and women, blacks and
whites, Protestants and Catholics, Hindus and Muslims, Jews and
gentiles—have identical averages on any trait one cares to measure.
Technically, those "base rates" could be plugged into actuarial formulas and guide predictions and policies pertaining to those groups. To
say that such profiling is fraught would be an understatement. We will
look at the morality of forbidden base rates in the discussion of Bayesian reasoning in chapter 5.

A second kind is the "taboo tradeoff." Resources are finite in life,
and tradeoffs unavoidable. Since not everyone values everything
equally, we can increase everyone's well-being by encouraging people to exchange something that is less valuable to them for something that is more valuable. But countering this economic fact is a
psychological one: people treat some resources as sacrosanct, and are
offended by the possibility that they may be traded for vulgar commodities like cash or convenience, even if everyone wins.

Organs for donation are an example.[35] No one needs both their kidneys, while a hundred thousand Americans desperately need one. That need is not filled either by posthumous donors (even when the state nudges them to consent by making donation the default) or by living altruists. If healthy donors were allowed to sell their kidneys (with the government providing vouchers to recipients who couldn't afford to pay), many people would be spared financial stress, many others would be spared disability and death, and no one would be worse off. Yet most people are not just opposed to this plan but offended by the very idea. Rather than providing arguments against it, they are insulted even to be asked. Switching the payoff from filthy lucre to wholesome vouchers (say, for education, health care, or retirement) softens the offense, but doesn't eliminate it. People are equally incensed when asked whether there should be subsidized markets for jury duty, military service, or children put up for adoption, ideas occasionally bruited by naughty libertarian economists.[36]

Taboo tradeoffs confront us not just in hypothetical policies but in everyday budgetary decisions. A dollar spent on health or safety—a pedestrian overpass, a toxic waste cleanup—is a dollar not spent on education or parks or museums or pensions. Yet editorialists are unembarrassed to make nonsensical proclamations like "No amount is too much to spend on X" or "We cannot put a price on Y" when it comes to sacred commodities like the environment, children, health care, or the arts, as if they were prepared to shut down schools to pay for sewage treatment plants or vice versa. Putting a dollar value on a human life is repugnant, but it's also unavoidable, because otherwise policymakers can spend profligate amounts on sentimental causes or pork-barrel projects that leave worse hazards untreated. When it comes to paying for safety, a human life in the United States is currently worth around $7–10 million (though planners are happy for

the price to be buried in dense technical documents). When it comes
to paying for health, the price is all over the map, one of the reasons
that America's health care system is so expensive and ineffective.

To show that merely *thinking* about taboo tradeoffs is perceived as
morally corrosive, Tetlock presented experimental participants with
the scenario of a hospital administrator faced with the choice of
spending a million dollars to save the life of a sick child or putting it
toward general hospital expenses. People condemned the administrator if he thought about it a lot rather than making a snap decision.
They made the opposite judgment, esteeming thought over reflex,
when the administrator grappled with a tragic rather than a taboo
tradeoff: whether to spend the money to save the life of one child or
the life of another.

The art of political rhetoric is to hide, euphemize, or reframe
taboo tradeoffs. Finance ministers can call attention to the lives a
budgetary decision will save and ignore the lives it costs. Reformers
can redescribe a transaction in a way that tucks the tit for tat in the
background: advocates for the women in red-light districts speak of
sex workers exercising their autonomy rather than prostitutes selling
their bodies; advertisers of life insurance (once taboo) describe the
policy as a breadwinner protecting a family rather than one spouse
betting that the other will die.[37]

Tetlock's third kind of taboo is the "heretical counterfactual."
Built into rationality is the ability to ponder what *would* happen if
some circumstance were *not* true. It's what allows us to think in abstract laws rather than the concrete present, to distinguish causation
from correlation (chapter 9). The reason we say the rooster does not
cause the sun to rise, even though one always follows the other, is
that if the rooster had *not* crowed, then the sun would still have risen.

Nonetheless, people often think it is immoral to let their minds
wander in certain make-believe worlds. Tetlock asked people, "What

if Joseph had abandoned Mary when Jesus was a child—would he have grown up as confident and charismatic?" Devout Christians refused to answer. Some devout Muslims are even touchier. When Salman Rushdie published *The Satanic Verses* in 1988, a novel containing a narrative that played out the life of Mohammad in a counterfactual world in which some of Allah's words really came from Satan, Iran's Ayatollah Khomeini issued a fatwa calling for his murder. Lest this mindset seem primitive and fanatical, try playing this game at your next dinner party: "Of course none of us would ever be unfaithful to our partners. But let's suppose, sheerly hypothetically, that we would. Who would be your adulterous paramour?" Or try this one: "Of course none of us is the least bit racist. But let's just say we were—which group would you be prejudiced against?" (A relative of mine was once dragged into this game and dumped her boyfriend after he replied, "Jews.")

How could it be rational to condemn the mere thinking of thoughts—an activity that cannot, by itself, impinge on the welfare of people in the world? Tetlock notes that we judge people not just by what they *do* but by who they *are*. A person who is capable of entertaining certain hypotheticals, even if the person has treated us well so far, might stab us in the back or sell us down the river were the temptation ever to arise. Imagine someone were to ask you: For how much money would you sell your child? Or your friendship, or citizenship, or sexual favors? The correct answer is to refuse to answer— better still, to be offended by the question. As with the rational handicaps in bargains, threats, and promises, a handicap in mental freedom can be an advantage. We trust those who are constitutionally incapable of betraying us or our values, not those who have merely chosen not to do so thus far.

Morality

One more realm that is sometimes excluded from the rational is the moral. Can we ever deduce what's right or wrong? Can we confirm it with data? It's not obvious how you could. Many people believe that "you can't get an *ought* from an *is*." The conclusion is sometimes attributed to Hume, with a rationale similar to his argument that reason must be a slave to the passions. "'Tis not contrary to reason," he famously wrote, "to prefer the destruction of the whole world to the scratching of my finger."[38] It's not that Hume was a callous sociopath. Turnabout being fair play, he continued, "'Tis not contrary to reason for me to chuse my total ruin, to prevent the least uneasiness of an Indian or person wholly unknown to me." Moral convictions would seem to depend on nonrational preferences, just like the other passions. This would jibe with the observation that what's considered moral and immoral varies across cultures, like vegetarianism, blasphemy, homosexuality, premarital sex, spanking, divorce, and polygamy. It also varies across historical periods within our own culture. In olden days a glimpse of stocking was looked on as something shocking.

Moral statements indeed must be distinguished from logical and empirical ones. Philosophers in the first half of the twentieth century took Hume's argument seriously and struggled with what moral statements could possibly mean if they are not about logic or empirical fact. Some concluded that "X is evil" means little more than "X is against the rules" or "I dislike X" or even "X, boo!"[39] Stoppard has fun with this in *Jumpers* when an inspector investigating a shooting is informed by the protagonist about a fellow philosopher's view that immoral acts are "not *sinful* but simply anti-social." The astonished inspector asks, "He thinks there's nothing *wrong* with killing people?" George replies, "Well, put like that, of course . . . But *philosophically*, he doesn't think it's actually, inherently wrong in itself, no."[40]

Like the incredulous inspector, many people are not ready to reduce morality to convention or taste. When we say "The Holocaust is bad," do our powers of reason leave us no way to differentiate that conviction from "I don't like the Holocaust" or "My culture disapproves of the Holocaust"? Is keeping slaves no more or less rational than wearing a turban or a yarmulke or a veil? If a child is deathly ill and we know of a drug that could save her, is administering the drug no more rational than withholding it?

Faced with this intolerable implication, some people hope to vest morality in a higher power. That's what religion is for, they say—even many scientists, like Stephen Jay Gould.[41] But Plato made short work of this argument 2,400 years ago in *Euthyphro*.[42] Is something moral because God commands it, or does God command some things because they are moral? If the former is true, and God had no reason for his commandments, why should we take his whims seriously? If God commanded you to torture and kill a child, would that make it right? "He would never do that!" you might object. But that flicks us onto the second horn of the dilemma. If God does have good reasons for his commandments, why don't we appeal to those reasons directly and skip the middleman? (As it happens, the God of the Old Testament did command people to slaughter children quite often.)[43]

In fact, it is not hard to ground morality in reason. Hume may have been technically correct when he wrote that it's not contrary to reason to prefer global genocide to a scratch on one's pinkie. But his grounds were very, very narrow. As he noted, it is *also* not contrary to reason to prefer bad things happening to oneself over good things—say, pain, sickness, poverty, and loneliness over pleasure, health, prosperity, and good company.[44] O-*kay*. But now let's just say—irrationally, whimsically, mulishly, for no good reason—that we prefer good things to happen to ourselves over bad things. Let's make a second wild and crazy assumption: that we are social animals who live with other people, rather than Robinson Crusoe on a desert island, so our well-being depends on what others do, like helping us when we are in need and not harming us for no good reason.

This changes everything. As soon as we start insisting to others, "You must not hurt me, or let me starve, or let my children drown," we cannot also maintain, "But I can hurt you, and let you starve, and let your children drown," and hope they will take us seriously. That is because as soon as I engage you in a rational discussion, I cannot insist that only my interests count just because I'm me and you're not, any more than I can insist that the spot I am standing on is a special place in the universe because I happen to be standing on it. The pronouns *I*, *me*, and *mine* have no logical heft—they flip with each turn in a conversation. And so any argument that privileges my well-being over yours or his or hers, all else being equal, is irrational.

When you combine self-interest and sociality with *impartiality*— the interchangeability of perspectives—you get the core of morality.[45] You get the Golden Rule, or the variants that take note of George Bernard Shaw's advice "Do not do unto others as you would have others do unto you; they may have different tastes." This sets up Rabbi Hillel's version, "What is hateful to you, do not do to your fellow." (That is the whole Torah, he said when challenged to explain it while the listener stood on one leg; the rest is commentary.) Versions of these rules have been independently discovered in Judaism, Christianity, Hinduism, Zoroastrianism, Buddhism, Confucianism, Islam, Baháí, and other religions and moral codes.[46] These include Spinoza's observation, "Those who are governed by reason desire nothing for themselves which they do not also desire for the rest of humankind." And Kant's Categorical Imperative: "Act only according to that maxim whereby you can at the same time will that it should become a universal law." And John Rawls's theory of justice: "The principles of justice are chosen behind a veil of ignorance" (about the particulars of one's life). For that matter the principle may be seen in the most fundamental statement of morality of all, the one we use to teach the concept to small children: "How would you like it if *he* did that to *you*?"

None of these statements depends on taste, custom, or religion. And though self-interest and sociality are not, strictly speaking, rational, they're hardly independent of rationality. How do rational agents come into existence in the first place? Unless you are talking about disembodied rational angels, they are products of evolution, with fragile, energy-hungry bodies and brains. To have remained alive long enough to enter into a rational discussion, they must have staved off injuries and starvation, goaded by pleasure and pain. Evolution, moreover, works on populations, not individuals, so a rational animal must be part of a community, with all the social ties that impel it to cooperate, protect itself, and mate. Reasoners in real life must be corporeal and communal, which means that self-interest and sociality are part of the package of rationality. And with self-interest and sociality comes the implication we call morality.

Impartiality, the main ingredient of morality, is not just a logical nicety, a matter of the interchangeability of pronouns. Practically speaking, it also makes everyone, on average, better off. Life presents many opportunities to help someone, or to refrain from hurting them, at a small cost to oneself (chapter 8). So if everyone signs on to helping and not hurting, everyone wins.[47] This does not, of course, mean that people are in fact perfectly moral, just that there's a rational argument as to why they should be.

Rationality about Rationality

Despite its lack of coolth, we should, and in many nonobvious ways do, follow reason. Merely asking why we should follow reason is confessing that we should. Pursuing our goals and desires is not the opposite of reason but ultimately the reason we have reason. We deploy reason to attain those goals, and also to prioritize them when they can't all be realized at once. Surrendering to desires in the moment is rational for a mortal being in an uncertain world, as long as future moments are not discounted too steeply or shortsightedly. When they are, our present rational self can outsmart a future, less rational self by restricting its choices, an example of the paradoxical rationality of ignorance, powerlessness, impetuousness, and taboo. And morality does not sit apart from reason but falls out of it as soon as the members of a self-interested social species impartially deal with the conflicting and overlapping desires among themselves.

All this rationalization of the apparently irrational may raise the worry that one could twist *any* quirk or perversity into revealing some hidden rationale. But the impression is untrue: sometimes the irrational is just the irrational. People can be mistaken or deluded about facts. They can lose sight of which goals are most important to them and how to realize them. They can reason fallaciously, or, more commonly, in pursuit of the wrong goal, like winning an argument rather than learning the truth. They can paint themselves into a corner, saw off the branch they're sitting on, shoot themselves in the foot, spend like a drunken sailor, play Chicken to the tragic end, stick their heads in the sand, cut off their nose to spite their face, and act as if they are the only ones in the world.

At the same time, the impression that reason always gets the last word is not unfounded. It's in the very nature of reason that it can always step back, look at how it is being applied or misapplied, and reason about that success or shortcoming. The linguist Noam Chomsky has argued that the essence of human language is *recursion*: a phrase can contain an example of itself without limit.[48] We can speak not only about my dog but about my mother's friend's husband's aunt's neighbor's dog; we can remark not only that she knows something, but that he knows that she knows it, and she knows that he knows that she knows it, ad infinitum. Recursive phrase structure is not just a way to show off. We would not have evolved the ability to speak phrases embedded in phrases if we did not have the ability to think thoughts embedded in thoughts.

And that is the power of reason: it can reason about itself. When something appears mad, we can look for a method to the madness. When a future self might act irrationally, a present self can outsmart it. When a rational argument slips into fallacy or sophistry, an even more rational argument exposes it. And if you disagree—if you think there is a flaw in this argument—it's reason that allows you to do so.

3

LOGIC AND CRITICAL THINKING

This modern type of the general reader may be known in conversation by the cordiality with which he assents to indistinct, blurred statements: say that black is black, he will shake his head and hardly think it; say that black is not so very black, he will reply, "Exactly." He has no hesitation ... to get up at a public meeting and express his conviction that at times, and within certain limits, the radii of a circle have a tendency to be equal; but, on the other hand, he would urge that the spirit of geometry may be carried a little too far.

—GEORGE ELIOT[1]

n the previous chapter, we asked why humans seem to be driven by what Mr. Spock called "foolish emotions." In this one we'll look at their irritating "illogic." The chapter is about logic, not in the loose sense of rationality itself but in the technical sense of inferring true statements (conclusions) from other true statements (premises). From the statements "All women are mortal" and "Xanthippe is a woman," for example, we can deduce "Xanthippe is mortal."

Deductive logic is a potent tool despite the fact that it can only draw out conclusions that are already contained in the premises (unlike inductive logic, the topic of chapter 5, which guides us in generalizing from evidence). Since people agree on many propositions—all women are mortal, the square of eight is sixty-four, rocks fall down and not up, murder is wrong—the goal of arriving at new, less obvious propositions is one we can all embrace. A tool with such power allows us to discover new truths about the world from the comfort of our armchairs, and to resolve disputes about the many things people don't agree on. The philosopher Gottfried Wilhelm Leibniz (1646–1716) fantasized that logic could bring about an epistemic utopia:

> The only way to rectify our reasonings is to make them as tangible as those of the Mathematicians, so that we can find our error at a glance, and when there are disputes among persons, we can simply say: Let us calculate, without further ado, to see who is right.[2]

You may have noticed that three centuries later we are still not resolving disputes by saying "Let us calculate." This chapter will explain why. One reason is that logic can be really hard, even for logicians, and it's easy to misapply the rules, leading to "formal fallacies." Another is that people often don't even try to play by the rules, and commit "informal fallacies." The goal of exposing these fallacies and coaxing people into renouncing them is called critical thinking. But a major reason we don't just calculate without further ado is that logic, like other normative models of rationality, is a tool that is suitable for seeking certain goals with certain kinds of knowledge, and it is unhelpful with others.

Formal Logic and Formal Fallacies

Logic is called "formal" because it deals not with the contents of statements but with their *forms*—the way they are assembled out of subjects, predicates, and logical words like AND, OR, NOT, ALL, SOME, IF, and THEN.[3] Often we apply logic to statements whose content we care about, such as "The President of the United States shall be removed from office on impeachment for, and conviction of, treason, bribery, or other high crimes and misdemeanors." We deduce that for a president to be removed, he must be not only impeached but also convicted, and that he need not be convicted of both treason and bribery; one is enough. But the laws of logic are general-purpose: they apply whether the content is topical, obscure, or even nonsensical. It was this point, and not mere whimsy, that led Lewis Carroll to create the "sillygisms" in his 1896 *Symbolic Logic* textbook, many of which are still used in logic courses today. For example, from the premises "A lame puppy would not say 'thank you' if you offered to lend it a skipping-rope" and "You offered to lend the lame puppy a skipping-rope," one may deduce "The puppy did not say 'thank you.'"[4]

Systems of logic are formalized as rules that allow one to deduce new statements from old statements by replacing some strings of symbols with others. The most elementary is called propositional calculus. *Calculus* is Latin for "pebble," and the term reminds us that logic consists of manipulating symbols mechanically, without pondering their content. Simple sentences are reduced to variables, like P and Q, which are assigned a truth value, TRUE or FALSE. Complex statements can be formed out of simple ones with the logical connectors AND, OR, NOT, and IF-THEN.

You don't even have to know what the connector words mean in

English. Their meaning consists only of rules that tell you whether a complex statement is true depending on whether the simple statements inside it are true. Those rules are stipulated in truth tables. The one on the left, which defines AND, can be interpreted line by line, like this: When P is TRUE and Q is TRUE, that means "P AND Q" is TRUE. When P is TRUE and Q is FALSE, that means "P AND Q" is FALSE. When P is FALSE . . . and so on for the last two lines.

P	Q	P AND Q	P	Q	P OR Q	P	NOT P
TRUE	TRUE	TRUE	TRUE	TRUE	TRUE	TRUE	FALSE
TRUE	FALSE	FALSE	TRUE	FALSE	TRUE	FALSE	TRUE
FALSE	TRUE	FALSE	FALSE	TRUE	TRUE		
FALSE	FALSE	FALSE	FALSE	FALSE	FALSE		

Let's take an example. In the meet-cute opening of the 1970 romantic tragedy *Love Story*, Jennifer Cavilleri explains to fellow Harvard student Oliver Barrett IV, whom she condescendingly called Preppy, why she assumes he went to prep school: "You look stupid and rich." Let's label "Oliver is stupid" as P and "Oliver is rich" as Q. The first line of the truth table for AND lays out the simple facts which have to be true for her conjunctive putdown to be true: that he's stupid, and that he's rich. He protests (not entirely honestly), "Actually, I'm smart and poor." Let's assume that "smart" means "NOT stupid" and "poor" means "NOT rich." We understand that Oliver is contradicting her by invoking the fourth line in the truth table: if he's not stupid and he's not rich, then he's not "stupid and rich." If all he wanted to do was contradict her, he could also have said, "Actually, I'm stupid and poor" (line 2) or "Actually, I'm smart and rich" (line 3). As it happens, Oliver is lying; he is not poor, which means that it's false for him to say he is "smart and poor."

Jenny replies, truthfully, "No, *I'm* smart and poor." Suppose we draw the cynical inference invited by the script that "Harvard students are rich OR smart." This inference is not a deduction but an induction—a fallible generalization from observation—but let's put aside how we got to that statement and look at the statement itself, asking what would make it true. It is a disjunction, a statement with an OR, and it may be verified by plugging our knowledge about the lovers-to-be into the truth table for OR (middle column), with P as "rich" and Q as "smart." Jenny is smart, even if she is not rich (line 3), and Oliver is rich, though he may or may not be smart (line 1 or 2), so the disjunctive statement about Harvard students, at least as it concerns these two, is true.

The badinage continues:

OLIVER: What makes you so smart?
JENNY: I wouldn't go for coffee with you.
OLIVER: I wouldn't ask you.
JENNY: That's what makes you stupid.

Let's fill out Jenny's answer as "If you asked me to have coffee, I would say 'no.'" Based on what we have been told, is the statement true? It is a *conditional*, a statement formed with an IF (the antecedent) and a THEN (the consequent). What is its truth table? Recall from the Wason selection task (chapter 1) that the only way for "IF P THEN Q" to be false is if P is true while Q is false. ("If a letter is labeled Express, it must have a ten-dollar stamp" means there can't be any Express letters without a ten-dollar stamp.) Here's the table:

P	Q	IF P THEN Q
TRUE	TRUE	TRUE
TRUE	FALSE	FALSE
FALSE	TRUE	TRUE
FALSE	FALSE	TRUE

If we take the students at their word, Oliver would not ask her. In other words, P is false, which in turn means that Jenny's IF-THEN statement is true (lines 3 and 4, third column). The truth table implies that her actual RSVP is irrelevant: as long as Oliver never asks, she's telling the truth. Now, as the flirtatious scene-closer suggests, Oliver eventually does ask her (P switches from FALSE to TRUE), and she does accept (Q is false). This means that her conditional IF P THEN Q was false, as playful banter often is.

The logical surprise we have just encountered—that as long as the antecedent of a conditional is false, the entire conditional is true (as long as Oliver never asks, she is telling the truth)—exposes a way in which a conditional in logic differs from a statement with an "if" and a "then" in ordinary conversation. Generally, we use a conditional to refer to a warranted prediction based on a testable causal law, like "If you drink coffee, you will stay awake." We aren't satisfied to judge the conditional as true just because it was never tested, like "If you drink turnip juice, you will stay awake," which would be logically true if you never drank turnip juice. We want there to be grounds for believing that in the counterfactual situations in which P *is* true (you do drink turnip juice), NOT Q (you fall asleep) would not happen. When the antecedent of a conditional is known to be false or necessarily false, we're tempted to say that the conditional was moot or irrelevant or speculative or even meaningless, not that it is true. But in the logical sense stipulated in the truth table, in which IF P THEN Q is just a synonym for NOT [P AND NOT Q], that is the strange outcome: "If pigs had wings, then $2 + 2 = 5$" is true, and so is "If $2 + 2 = 3$, then $2 + 2 = 5$." For this reason, logicians use a technical term to refer to the conditional in the truth-table sense, calling it a "material conditional."

Here is a real-life example of why the difference matters. Suppose we want to score pundits on the accuracy of their predictions. How should we evaluate the conditional prediction, made in 2008, "If Sarah Palin were to become president, she would outlaw all abortions"? Does the pundit get credit because the statement is, logically speaking, true? Or should it not count either way? In the real forecasting competition from which the example was drawn, the scorers had to decide what to do about such predictions, and decided not to count it as a true prediction: they chose to interpret the conditional in its everyday sense, not as a material conditional in the logical sense.[5]

The difference between "if" in everyday English and IF in logic is just one example of how the mnemonic symbols we use for connectors in formal logic are not synonymous with the ways they are used in conversation, where, like all words, they have multiple meanings that are disambiguated in context.[6] When we hear "He sat down and told me his life story," we interpret the "and" as implying that he first did one and then the other, though logically it could have been the other way around (as in the wisecrack from another era, "They got married and had a baby, but not in that order"). When the mugger says "Your money or your life," it is technically accurate that you could keep both your money and your life because P OR Q embraces the case where P is true and Q is true. But you would be ill advised to press that argument with him; everyone interprets the "or" in context as the logical connector XOR, "exclusive or," P OR Q AND NOT

[P AND Q]. It's also why when the menu offers "soup or salad," we don't argue with the waiter that we are logically entitled to both. And technically speaking, propositions like "Boys will be boys," "A deal is a deal," "It is what it is," and "Sometimes a cigar is just a cigar" are empty tautologies, necessarily true by virtue of their form and therefore devoid of content. But we interpret them as having a meaning; in the last example (attributed to Sigmund Freud), that a cigar is not always a phallic symbol.

EVEN WHEN THE WORDS are pinned down to their strict logical meanings, logic would be a minor exercise if it consisted only in verifying whether statements containing logical terms are true or false. Its power comes from rules of valid *inference*: little algorithms that allow you to leap from true premises to a true conclusion. The most famous is called "affirming the antecedent" or *modus ponens* (the premises are written above the line, the conclusion below):

IF P THEN Q

P
———
Q

"If someone is a woman, then she is mortal. Xanthippe is a woman. Therefore, Xanthippe is mortal." Another valid rule of inference is called "denying the consequent," the law of contraposition, or *modus tollens*:

IF P THEN Q

NOT Q
———
NOT P

"If someone is a woman, then she is mortal. Stheno the Gorgon is immortal. Therefore, Stheno the Gorgon is not a woman."

These are the most famous but by no means the only valid rules of inference. From the time Aristotle first formalized logic until the late nineteenth century, when it began to be mathematized, logic was basically a taxonomy of the various ways one may or may not deduce conclusions from various collections of premises. For example, there is the valid (but mostly useless) disjunctive addition:

$$\frac{P}{P \text{ OR } Q}$$

"Paris is in France. Therefore, Paris is in France or unicorns exist." And there is the more useful disjunctive syllogism or process of elimination:

$$\frac{P \text{ OR } Q}{\text{NOT } P}$$
$$\frac{}{Q}$$

"The victim was killed with a lead pipe or a candlestick. The victim was not killed with a lead pipe. Therefore, the victim was killed with a candlestick." According to a story, the logician Sidney Morgenbesser and his girlfriend underwent couples counseling during which the bickering pair endlessly aired their grievances about each other. The exasperated counselor finally said to them, "Look, someone's got to change." Morgenbesser replied, "Well, I'm not going to change. And she's not going to change. So *you're* going to have to change."

More interesting still is the Principle of Explosion, also known as "From contradiction, anything follows."

P

NOT P
———
Q

Suppose you believe P, "Hextable is in England." Suppose you also believe NOT P, "Hextable is not in England." By disjunctive addition, you can go from P to P OR Q, "Hextable is in England or unicorns exist." Then, by the disjunctive syllogism, you can go from P OR Q and NOT P to Q: "Hextable is not in England. Therefore, unicorns exist." Congratulations! You just logically proved that unicorns exist. People often misquote Ralph Waldo Emerson as saying, "Consistency is the hobgoblin of little minds." In fact he wrote about a *foolish* consistency, which he advised "great souls" to transcend, but either way, the putdown is dubious.[7] If your belief system contains a contradiction, you can believe anything. (Morgenbesser once said of a philosopher he didn't care for, "There's a guy who asserted both P and not P, and then drew out all the consequences.")[8]

The way that valid rules of inference can yield absurd conclusions exposes an important point about logical arguments. A *valid* argument correctly applies rules of inference to the premises. It only tells us that *if* the premises are true, then the conclusion must be true. It makes no commitment as to whether the premises *are* true, and thus says nothing about the truth of the conclusion. This may be contrasted with a *sound* argument, one that applies the rules correctly to *true* premises and thus yields a true conclusion. Here is a valid argument: "If Hillary Clinton wins the 2016 election, then in 2017 Tim Kaine is the vice president. Hillary Clinton wins the 2016 election. Therefore, in 2017 Tim Kaine is the vice president." It is not a sound argument, because Clinton did not in fact win the election. "If Donald Trump wins the 2016 election, then in 2017 Mike Pence is the vice president. Donald

Trump wins the 2016 election. Therefore, in 2017 Mike Pence is the vice president." This argument is both valid and sound.

Presenting a valid argument as if it were sound is a common fallacy. A politician promises, "If we eliminate waste and fraud from the bureaucracy, we can lower taxes, increase benefits, and balance the budget. I will eliminate waste and fraud. Therefore, vote for me and everything will be better." Fortunately, people can often spot a lack of soundness, and we have a family of retorts to the sophist who draws plausible conclusions from dubious premises: "That's a big if." "If wishes were horses, beggars would ride." "Assume a spherical cow" (among scientists, from a joke about a physicist recruited by a farmer to increase milk production). And then there's my favorite, the Yiddish *As di bubbe volt gehat beytsim volt zi gevain mayn zaidah*, "If my grandmother had balls, she'd be my grandfather."

Of course, many inferences are not even valid. The classical logicians also collected a list of invalid inferences or formal fallacies, sequences of statements in which the conclusions may seem to follow from the premises but in fact do not. The most famous of these is *affirming the consequent*: "IF P THEN Q. Q. Therefore, P." If it rains, then the streets are wet. The streets are wet. Therefore, it rained. The argument is not valid: a street-cleaning truck could have just gone by. An equivalent fallacy is *denying the antecedent*: "IF P THEN Q. NOT P. Therefore, NOT Q." It didn't rain, therefore the streets are not wet. It's also not valid, and for the same reason. A different way of putting it is that the statement IF P THEN Q does not entail its converse, IF Q THEN P, or its inverse, IF NOT P THEN NOT Q.

But people are prone to affirming the consequent, confusing "P implies Q" with "Q implies P." That's why in the Wason selection task, so many people who were asked to verify "If D then 3" turn over the 3 card. It's why conservative politicians encourage voters to slide from "If someone is a socialist, he probably is a Democrat" to

"If someone is a Democrat, he probably is a socialist." It's why crackpots proclaim that all of history's great geniuses were laughed at in their era, forgetting that "If genius, then laughed at" does not imply "If laughed at, then genius." It should be kept in mind by the slackers who note that the most successful tech companies were started by college dropouts.

Fortunately, people often spot the fallacy. Many of us who grew up in the 1960s still snicker at the drug warriors of the era who said that every heroin user started with marijuana, therefore marijuana is a gateway drug to heroin. And then there's Irwin, the hypochondriac who told his doctor, "I'm sure I have liver disease." "That's impossible," replied the doctor. "If you had liver disease you'd never know it—there's no discomfort of any kind." Irwin replies, "Those are my symptoms exactly!"

Incidentally, if you've been paying close attention to the wording of the examples, you will have noticed that I did not consistently mind my Ps and Qs, as I should have if logic consists of manipulating symbols. Instead, I sometimes altered their subjects, tenses, numbers, and auxiliaries. "Someone is a woman" became "Xanthippe is a woman"; "you asked" alternated with "Oliver does ask"; "you must wear a helmet" switched with "the child is wearing a helmet." These kinds of edits matter: "you must wear a helmet" in that context does not literally contradict "a child without a helmet." That's why logicians have developed more powerful logics that break the Ps and Qs of propositional calculus into finer pieces. These include predicate calculus, which distinguishes subjects from predicates and ALL from SOME; modal logic, which distinguishes statements that happen to be true in this world, like "Paris is the capital of France," from those that are necessarily true in all worlds, like "2 + 2 = 4"; temporal logic, which distinguishes past, present, and future; and deontic logic, which worries about permission, obligation, and duty.[9]

Formal Reconstruction

What earthly use is it to be able to identify the various kinds of valid and invalid arguments? Often they can expose fallacious reasoning in everyday life. Rational argumentation consists in laying out a common ground of premises that everyone accepts as true, together with conditional statements that everyone agrees make one proposition follow from another, and then cranking through valid rules of inference that yield the logical, and only the logical, implications of the premises. Often an argument falls short of this ideal: it uses a fallacious rule of inference, like affirming the consequent, or it depends on a premise that was never explicitly stated, turning the syllogism into what logicians call an enthymeme. Now, no mortal has the time or attention span to lay out every last premise and implication in an argument, so in practice almost all arguments are enthymemes. Still, it can be instructive to unpack the logic of an argument as a set of premises and conditionals, the better to spot the fallacies and missing assumptions. It's called formal reconstruction, and philosophy professors sometimes assign it to their students to sharpen their reasoning.

Here's an example. A candidate in the 2020 Democratic presidential primary, Andrew Yang, ran on a platform of implementing a universal basic income (UBI). Here is an excerpt from his website in which he justifies the policy (I have numbered the statements):

(1) The smartest people in the world now predict that ⅓ of Americans will lose their job to automation in 12 years.
(2) Our current policies are not equipped to handle this crisis.
(3) If Americans have no source of income, the future could be very dark. (4) A $1,000/month UBI—funded by a Value

Added Tax—would guarantee that all Americans benefit from automation.[10]

Statements (1) and (2) are factual premises; let's assume they are true. (3) is a conditional, and is uncontroversial. There is a leap from (3) to (4), but it can be bridged in two steps. There is a missing (but reasonable) conditional, (2a) "If Americans lose their jobs, they will have no source of income," and there is the (valid) denial of the consequent of (3), yielding "If the future is not to be dark, Americans must have a source of income." However, upon close examination we discover that the antecedent of (2a), "Americans will lose their jobs," was never stated. All we have is (1)—the smartest people in the world *predict* they will lose their jobs. To get from (1) to the antecedent of (2a), we need to add another conditional, (1a) "If the smartest people in the world predict something, it will come true." But we know that this conditional is false. Einstein, for example, announced in 1952 that only the creation of a world government, P, would prevent the impending self-destruction of mankind, Q (IF NOT P THEN Q), yet no world government was created (NOT P) and mankind did not destroy itself (NOT Q; at least if "impending" means "within several decades"). Conversely, some things may come true that are predicted by people who are not the smartest in the world but are experts in the relevant subject, in this case, the history of automation. Some of those experts predict that for every job lost to automation, a new one will materialize that we cannot anticipate: the unemployed forklift operators will retrain as tattoo removal technicians and video game costume designers and social media content moderators and pet psychiatrists. In that case the argument would fail—a third of Americans will not necessarily lose their jobs, and a UBI would be premature, averting a nonexistent crisis.

The point of this exercise is not to criticize Yang, who was admirably explicit in his platform, nor to suggest that we diagram a logic chart for every argument we consider, which would be unbearably tedious. But the habit of formal reconstruction, even if carried out partway, can often expose fallacious inferences and unstated premises which would otherwise lie hidden in any argument, and is well worth cultivating.

Critical Thinking and Informal Fallacies

Though formal fallacies such as denying the antecedent may be exposed when an argument is formally reconstructed, the more common errors in reasoning can't be pigeonholed in this way. Rather than crisply violating an argument form in the propositional calculus, arguers exploit some psychologically compelling but intellectually spurious lure. They are called *informal* fallacies, and fans of rationality have given them names, collected them by the dozens, and arranged them (together with the formal fallacies) into web pages, posters, flash cards, and the syllabuses of freshman courses on "critical thinking."[11] (I couldn't resist; see the index.)

Many informal fallacies come out of a feature of human reasoning which lies so deep in us that, according to the cognitive scientists Dan Sperber and Hugo Mercier, it was the selective pressure that allowed reasoning to evolve. We like to win arguments.[12] In an ideal forum, the winner of an argument is the one with the most cogent position. But few people have the rabbinical patience to formally reconstruct an argument and evaluate its correctness. Ordinary conversation is held together by intuitive links that allow us to connect the dots even as the discussion falls short of Talmudic explicitness.

Skilled debaters can exploit these habits to create the illusion that they have grounded a proposition in a sound logical foundation when in reality it is levitating in midair.

Foremost among informal fallacies is the *straw man*, the effigy of an opponent that is easier to knock over than the real thing. "Noam Chomsky claims that children are born talking." "Kahneman and Tversky say that humans are imbeciles." It has a real-time variant practiced by aggressive interviewers, the *so-what-you're-saying-is* tactic. "Dominance hierarchies are common in the animal kingdom, even in creatures as simple as lobsters." "So what you're saying is that we should organize our societies along the lines of lobsters."[13]

Just as arguers can stealthily replace an opponent's proposition by one that is easier to attack, they can replace their own proposition with one that is easier to defend. They can engage in *special pleading*, such as explaining that ESP fails in experimental tests because it is disrupted by the negative vibes of skeptics. Or that democracies never start wars, except for ancient Greece, but it had slaves, and Georgian England, but the commoners couldn't vote, and nineteenth-century America, but its women lacked the franchise, and India and Pakistan, but they were fledgling states. They can *move the goalposts*, demanding that we "defund the police" but then explaining that they only mean reallocating part of its budget to emergency responders. (Rationality cognoscenti call it the *motte-and-bailey* fallacy, after the medieval castle with a cramped but impregnable tower into which one can retreat when invaders attack the more desirable but less defensible courtyard.)[14] They can claim that no Scotsman puts sugar on his porridge, and when confronted with Angus, who puts sugar on his porridge, say this shows that Angus is not a true Scotsman. The *no true Scotsman* fallacy also explains why no true Christian ever kills, no true communist state is repressive, and no true Trump supporter endorses violence.

These tactics shade into *begging the question*, a phrase that philosophers beg people not to use as a malaprop for "raising the question" but to reserve for the informal fallacy of assuming what you're trying to prove. It includes circular explanations, as in Molière's *virtus dormitiva* (his doctor's explanation for why opium puts people to sleep), and tendentious presuppositions, as in the classic "When did you stop beating your wife?" In one joke, a man boasts about the mellifluous cantor in his synagogue, and another retorts, "Ha! If I had his voice, I'd be just as good."

One can always maintain a belief, no matter what it is, by saying that the *burden of proof* is on those who disagree. Bertrand Russell responded to this fallacy when he was challenged to explain why he was an atheist rather than an agnostic, since he could not prove that God does not exist. He replied, "Nobody can prove that there is not between the Earth and Mars a china teapot revolving in an elliptic orbit."[15] Sometimes both sides pursue the fallacy, leading to the style of debate called burden tennis. ("The burden of proof is on you." "No, the burden of proof is on *you*.") In reality, since we start out ignorant about everything, the burden of proof is on anyone who wants to show anything. (As we will see in chapter 5, Bayesian reasoning offers a principled way to reason about who should carry the burden as knowledge accumulates.)

Another diversionary tactic is called *tu quoque*, Latin for "you too," also known as *what-aboutery*. It was a favorite of the apologists for the Soviet Union in the twentieth century, who presented the following defense of its totalitarian repression: "What about the way the United States treats its Negroes?" In another joke, a woman comes home from work early to find her husband in bed with her best friend. The startled man says, "What are you doing home so early?" She replies, "What are *you* doing in bed with my best friend!?" He snaps, "Don't change the subject!"

The "smartest people in the world" claim from the Yang Gang is a mild example of the *argument from authority*. The authority being deferred to is often religious, as in the gospel song and bumper sticker "God said it, I believe it, that settles it." But it can also be political or academic. Intellectual cliques often revolve around a guru whose pronouncements become secular gospel. Many academic disquisitions begin, "As Derrida has taught us . . ."—or Foucault, or Butler, or Marx, or Freud, or Chomsky. Good scientists disavow this way of talking, but they are sometimes raised up as authorities by others. I often get letters taking me to task for worrying about human-caused climate change because, they note, this brilliant physicist or that Nobel laureate denies it. But Einstein was not the only scientific authority whose opinions outside his area of expertise were less than authoritative. In their article "The Nobel Disease: When Intelligence Fails to Protect against Irrationality," Scott Lilienfeld and his colleagues list the flaky beliefs of a dozen science laureates, including eugenics, megavitamins, telepathy, homeopathy, astrology, herbalism, synchronicity, race pseudoscience, cold fusion, crank autism treatments, and denying that AIDS is caused by HIV.[16]

Like the argument from authority, the *bandwagon* fallacy exploits the fact that we are social, hierarchical primates. "Most people I know think astrology is scientific, so there must be something to it." While it may not be true that "the majority is always wrong," it certainly is not always right.[17] The history books are filled with manias, bubbles, witch hunts, and other extraordinary popular delusions and madnesses of crowds.

Another contamination of the intellectual by the social is the attempt to rebut an idea by insulting the character, motives, talents, values, or politics of the person who holds it. The fallacy is called arguing *ad hominem*, against the person. A crude but common version is endorsed by Wally in *Dilbert*:

Often the expression is more genteel but no less fallacious. "We don't have to take Smith's argument seriously; he is a straight white male and teaches at a business school." "The only reason Jones argues that climate change is happening is that it gets her grants and fellowships and invitations to give TED talks." A related tactic is the *genetic* fallacy, which has nothing to do with DNA but is related to the words "genesis" and "generate." It refers to evaluating an idea not by its truth but by its origins. "Brown got his data from the *CIA World Factbook*, and the CIA overthrew democratic governments in Guatemala and Iran." "Johnson cited a study funded by a foundation that used to support eugenics."

Sometimes the ad hominem and genetic fallacies are combined to forge chains of *guilt by association*: "Williams's theory must be repudiated, because he spoke at a conference organized by someone who published a volume containing a chapter written by someone who said something racist." Though no one can gainsay the pleasure of ganging up on an evildoer, the ad hominem and genetic fallacies are genuinely fallacious: good people can hold bad beliefs and vice versa. To take a pointed example, lifesaving knowledge in public health, including the carcinogenicity of tobacco smoke, was originally discovered by Nazi scientists, and tobacco companies were all too happy to reject the smoking–cancer link because it was "Nazi science."[18]

Then there are arguments directly aimed at the limbic system rather than the cerebral cortex. These include the *appeal to emotion*: "How can anyone look at this photo of the grieving parents of a dead child and say that war deaths have declined?" And the increasingly popular *affective* fallacy, in which a statement may be rejected if it is "hurtful" or "harmful" or may cause "discomfort." Here we see a perpetrator of the affective fallacy as a child:

"It may be wrong, but it's how I feel."

David Sipress/The New Yorker Collection/The Cartoon Bank

Many facts, of course, are hurtful: the racial history of the United States, global warming, a cancer diagnosis, Donald Trump. Yet they are facts for all that, and we must know them, the better to deal with them.

The ad hominem, genetic, and affective fallacies used to be treated as forehead-slapping blunders or dirty rotten tricks. Critical-thinking teachers and high school debate coaches would teach their students how to spot and refute them. Yet in one of the ironies of modern intellectual life, they are becoming the coin of the realm. In large swaths of academia and journalism the fallacies are applied with gusto, with ideas attacked or suppressed because their proponents, sometimes from centuries past, bear unpleasant odors and stains.[19] It reflects a shift in one's conception of the nature of beliefs: from ideas that may be true or false to expressions of a person's moral and cultural identity. It also bespeaks a change in how scholars and critics conceive of their mission: from seeking knowledge to advancing social justice and other moral and political causes.[20]

To be sure, sometimes the context of a statement really is relevant to evaluating its truth. This can leave the misimpression that informal fallacies are OK after all. One can be skeptical of a study showing the efficacy of a drug carried out by someone who stands to profit from the drug, but noting a conflict of interest is not an ad hominem fallacy. One can dismiss a claim that was based on divine inspiration or exegesis of ancient texts or interpreting goat entrails; this debunking is not the genetic fallacy. One can take note of a near consensus among scientists to counter the assertion that we must be agnostic about some issue because the experts disagree; this is not the bandwagon fallacy. And we can impose higher standards of evidence for a hypothesis that would call for drastic measures if it were true; this is not the affective fallacy. The difference is that in the legitimate arguments, one can give *reasons* for why the context of a statement should affect our credence in whether it is true or how we should act on it, such as indicating the degree to which the evidence is trustworthy. With the fallacies, one is surrendering to feelings that have no bearing on the truth of the claim.

So with all these formal and informal fallacies waiting to entrap us (*Wikipedia* lists more than a hundred), why can't we do away with this jibber-jabber once and for all and implement Leibniz's plan for logical discourse? Why can't we make our reasonings as tangible as those of the mathematicians so that we can find our errors at a glance? Why, in the twenty-first century, do we still have barroom arguments, Twitter wars, couples counseling, presidential debates? Why don't we say "Let us calculate" and see who is right? We are not living in Leibniz's utopia, and, as with other utopias, we never will. There are at least three reasons.

Logical versus Empirical Truths

One reason logic will never rule the world is the fundamental distinction between *logical* propositions and *empirical* ones, which Hume called "relations of ideas" and "matters of fact," and philosophers call analytic and synthetic. To determine whether "All bachelors are unmarried" is true, you just need to know what the words mean (replacing *bachelor* with the phrase "male AND adult AND NOT married") and check the truth table. But to determine whether "All swans are white" is true, you have to get out of your armchair and look. If you visit New Zealand, you will discover the proposition is false, because the swans there are black.

It's often said that the Scientific Revolution of the seventeenth century was launched when people first appreciated that statements about the physical world are empirical and can be established only by observation, not scholastic argumentation. There is a lovely story attributed to Francis Bacon:

> In the year of our Lord 1432, there arose a grievous quarrel among the brethren over the number of teeth in the mouth of a horse. For thirteen days the disputation raged without ceasing. All the ancient books and chronicles were

fetched out, and wonderful and ponderous erudition such as was never before heard of in this region was made manifest. At the beginning of the fourteenth day, a youthful friar of goodly bearing asked his learned superiors for permission to add a word, and straightway, to the wonderment of the disputants, whose deep wisdom he sore vexed, he beseeched them to unbend in a manner coarse and unheard-of and to look in the open mouth of a horse and find answer to their questionings. At this, their dignity being grievously hurt, they waxed exceeding wroth; and, joining in a mighty uproar, they flew upon him and smote him, hip and thigh, and cast him out forthwith. For, said they, surely Satan hath tempted this bold neophyte to declare unholy and unheard-of ways of finding truth, contrary to all the teachings of the fathers.

Now, this event almost certainly never happened, and it's doubtful that Bacon said it did.[21] But the story captures one reason that we will never resolve our uncertainties by sitting down and calculating.

Formal versus Ecological Rationality

A second reason Leibniz's dream will never come true lies in the nature of formal logic: it is *formal*, blinkered from seeing anything but the symbols and their arrangement as they are laid out in front of the reasoner. It is blind to the *content* of the proposition—what those symbols mean, and the context and background knowledge that might be mixed into the deliberation. Logical reasoning in the strict sense means forgetting everything you know. A student taking a test in Euclidean geometry gets no credit for pulling out a ruler and measuring the two sides of the triangle with equal angles, sensible as that might be in everyday life, but rather is required to prove it. In the same way, students doing the logic exercises in Carroll's textbook must not be distracted by their irrelevant knowledge that puppies can't talk. The only legitimate reason to conclude that the lame one failed to say "Thank you" is that that's what is stipulated in the consequent of a conditional whose antecedent is true.

Logic, in this sense, is not rational. In the world in which we evolved and most of the world in which we spend our days, it makes no sense to ignore everything you know.[22] It does make sense in certain unnatural worlds—logic courses, brainteasers, computer programming, legal proceedings, the application of science and math to areas in which common sense is silent or misleading. But in the natural world, people do pretty well by commingling their logical abilities with their encyclopedic knowledge, as we saw in chapter 1 with the San. We also saw that when we add certain kinds of verisimilitude to the brainteasers, people recruit their subject knowledge and no longer embarrass themselves. True, when they are asked to verify "If a card has a D on one side it must have a 3 on the other," they mistakenly turn over the "3" and neglect to turn over the "7." But when they are asked to imagine themselves as bouncers in a bar and verify "If a patron is drinking alcohol he must be over twenty-one," they know to check the beverages in front of the teenagers and to card anyone drinking beer.[23]

The contrast between the *ecological* rationality that allows us to thrive in a natural environment and the *logical* rationality demanded by formal systems is one of the defining features of modernity.[24] Studies of unlettered peoples by cultural psychologists and anthropologists have shown that they are rooted in the rich texture of reality and have little patience for the make-believe worlds familiar to graduates of Western schooling. Here Michael Cole interviews a member of the Kpelle people in Liberia:

> **Q:** Flumo and Yakpalo always drink rum together. Flumo is drinking rum. Is Yakpalo drinking rum?
>
> **A:** Flumo and Yakpalo drink rum together, but the time Flumo was drinking the first one Yakpalo was not there on that day.
>
> **Q:** But I told you that Flumo and Yakpalo always drink rum together. One day Flumo was drinking rum. Was Yakpalo drinking rum?
>
> **A:** The day Flumo was drinking the rum Yakpalo was not there on that day.
>
> **Q:** What is the reason?
>
> **A:** The reason is that Yakpalo went to his farm on that day and Flumo remained in town on that day.[25]

The Kpelle man treats the question as a sincere inquiry, not a logic puzzle. His response, though it would count as an error on a test, is by no means irrational: it uses relevant information to come up with the correct answer. Educated Westerners have learned how to play the game of forgetting what they know and fixating on the premises of a problem—though even they have trouble separating their factual knowledge from their logical reasoning. Many people will insist, for example, that the following argument is logically invalid: "All things made of plants are healthy. Cigarettes are made of plants. Therefore, cigarettes are healthy."[26] Change "cigarettes" to "salads" and they confirm that it's fine. Philosophy professors who present students with contrived thought experiments, like whether it is permissible to throw a fat man over a bridge to stop a runaway trolley which threatens five workers on the track, often get frustrated when students look for loopholes, like shouting at the workers to get out of the way. Yet that is exactly the rational thing one would do in real life.

The zones in which we play formal, rule-governed games—law, science, digital devices, bureaucracy—have expanded in the modern world with the invention of powerful, content-blind formulas and rules. But they still fall short of life in all its plentitude. Leibniz's logical utopia, which requires self-inflicted amnesia for background knowledge, not only runs against the grain of human cognition but is ill suited for a world in which not every relevant fact can be laid out as a premise.

Classical versus Family Resemblance Categories

A third reason that rationality will never be reduced to logic is that the concepts that people care about differ in a crucial way from the predicates of classical logic. Take the predicate "even number," which can be defined by the biconditional "If an integer is even, it can be divided by 2 without remainder, and vice versa." The biconditional is true, as is the proposition "8 can be divided by 2 without remainder," and from these true premises we can deduce the true conclusion "8 is even." Likewise with "If a person is a grandmother, she is female and the mother of a parent, and vice versa" and "If a person is a bachelor, he is male and adult and not married, and vice versa." We might suppose that with enough effort every human concept can be defined in this way, by laying out necessary conditions for it to be true (the first IF-THEN in the biconditional) and sufficient conditions for it to be true (the "vice versa" converse).

This reverie was famously punctured by the philosopher Ludwig Wittgenstein (1889–1951).[27] Just *try*, he said, to find necessary and sufficient conditions for any of our everyday concepts. What is the common denominator across all the pastimes we call "games"? Physical activity? Not with board games. Gaiety? Not in chess. Competitors? Not solitaire. Winning and losing? Not ring-around-the-rosy or a child throwing a ball against a wall. Skill? Not bingo. Chance? Not crossword puzzles. And Wittgenstein did not live to see mixed martial arts, Pokémon GO, or *Let's Make a Deal*.[28]

The problem is not that no two games have anything in common. Some are merry, like tag and charades; some have winners, like Monopoly and football; some involve projectiles, like baseball and tiddlywinks. Wittgenstein's point was that the concept of "game" has no common thread running through it, no necessary and sufficient features that could be turned into a definition. Instead, various characteristic features run through different subsets of the category, the same way that physical features may be found in different combinations in the members of a family. Not every scion of Robert Kardashian and Kristen Mary Jenner has the pouty Kardashian lips or the raven Kardashian hair or the caramel Kardashian skin or the ample Kardashian derrière. But most of the sisters have some of them, so we can recognize a Kardashian when we see one, even if there is no true proposition "If someone has features X and Y and Z, that person is a Kardashian." Wittgenstein concluded that it is family resemblance, not necessary and sufficient features, that holds the members of a category together.

Most of our everyday concepts turn out to be family resemblance categories, not the "classical" or "Aristotelian" categories that are easily stipulated in logic.[29] These categories often have stereotypes, like the little picture of a bird in a dictionary next to the definition of *bird*, but the definition itself fails to embrace all and only the exemplars. The category "chairs," for example, includes wheelchairs with no legs, rolling stools with no back, beanbag chairs with no seat, and the exploding props used in Hollywood fight scenes which cannot support a sitter. Even the ostensibly classical categories that professors used to cite to illustrate the concept turn out to be riddled with exceptions. Is there a definition of "mother" that embraces adoptive mothers, surrogates, and egg donors? If a "bachelor" is an unmarried man, is the pope a bachelor? What about the male half of a monogamous couple who never bothered to get the piece of paper from city hall? And you can get in a lot of trouble these days if you try to lay out necessary and sufficient conditions for "woman."

As if this weren't bad enough for the dream of a universal logic, the fact that concepts are defined by family resemblance rather than necessary and sufficient conditions means that propositions can't even be given the values TRUE or FALSE. Their predicates may be truer of some subjects than others, depending on how stereotypical the subject is—in other words, how many of the family's typical features it has. Everyone agrees that "Football is a sport" is true, but many feel that "Synchronized swimming is a sport" is, at best, truthy. The same holds for "Parsley is a vegetable," "A parking violation is a crime," "Stroke is a disease," and "Scorpions are bugs." In everyday judgments, truth can be fuzzy.

It's not that *all* concepts are fuzzy family resemblance categories.[30] People are perfectly capable of putting things in little boxes. Everyone understands that a number is either even or odd, with no in-betweens. We joke that you can't be a little bit pregnant or a little bit married. We understand laws that preempt endless disputation over borderline cases by drawing red lines around concepts like "adult," "citizen," "owner," "spouse," and other consequential categories.

Indeed, an entire category of informal fallacies arises from people being all too eager to think in black and white. There's the *false dichotomy*: "Nature versus nurture"; "America—love it or leave it"; "You're either with us or with the terrorists"; "Either you're part of the solution or you're part of the problem." There's the *slippery slope* fallacy: if we legalize abortion, soon we'll legalize infanticide; if we allow people to marry an individual who is not of the opposite sex, we will have to allow people to marry an individual who is not of the same species. And the *paradox of the heap* begins with the truth that if something is a heap, then it's still a heap if you remove one grain. But when you remove another, and then another, you reach a point where it's no longer a heap, which implies that there's no such thing as a heap. By the same logic, the job will get done even if I put it off by just one more day (the mañana fallacy), and I can't get fat by eating just one more French fry (the dieter's fallacy).

Wittgenstein's response to Leibniz and Aristotle is not just a debating point for philosophy seminars. Many of our fiercest controversies involve decisions on how to reconcile fuzzy family resemblance concepts with the classical categories demanded by logic and law. Is a fertilized ovum a "person"? Did Bill and Monica have "sex"? Is a sport utility vehicle a "car" or a "truck"? (The latter classification put tens of millions of vehicles on American roads that met laxer standards for safety and emissions.) And not long ago I received the following email from the Democratic Party:

> House Republicans are ramming through legislation this week to classify pizza as a "vegetable" for the purpose of school lunches. Why? Because a massive lobbying effort of Republican lawmakers by the frozen pizza industry is underway. . . .
>
> In this Republican Congress, almost anything is up for sale to the most powerful lobbyists—including the literal definition of the word "vegetable"—and this time, it's coming at the expense of our kids' health.
>
> Sign this petition and spread the word: Pizza isn't a vegetable.

Logical Computation versus
Pattern Association

If many of our judgments are too squishy to be captured in logic, how do we think at all? Without the guardrails of necessary and sufficient conditions, how do we come to agree that football is a sport and Kris Jenner is a mother and, House Republicans notwithstanding, pizza isn't a vegetable? If rationality is not implemented in the mind as a list of propositions and a chain of logical rules, how is it implemented?

One answer may be found in the family of cognitive models called pattern associators, perceptrons, connectionist nets, parallel distributed processing models, artificial neural networks, and deep learning systems.[31] The key idea is that rather than manipulating strings of symbols with rules, an intelligent system can aggregate tens, thousands, or millions of graded signals, each capturing the degree to which a property is present.

Take the surprisingly contentious concept "vegetable." It's clearly a family resemblance category. There is no Linnaean taxon that includes carrots, fiddleheads, and mushrooms; no type of plant organ that embraces broccoli, spinach, potatoes, celery, peas, and eggplant; not even a distinctive taste or color or texture. But as with the Kardashians, we tend to know vegetables when we see them, because overlapping traits run through the different members of the family. Lettuce is green and crisp and leafy, spinach is green and leafy, celery is green and crisp, red cabbage is red and leafy. The greater the number of veggie-like traits something has, and the more definitively it has them, the more apt we are to call it a vegetable. Lettuce is a vegetable par excellence; parsley, not so much; garlic, still less.

Conversely, certain traits militate against something being a vegetable. While some vegetables verge on sweetness, like acorn squash, once a plant part gets too sweet, like a cantaloupe, we call it a fruit instead. And though portobello mushrooms are meaty and spaghetti squash is pasta-like, anything made from animal flesh or flour dough is blackballed. (Bye-bye, pizza.)

This means we can capture vegetableness in a complicated statistical formula. Each of an item's traits (its greenness, its crunchiness, its sweetness, its doughiness) is quantified, and then multiplied by a numerical weight that reflects how diagnostic that trait is of the category: high positive for greenness, lower positive for crispness, low negative for sweetness, high negative for doughiness. Then the weighted values are added up, and if the sum exceeds a threshold, we say it's a vegetable, with higher numbers indicating better examples.

Now, no one thinks we make our fuzzy judgments by literally carrying out chains of multiplication and addition in our heads. But the equivalent can be done by networks of neuron-like units which can "fire" at varying rates, representing the fuzzy truth value. A toy version is shown on the next page. At the bottom we have a bank of input neurons fed by the sense organs, which respond to simple traits like "green" and "crisp." At the top we have the output neurons, which display the network's guess of the category. Every input neuron is connected to every output neuron via a "synapse" of varying strength, both excitatory (implementing the positive multipliers) and inhibitory (implementing the negative ones). The activated input units propagate signals, weighted by the synapse strengths, to the output units, each of which adds up the weighted set of incoming signals and fires accordingly. In the diagram, the excitatory connections are shown with arrows and the inhibitory ones with dots, and

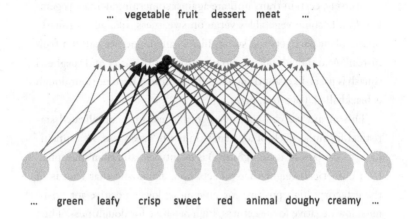

the thickness of the lines represents the strengths of synapses (shown only for the vegetable output, to keep it simple).

Who, you may ask, programmed in the all-important connection weights? The answer is no one; they are learned from experience. The network is *trained* by presenting it with many examples of different foods, together with the correct category provided by a teacher. The neonate network, born with small random weights, offers random feeble guesses. But it has a learning mechanism that works by a getting-warmer/getting-colder rule. It compares each node's output with the correct value supplied by the teacher, and nudges the weight up or down to close the gap. After hundreds of thousands of training examples, the connection weights settle into the best values, and the networks can get pretty good at classifying things.

But that's true only when the input features indicate the output categories in a linear, more-is-better, add-'em-up way. It works for categories where the whole is the (weighted) sum of its parts, but it fails when a category is defined by tradeoffs, sweet spots, winning combinations, poison pills, deal-killers, perfect storms, or too much of a good thing. Even the simple logical connector XOR (exclusive or), "*x* or *y* but not both," is beyond the powers of a two-layer neural network, because *x*-ness has to boost the output, and *y*-ness has to boost the output, but in combination they have to squelch it. So while a simple network can learn to recognize carrots and cats, it may fail with an unruly category like "vegetable." An item that's red and round is likely to be a fruit if it is crunchy and has a stem (like an apple), but a vegetable if it is crunchy and has roots (like a beet) or if it is fleshy and has a stem (like a tomato). And what combination of colors and shapes and textures could possibly rope in mushrooms, spinach, cauliflower, carrots, and beefsteak tomatoes? A two-layer network gets confused by the crisscrossing patterns, jerking its weights up and down with each training example and never settling on values that consistently separate the members from the nonmembers.

The problem may be tamed by inserting a "hidden" layer of neurons between the input and the output, as shown on the next page. This changes the network from a stimulus-response creature to one with internal representations—concepts, if you will. Here they might stand for cohesive intermediate categories like "cabbage-like," "savory fruits," "squashes and gourds," "greens," "fungi," and "roots and tubers," each with a set of input weights that allow it to pick out the corresponding stereotype, and strong weights to "vegetable" in the output layer.

The challenge in getting these networks to work is how to train them. The problem is with the connections from the input layer to the hidden layer: since the units are hidden from the environment, their guesses cannot be matched against "correct" values supplied by the teacher. But a breakthrough in the 1980s, the error back-propagation learning algorithm, cracked the problem.[32] First, the mismatch between each output unit's guess and the correct answer is used to tweak the weights of the hidden-to-output connections in the top layer, just

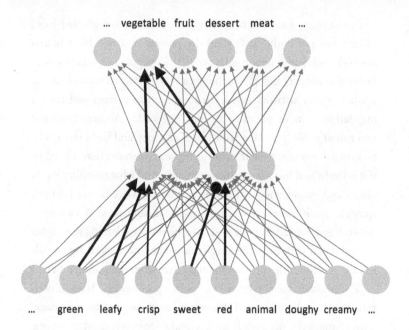

... vegetable fruit dessert meat ...

... green leafy crisp sweet red animal doughy creamy ...

like in the simple networks. Then the sum of all these errors is propagated backwards to each hidden unit to tweak the input-to-hidden connections in the middle layer. It sounds like it could never work, but with millions of training examples the two layers of connections settle into values that allow the network to sort the sheep from the goats. Just as amazingly, the hidden units can spontaneously discover abstract categories like "fungi" and "roots and tubers," if that's what helps them with the classifying. But more often the hidden units don't stand for anything we have names for. They implement whichever complex formulas get the job done: "a teensy bit of this feature, but not too much of that feature, unless there's really a lot of this other feature."

In the second decade of the twenty-first century, computer power skyrocketed with the development of graphics processing units, and data got bigger and bigger as millions of users uploaded text and images to the web. Computer scientists could put multilayer networks on megavitamins, giving them two, fifteen, even a thousand hidden layers, and training them on billions or even trillions of examples. The networks are called deep learning systems because of the number of layers between the input and the output (they're not deep in the sense of understanding anything). These networks are powering "the great AI awakening" we are living through, which is giving us the first serviceable products for speech and image recognition, question-answering, translation, and other humanlike feats.[33]

Deep learning networks often outperform GOFAI (good old-fashioned artificial intelligence), which executes logic-like deductions on hand-coded propositions and rules.[34] The contrast in the way they work is stark: unlike logical inference, the inner workings of a neural network are inscrutable. Most of the millions of hidden units don't stand for any coherent concept that we can make sense of, and the computer scientists who train them can't explain how they arrive at any particular answer. That is why many technology critics fear that as AI systems are entrusted with decisions about the fates of people, they could perpetuate biases that no one can identify and uproot.[35] In 2018 Henry Kissinger warned that since deep learning systems don't work on propositions we can examine and justify, they portend the end of the Enlightenment.[36] That is a stretch, but the contrast between logic and neural computation is clear.

Is the human brain a big deep learning network? Certainly not, for many reasons, but the similarities are illuminating. The brain has around a hundred billion neurons connected by a hundred trillion synapses, and by the time we are eighteen we have been absorbing examples from our environments for more than three hundred million waking seconds. So we are prepared to do a lot of pattern-matching and associating, just like these networks. The networks are tailor-made for the fuzzy family resemblance categories that make up so much of our conceptual repertoire. Neural networks thus provide clues about the portion of human cognition that is rational but not, technically speaking, logical. They demystify the inarticulate yet sometimes uncanny mental power we call intuition, instinct, inklings, gut feelings, and the sixth sense.

FOR ALL THE CONVENIENCE that Siri and Google Translate bring to our lives, we must not think that neural networks have made logic obsolete. These systems, driven by fuzzy associations and incapable of parsing syntax or consulting rules, can be stunningly stupid.[37] If you ask Google for "fast-food restaurants near me that are not McDonald's," it will give you a list of all the McDonald's within a fifty-mile radius. Ask Siri, "Did George Washington use a computer?" and she will direct you to a computer reconstruction of George Washington's face and the Computing System Services of George Washington University. The vision modules that someday will drive our cars are today apt to confuse road signs with refrigerators, and overturned vehicles with punching bags, fireboats, and bobsleds.

Human rationality is a hybrid system.[38] The brain contains pattern associators that soak up family resemblances and aggregate large numbers of statistical clues. But it also contains a logical symbol manipulator that can assemble concepts into propositions and draw out their implications. Call it System 2, or recursive cognition, or rule-based reasoning. Formal logic is a tool that can purify and extend this mode of thinking, freeing it from the bugs that come with being a social and emotional animal.

Because our propositional reasoning frees us from similarity and stereotypes, it enables the highest achievements of human rationality, such as science, morality, and law.[39] Though porpoises fit into the family resemblance among fishes, the rules that define membership in Linnaean classes (like "IF an animal suckles its young, THEN it is a mammal") tell us they are not in fact members. Through chains of categorical reasoning like this, we can be convinced that humans are apes, the sun is a star, and solid objects are mostly empty space. In the social sphere, our pattern-finders easily see the ways in which people differ: some individuals are richer, smarter, stronger, swifter, better-looking, and more like us than others. But when we embrace the proposition that all humans are created equal ("IF X is human, THEN X has rights"), we can sequester these impressions from our legal and moral decision making, and treat all people equally.

4

PROBABILITY AND RANDOMNESS

A thousand stories which the ignorant tell, and believe,
die away at once, when the computist takes them in his
gripe.

—SAMUEL JOHNSON[1]

Though Albert Einstein never said most of the things he supposedly said, he did say, in several variations, "I shall never believe that God plays dice with the world."[2] Whether or not he was right about the subatomic world, the world we live in certainly *looks* like a game of dice, with unpredictability at every scale. The race is not always to the swift, nor the battle to the strong, nor bread to the wise, nor favor to those of skill, but time and chance happen to them all. An essential part of rationality is dealing with randomness in our lives and uncertainty in our knowledge.

What Is Randomness? Where Does It Come From?

In the strip below, Dilbert's question awakens us to the fact that the word "random" in common parlance refers to *two* concepts: a lack of patterning in data, and a lack of predictability in a process. When he doubts that the consecutive nines produced by the troll are truly random, he's referring to their patterning.

Dilbert's impression that there's a pattern in the sequence is not a figment of his imagination, like seeing butterflies in inkblots. Non-random patterning can be quantified. Brevity is the soul of pattern: we say that a dataset is nonrandom when its shortest possible description is shorter than the dataset itself.[3] The description "6 9s" is two characters long (in an efficient shorthand for the descriptions), whereas the dataset itself, "999999," is six characters long. Other strings we feel to be nonrandom also submit to compression: "123456" boils down to "1st 6"; "505050" squashes down to "3 50s." In contrast, data we feel to be random, like "634579," can't be abridged into anything more concise; they must be rendered verbatim.

The troll's answer captures the second sense of randomness: an anarchic, unpredictable generating process. The troll is correct that a random *process* can generate nonrandom *patterns*, at least for a while— for six digits' worth of output, in this case. After all, if there's no rhyme or reason to the generator, what's to stop it from producing six 9s or any other nonrandom pattern, at least occasionally? As the generator continues and the sequence gets longer, we can expect random patterning to reassert itself, because the freakish run is unlikely to continue.

The troll's punch line is profound. As we shall see, mistaking a nonrandom *pattern* for a nonrandom *process* is one of the thickest chapters in the annals of human folly, and knowing the difference between them is one of the greatest gifts of rationality that education can confer.

All this raises the question of what kinds of physical mechanism can generate random events. Einstein notwithstanding, most physicists believe there is irreducible randomness in the subatomic realm of quantum mechanics, like the decay of an atomic nucleus or the emission of a photon when an electron jumps from one energy state to another. It's possible for this quantum uncertainty to be amplified to scales that impinge on our lives. When I was a research assistant in an animal behavior lab, the refrigerator-sized minicomputers of the day were too slow to generate random-looking numbers in real time, and my supervisor had invented a gadget with a capsule filled with a radioactive isotope and a teensy-weensy Geiger counter that detected the intermittent particle spray and tripped a switch that fed the pigeon.[4] But in most of the intermediate-sized realm in which we spend our days, quantum effects cancel out and may as well not exist.

So how could randomness arise in a world of billiard balls obeying Newton's equations? As the 1970s poster proclaimed (satirizing billboards about the speed limit), "Gravity. It isn't just a good idea. It's the law."[5] In theory, couldn't the demon imagined by Pierre-Simon Laplace in 1814, who knew the position and momentum of every particle in the universe, plug them into equations for the laws of physics and predict the future perfectly?

In reality, there are two ways in which a law-governed world can generate events that for all intents and purposes are random. One of them is familiar to popular science readers: the butterfly effect, named after the possibility that the flapping of a butterfly's wings in Brazil could trigger a tornado in Texas. Butterfly effects can arise in deterministic nonlinear dynamical systems, also known as "chaos," where minuscule differences in initial conditions, too small for any instrument to measure, can feed on themselves and blow up into gargantuan effects.

The other way in which a deterministic system can appear to be random from a human vantage point also has a familiar name: the coin flip. The fate of a tossed coin is not literally random; a skilled magician can flick one just so to get a head or a tail on demand. But when an outcome depends on a large number of tiny causes that are impractical to keep track of, like the angles and forces that launched the penny and the wind currents buffeting it in midair, it might as well be random.

What Does "Probability" Mean?

When the TV meteorologist says there's a 30 percent chance of rain in the area tomorrow, what does she mean? Most people are foggy about the answer. Some think it means it will rain in 30 percent of the area. Others think it means it will rain 30 percent of the time. A few think it means that 30 percent of meteorologists think it will rain. And some think it means it will rain somewhere in the area on 30 percent of the days in which a prediction like this is made. (The last of these is in fact closest to what the meteorologist had in mind.)[6]

Weather-watchers are not the only ones who are confused. In 1929, Bertrand Russell noted that "probability is the most important concept in modern science, especially as nobody has the slightest notion what it means."[7] More accurately, different people have different notions of what it means, as we saw in chapter 1 with the Monty Hall and Linda problems.[8]

There is the *classical* definition of probability, which goes back to the origins of probability theory as a way of understanding games of chance. You lay out the possible outcomes of a process that have an equal chance of occurring, add up the ones that count as examples of the event, and divide by the number of possibilities. A die can land on any of six sides. An "even number" corresponds to its landing on the sides with two dots, four dots, or six dots showing. With three ways it can land "even" out of the six possibilities in all, we say that the classical probability it will roll "even" is three out of six, or .5. (In chapter 1 I used the classical definition to explain the correct strategy in the Monty Hall dilemma, and noted that miscounting the possible outcomes was what lured some of the overconfident experts to the incorrect strategy.)

But why did we think that the outcome of landing on each face had an equal chance of happening in the first place? We assessed the die's *propensity*, its physical disposition to do various things. This includes the symmetry of the six faces, the haphazard way the shooter releases it, and the physics of tumbling.

Closely related is a third, *subjectivist* interpretation. Before you fling the die, based on everything you know, how would you quantify, on a scale from 0 to 1, your belief that it will land even? This credence estimate is sometimes called the Bayesian interpretation of probability (a bit misleadingly, as we'll see in the next chapter).

Then there is the *evidential* interpretation: the degree to which you believe the information presented warrants the conclusion. Think of a court of law, where in judging the probability that the defendant is guilty, you ignore inadmissible and prejudicial background information and consider only the strength of the prosecutor's case. It was the evidential interpretation that made it rational to judge that Linda, having been presented as a social justice warrior, was likelier to be a feminist bank teller than a bank teller.

Finally there is the *frequentist* interpretation: if you did toss the die many times, say, a thousand, and counted the outcomes, you'd find that the result was even in around five hundred of the tosses, or half of them.

Ordinarily the five interpretations are aligned. In the case of a coin toss, the penny is symmetrical; coming up heads comprises exactly one out of the two possible outcomes; your gut feeling is halfway between "heads for sure" and "tails for sure"; the argument for heads is as strong as the argument for tails; and in the long run half the tosses you'll see are heads. The probability of heads is .5 in every case. But the interpretations don't mean the same thing, and sometimes they part company. When they do, statements about probabilities can result in confusion, controversy, even tragedy.

Most dramatically, the first four interpretations apply to the vaguely mystical notion of the probability of a single instance. What is the probability that you are over fifty? That the next pope will be Bono? That Britney Spears and Katy Perry are the same person? That there is life on Enceladus, one of the moons of Saturn? You might object that the questions are meaningless: either you are over fifty or you aren't, and "probability" has nothing to do with it. But in the subjectivist interpretation, I can put a number on my ignorance. This offends some statisticians, who want to reserve the concept of probability for relative frequency in a set of events, which are really real and can be counted. One quipped that single-event probabilities belong not in mathematics but in psychoanalysis.[9]

Laypeople, too, can have trouble wrapping their minds around the concept of the numerical probability of a single event. They are mad at the weatherman after getting soaked on a day when she had predicted a 10 percent chance of rain, and they laugh at the poll aggregator who predicted that Hillary Clinton had a 60 percent chance of winning the 2016 presidential election. These soothsayers defend themselves by invoking a frequentist interpretation of their probabilities: on one out of ten days in which she makes such a prediction, it rains; in six of ten elections with those polling numbers, the leading candidate wins. In this strip, Dilbert's boss illustrates a common fallacy:

DILBERT   **BY SCOTT ADAMS**

As we saw in chapter 1 with Linda and will see again in the next chapter, reframing a probability from credence in a single event to frequency in a set of events can recalibrate people's intuitions. A prosecutor in a big city who says "The probability that the DNA on the victim's clothing would match the DNA on the suspect if he were innocent is one in a hundred thousand" is likelier to win a conviction than one who says "Out of every hundred thousand innocent people in this city, one will show a match." The first feels like an estimate of subjective doubt that is indistinguishable from zero; the second invites us to imagine that falsely accused fellow, together with the many others living in the metropolis.

People also confuse probability in the frequentist sense with propensity. Gerd Gigerenzer recounts a tour of an aerospace factory in which the guide told the visitors that its Ariane rockets had a 99.6 percent security factor.[10] They were standing in front of a poster depicting the ninety-four rockets and their histories, eight of which crashed or blew up. When Gigerenzer asked how a rocket with a 99.6 percent security factor could fail almost 9 percent of the time, the guide explained that the factor was calculated from the reliabilities of the individual parts, and the failures were the result of human error. Of course what we ultimately care about is how often the rocket slips the surly bonds of earth or buys the farm, regardless of the causes, so the only probability that matters is the overall frequency. By the same misunderstanding, people sometimes wonder why a popular candidate who is miles ahead in the polls is given only a 60 percent chance of winning the election, when nothing but a last-minute shocker could derail him. The answer is that the probability estimate takes into account last-minute shockers.

Probability versus Availability

Despite the difference in interpretations, probability is intimately tied to events as a proportion of opportunities, whether directly, in the classical and frequentist definitions, or indirectly, with the other judgments. Surely whenever we say that one event is more probable than another, we believe it will occur more often given the opportunity. To estimate risk, we should tally the number of instances of an event and mentally divide it by the number of occasions on which it could have taken place.

Yet one of the signature findings in the science of human judgment is that this is not how human probability estimation generally works. Instead, people judge the probability of events by the ease with which instances come into mind, a habit that Tversky and Kahneman called the *availability heuristic*.[11] We use the ranking from our brain's search engine—the images, anecdotes, and mental videos it coughs up—as our best guess of the probabilities. The heuristic exploits a feature of human memory, namely that recall is affected by frequency: the more often we encounter something, the stronger the trace it leaves in our brains. So working backwards and estimating frequency from recallability often works serviceably well. When asked which birds are most common in a city, you would not do badly by tapping your memory and guessing pigeons and sparrows rather than waxwings and flycatchers, instead of going to the trouble of consulting a bird census.

For most of human existence, availability and hearsay were the *only* ways to estimate frequency. Statistical databases were kept by some governments, but they were considered state secrets and divulged only to administrative elites. With the rise of liberal democracies in the nineteenth century, data came to be considered a public good.[12] Even today, when data on just about everything is a few clicks away, not many people avail themselves of it. We instinctively draw on our impressions, which distort our understanding whenever the strengths of those impressions don't mirror frequencies in the world. That can happen when our experiences are a biased sample of those events, or when the impressions are promoted or demoted in our mental search results by psychological amplifiers such as recency, vividness, or emotional poignancy. The effects on human affairs are sweeping.

Outside our immediate experience, we learn about the world through the media. Media coverage thus drives people's sense of frequency and risk: they think they are likelier to be killed by a tornado than by asthma, despite asthma being eighty times deadlier, presumably because tornadoes are more photogenic.[13] For similar reasons the kinds of people who can't stay out of the news tend to be overrepresented in our mental censuses. What percentage of teenage girls give birth each year, worldwide? People guess 20 percent, around ten times too many. What proportion of Americans are immigrants? Around 28 percent, say survey respondents; the correct answer is 12 percent. Gay? Americans guess 24 percent; polls indicate 4.5 percent.[14] African Americans? About a third, people say, around two and half times higher than the real figure, 12.7 percent. That's still more accurate than their estimate for another conspicuous minority, Jews, where respondents are off by a factor of nine (18 versus 2 percent).[15]

The availability heuristic is a major driver of world events, often in irrational directions. Other than disease, the most lethal risk to life and limb is accidents, which kill about five million people a year (out of 56 million deaths in all), about a quarter of them in traffic accidents.[16] But except when they take the life of a photogenic celebrity, car crashes seldom make the news, and people are insouciant about the carnage. Plane crashes, in contrast, get lavish coverage, but they kill only about 250 people a year worldwide, making planes about a thousand times safer per passenger mile than cars.[17] Yet we all know people with a fear of flying but no one with a fear of driving, and a gory plane crash can scare airline passengers onto the highways for months afterward, where thousands more die.[18] The *SMBC* cartoon makes a similar point.

Among the most vivid and ghastly deaths imaginable is the one described in the song from *The Threepenny Opera*: "When that shark bites with his teeth, babe, scarlet billows start to spread."[19] In 2019, after a Cape Cod surfer became the first shark fatality in Massachusetts in more than eighty years, the towns equipped every beach with menacing *Jaws*-like warning billboards and hemorrhage-control kits, and commissioned studies on towers, drones, planes,

Used by permission of Zach Weinersmith

balloons, sonar, acoustic buoys, and electromagnetic and odorant repellents. Yet every year on Cape Cod between fifteen and twenty people die in car crashes, and cheap improvements in signage, barriers, and traffic law enforcement could save many more lives at a fraction of the cost.[20]

The availability bias may affect the fate of the planet. Several eminent climate scientists, having crunched the numbers, warn that "there is no credible path to climate stabilization that does not include a substantial role for nuclear power."[21] Nuclear power is the safest form of energy humanity has ever used. Mining accidents, hydroelectric dam failures, natural gas explosions, and oil train crashes all kill people, sometimes in large numbers, and smoke from burning coal kills them in enormous numbers, more than half a million per year. Yet nuclear power has stalled for decades in the United States and is being pushed back in Europe, often replaced by dirty and dangerous coal. In large part the opposition is driven by memories of three accidents: Three Mile Island in 1979, which killed no one;

Fukushima in 2011, which killed one worker years later (the other deaths were caused by the tsunami and from a panicked evacuation); and the Soviet-bungled Chernobyl in 1986, which killed 31 in the accident and perhaps several thousand from cancer, around the same number killed by coal emissions *every day*.[22]

Availability, to be sure, is not the only distorter of risk perception. Paul Slovic, a collaborator of Tversky and Kahneman, showed that people also overestimate the danger from threats that are novel (the devil they don't know instead of the devil they do), out of their control (as if they can drive more safely than a pilot can fly), human-made (so they avoid genetically modified foods but swallow the many toxins that evolved naturally in plants), and inequitable (when they feel they assume a risk for another's gain).[23] When these bugbears combine with the prospect of a disaster that kills many people at once, the sum of all fears becomes a *dread risk*. Plane crashes, nuclear meltdowns, and terrorist attacks are prime examples.

TERRORISM, LIKE OTHER LOSSES of life with malice aforethought, brews up a different chemistry of fear. Body-counting data scientists are often perplexed at the way that highly publicized but low-casualty killings can lead to epochal societal reactions. The worst terrorist attack in history by far was 9/11, and it claimed 3,000 lives; in most bad years, the United States suffers a few dozen terrorist deaths, a rounding error in the tally of homicides and accidents. (The annual toll is lower, for example, than the number of people killed by lightning, bee stings, or drowning in bathtubs.) Yet 9/11 led to the creation of a new federal department, massive surveillance of citizens and hardening of public facilities, and two wars which killed more than twice as many Americans as the number who died in 2001, together with hundreds of thousands of Iraqis and Afghans.[24]

To take another low-death/high-fear hazard, rampage killings in American schools claim around 35 victims a year, compared with about 16,000 routine police-blotter homicides.[25] Yet American schools have implemented billions of dollars of dubious safety measures, like installing bulletproof whiteboards and arming teachers with pepper-ball guns, while traumatizing children with terrifying active-shooter drills. In 2020 the brutal murder of George Floyd, an unarmed African American man, by a white police officer led to massive protests and the sudden adoption of a radical academic doctrine, Critical Race Theory, by universities, newspapers, and corporations. These upheavals were driven by the impression that African Americans are at serious risk of being killed by the police. Yet as with terrorism and school shootings, the numbers are surprising. A total of 65 unarmed Americans of all races are killed by the police in an average year, of which 23 are African American, which is around three tenths of one percent of the 7,500 African American homicide victims.[26]

It would be psychologically obtuse to explain the outsize reaction to publicized killings solely by availability-inflated fear. As with many signs of apparent irrationality, there are other logics at work, in the service of goals other than accurate probabilities.

Our disproportionate reaction to murder most foul may be irrational in the framework of probability theory but rational in the framework of game theory (chapter 8). Homicide is not like other lethal hazards. A hurricane or shark doesn't care how we will respond to the harm they have in store for us, but a human killer might. So when people react to a killing with public shock and anger, and redouble their commitment to self-defense, justice, or revenge, it sends a signal to the premeditating killers out there, possibly giving them second thoughts.

Game theory may also explain the frenzy set off by a special kind of event that Thomas Schelling described in 1960, which may be called a communal outrage.[27] A communal outrage is a flagrant, widely witnessed attack upon a member or symbol of a collective. It is felt to be an intolerable affront and incites the collective to rise up and righteously avenge it. Examples include the explosion of the USS *Maine* in 1898, leading to the Spanish-American War; the sinking of the RMS *Lusitania* in 1915, tipping the United States toward entering World War I; the Reichstag fire of 1933, enabling the establishment of the Nazi regime; Pearl Harbor in 1941, sending America into World War II; 9/11, which licensed the invasions of Afghanistan and Iraq; and the harassment of a produce peddler in Tunisia in 2010, whose self-immolation set off the Tunisian Revolution and Arab Spring. The logic of these reactions is *common knowledge* in the technical sense of something that everyone knows that everyone knows that everyone knows.[28] Common knowledge is necessary for *coordination*, in which several parties act in the expectation that each of the others will too. Common knowledge can be generated by *focal points*, public happenings which people see other people seeing. A public outrage can be the common knowledge that solves the problem of getting everyone to act in concert when a vexation has built up gradually and the right moment to deal with it never seems to arrive. The unignorable atrocity can trigger simultaneous indignation in a dispersed constituency and forge them into a resolute collective. The amount of harm inflicted by the attack is beside the point.

Not just beside the point but taboo. A communal outrage inspires what the psychologist Roy Baumeister calls a victim narrative: a moralized allegory in which a harmful act is sanctified, the damage consecrated as irreparable and unforgivable.[29] The goal of the narrative is not accuracy but solidarity. Picking nits about what actually happened is seen as not just irrelevant but sacrilegious or treasonous.[30]

At best, a public outrage can mobilize overdue action against a long-simmering trouble, as is happening in the grappling with systemic racism in response to the Floyd killing. Thoughtful leadership can channel an outrage into responsible reform, captured in the politician's saying "Never let a crisis go to waste."[31] But the history of public outrages suggests they can also empower demagogues and egg impassioned mobs into quagmires and disasters. Overall, I suspect that more good comes from cooler heads assessing harms accurately and responding to them proportionately.[32]

OUTRAGES CANNOT BECOME PUBLIC without media coverage. It was in the aftermath of the *Maine* explosion that the term "yellow journalism" came into common usage. Even when journalists don't whip readers into a jingoistic lather, intemperate public reactions are a built-in hazard. I believe journalists have not given enough thought to the way that media coverage can activate our cognitive biases and distort our understanding. Cynics might respond that the journalists couldn't care less, since the only thing that matters to them is clicks and eyeballs. But in my experience most journalists are idealists, who feel they answer to the higher calling of informing the public.

The press is an availability machine. It serves up anecdotes which feed our impression of what's common in a way that is guaranteed to mislead. Since news is what happens, not what doesn't happen, the denominator in the fraction corresponding to the true probability of an event—all the opportunities for the event to occur, including those in which it doesn't—is invisible, leaving us in the dark about how prevalent something really is.

The distortions, moreover, are not haphazard, but misdirect us toward the morbid. Things that happen suddenly are usually bad—a war, a shooting, a famine, a financial collapse—but good things may consist of nothing happening, like a boring country at peace or a for-gettable region that is healthy and well fed. And when progress takes place, it isn't built in a day; it creeps up a few percentage points a year, transforming the world by stealth. As the economist Max Roser points out, news sites could have run the headline 137,000 PEOPLE ESCAPED EXTREME POVERTY YESTERDAY every day for the past twenty-five years.[33] But they never ran the headline, because there was never a Thursday in October in which it suddenly happened. So one of the greatest developments in human history—a billion and a quarter people escaping from squalor—has gone unnoticed.

The ignorance is measurable. Pollsters repeatedly find that while people tend to be too optimistic about their own lives, they are too pessimistic about their societies. For instance, in most years between 1992 and 2015, an era that criminologists call the Great American Crime Decline, a majority of Americans believed that crime was ris-ing.[34] In their "Ignorance Project," Hans and Ola Rosling and Anna Rosling-Rönnlund have shown that the understanding of global trends in most educated people is exactly backwards: they think that longev-ity, literacy, and extreme poverty are worsening, whereas all have dra-matically improved.[35] (The Covid-19 pandemic set these trends back in 2020, almost certainly temporarily.)

Availability-driven ignorance can be corrosive. A looping mental newsreel of catastrophes and failures can breed cynicism about the ability of science, liberal democracy, and institutions of global coop-eration to improve the human condition. The result can be a para-lyzing fatalism or a reckless radicalism: a call to smash the machine, drain the swamp, or empower a demagogue who promises "I alone can fix it."[36] Calamity-peddling journalism also sets up perverse incentives for terrorists and rampage shooters, who can game the system and win instant notoriety.[37] And a special place in Journalist Hell is reserved for the scribes who in 2021, during the rollout of

Covid vaccines known to have a 95 percent efficacy rate, wrote stories on the vaccinated people who came down with the disease—by definition not news (since it was always certain there would be some) and guaranteed to scare thousands from this lifesaving treatment.

How can we recognize the genuine dangers in the world while calibrating our understanding to reality? Consumers of news should be aware of its built-in bias and adjust their information diet to include sources that present the bigger statistical picture: less Facebook News Feed, more *Our World in Data*.[38] Journalists should put lurid events in context. A killing or plane crash or shark attack should be accompanied by the annual rate, which takes into account the denominator of the probability, not just the numerator. A setback or spate of misfortunes should be put into the context of the longer-term trend. News sources might include a dashboard of national and global indicators—the homicide rate, CO_2 emissions, war deaths, democracies, hate crimes, violence against women, poverty, and so on—so readers can see the trends for themselves and get a sense of which policies move the needle in the right direction. Though editors have told me that readers hate math and will never put up with numbers spoiling their stories and pictures, their own media belie this condescension. People avidly consume data in the weather, business, and sports pages, so why not the news?

Conjunctive, Disjunctive, and Conditional Probabilities

A TV meteorologist announces there is a 50 percent chance of rain on Saturday and a 50 percent chance of rain on Sunday and concludes there is a 100 percent chance of rain over the weekend.[39] In an old joke, a man carries a bomb onto a plane for his own safety because, he figures, what are the chances that a plane will have *two* bombs on it? And then there's the argument that the pope is almost certainly a space alien. The probability that a randomly selected person on earth is the pope is tiny: one out of 7.8 billion, or .00000000013. Francis is the pope. Therefore, Francis is probably not a human being.[40]

In reasoning about probability, it's easy to go off the rails. These clangers come from misapplying the next step in understanding probability: how to calculate the probabilities of a conjunction, a disjunction, a complement, and a conditional. If these terms sound familiar, it's because they are the probabilistic equivalents of AND, OR, NOT, and IF-THEN from the previous chapter. Though the formulas are simple, each lays a trap, and tripping them is what gives rise to the probability gaffes.[41]

The probability of a *conjunction* of two independent events, prob(A AND B), is the product of the probabilities of each: prob(A) × prob(B). If the Greens have two children, what is the probability that both are girls? It's the probability that the first one is a girl, .5, times the probability that the second is a girl, also .5, or .25. Translating from single-event to frequentist language, we will find that in all the two-children families we look at, a quarter will be all-girl. More intuitive still, the classical definition of probability advises us to lay out the logical possibilities: Boy-Boy, Boy-Girl, Girl-Boy, Girl-Girl. One of these four is all-girl.

The trap in the conjunction formula lies in the proviso *independent*. Events are independent when they are disconnected: the chance of seeing one has no bearing on the chance of seeing the other. Imagine a society, perhaps not far off, in which people can choose the sex of their children. Imagine for the sake of the example that parents are gender chauvinists, half wanting only boys and the other only girls. If the first child is a girl, it tips us off that the parents preferred a girl, which means they would opt for a girl again, and vice versa if the first child is a boy. The events are not independent, and multiplication fails. If the preferences were absolute and the technology perfect, every family would have only sons or only daughters, and the probability that a two-child family is all-girl would be .5, not .25.

Failing to think about whether events are independent can lead to big boo-boos. When a streak of rare occurrences pops up in entities that are not quarantined from one another—the occupants of a building, who give each other colds, or the members of a peer group, who copy each other's fashions, or the survey answers from a single respondent, who retains his biases from question to question, or the measurements of anything on successive days or months or years, which may show inertia—then the set of observations is in effect a single event, not a freakish run of events, and their probabilities may not be multiplied. For example, if the crime rate was below average in each of the twelve months after Neighborhood Watch signs were posted in a city, it would be a mistake to conclude that the run must be due to the signage rather than chance. Crime rates change slowly, with the patterns in one month carrying over to the next, so the outcome is closer to a single coin flip than a run of twelve coin flips.

In the legal arena, misapplying the formula for a conjunction is not just a math error but a miscarriage of justice. A notorious example is the bogus "Meadow's Law," named after a British pediatrician who declared that when crib deaths within a family are examined, "one is a tragedy, two is suspicious and three is murder unless there is proof to the contrary." In the 1999 case of the attorney Sally Clark, who had lost two infant sons, the doctor testified that since the probability of a crib death in an affluent nonsmoking family is 1 in 8,500, the probability of two crib deaths is the square of that number, 1 in 72 million. Clark was sentenced to life imprisonment for murder.

Appalled statisticians pointed out the mistake: crib deaths within
a family are not independent, because siblings may share a genetic
predisposition, the home may have elevated risk factors, and the parents may have reacted to the first tragedy by taking misguided pre-
cautions that increased the chance of a second. Clark was released
after a second appeal (on different grounds), and in the following
years hundreds of cases based on similar errors had to be reviewed.[42]

Another howler in calculating conjunctions had a cameo in the
bizarre attempt by Donald Trump and his supporters to overturn
the results of the 2020 presidential election based on baseless claims
of voter fraud. In a motion filed with the US Supreme Court, the
Texas attorney general Ken Paxton wrote: "The probability of former
Vice President Biden winning the popular vote in the four Defendant States—Georgia, Michigan, Pennsylvania, and Wisconsin—
independently given President Trump's early lead in those States as
of 3 a.m. on November 4, 2020, is less than one in a quadrillion, or 1
in 1,000,000,000,000,000. For former Vice President Biden to win
these four States collectively, the odds of that event happening decrease to less than one in a quadrillion to the fourth power." Paxton's
jaw-dropping math assumed that the votes being tallied over the
course of the counting were statistically independent, like repeated
rolls of a die. But urbanites tend to vote differently from suburban-
ites, who in turn vote differently from country folk, and in-person
voters differ from those who mail in their ballots (particularly in
2020, when Trump discouraged his supporters from voting by mail).
Within each sector, the votes are not independent, and the base rates
differ from sector to sector. Since the results from each precinct
are announced as they become available, and the mail-in ballots are
counted later still, then as the different tranches are added up, the
running tally favoring each candidate can rise or fall, and the final
result cannot be extrapolated from the interim ones. The flapdoodle

was raised to the fourth power when Paxton multiplied the bogus probabilities from the four states, whose votes are not independent either: whatever sways voters in the Great Lake State is also likely to sway them in America's Dairyland.[43]

STATISTICAL INDEPENDENCE IS TIED to the concept of causation: if one event affects another, they are not statistically independent (though, as we shall see, the converse isn't true: events that are causally isolated may be statistically dependent). That is why the gambler's fallacy is a fallacy. One spin of a roulette wheel cannot impinge on the next, so the high roller who expects a run of blacks to set up a red will lose his shirt: the probability is always a bit less than .5 (because of the green slots with 0 and 00). This shows that fallacies of statistical independence can go both ways: falsely assuming independence (as in Meadow's fallacy) and falsely assuming dependence (as in the gambler's fallacy).

Whether events are independent is not always obvious. Among the most famous applications of research on cognitive biases to everyday life was Tversky's analysis (with the social psychologist Tom Gilovich) of the "hot hand" in basketball.[44] Every hoops fan knows that from time to time a player can be "on fire," "in the zone," or "unconscious," especially "streak shooters" like Vinnie "The Microwave" Johnson, the 1980s Detroit Pistons guard who earned his sobriquet because he heats up in a hurry. In the teeth of incredulity from every fan, coach, player, and sportswriter, Tversky and Gilovich claimed that the hot hand was an illusion, a gambler's fallacy in reverse. The data they analyzed suggested that the outcome of every attempt is statistically independent of the preceding run of attempts.

Now, before looking at the data, one cannot blow off the possibility of a hot hand on the grounds of causal plausibility in the way one can blow off the gambler's fallacy. Unlike a roulette wheel, a player's body and brain do have a memory, and it is far from superstitious to think that a spurt of energy or confidence might persist over a span of minutes. So it was not a breach of the scientific worldview when other statisticians took a second look at the data and concluded that the boffins were wrong and the jocks were right: there *is* a hot hand in basketball. The economists Joshua Miller and Adam Sanjurjo showed that when you select streaks of hits or misses from a long run of data, the outcome of the very next attempt is not statistically independent of that streak. The reason is that if the attempt had happened to be successful and continue the streak, it might have been counted as part of that streak in the first place. Any attempt which is singled out because it took place following a streak is biased to be an unsuccessful attempt: one that had no chance of being defined as part of the streak itself. That throws off the calculations of what one should expect by chance, which in turn throws off the conclusion that basketball players are no streakier than roulette wheels.[45]

The hot hand fallacy fallacy has three lessons. First, events can be statistically dependent not only when one event causally impinges on the other but when it affects which event is selected for comparison. Second, the gambler's fallacy may arise from a not-so-irrational feature of perception: when we look for streaks in a long run of events, a streak of a given length really is likelier to be reversed than to continue. Third, probability can be truly, deeply unintuitive: even the mavens can mess up the math.

LET'S TURN TO THE PROBABILITY of a *disjunction* of events, prob(A OR B). It is the probability of A plus the probability of B minus the probability of both A and B. If the Browns have two children, the probability that at least one is a girl—that is, that the first is a girl or the second is a girl—is .5 + .5 − .25, or .75. You can arrive at the same result by counting the combinations: Boy-Girl + Girl-Girl + Girl-Boy (three possibilities) out of Boy-Girl + Boy-Boy + Girl-Boy + Girl-Girl (four opportunities). Or by tallying frequencies: in a large set of families with two children, you'll find that three fourths have at least one daughter.

The arithmetic of OR shows us what went wrong with the weathercaster who said it was certain to rain over the weekend because there was a 50 percent chance of rain on each day: by simply adding the two probabilities, he inadvertently double-counted the weekends on which it would rain on *both* days, neglecting to subtract .25 for the conjunction. He applied a rule that works for exclusive-OR (XOR), namely A or B but not both. The probabilities of mutually exclusive events *can* be added to get the disjunction, and the sum of all of them is 1, certainty. The probability that a child is either a boy (.5) or a girl (.5) is their sum, 1, since the child must be either one or the other (this being an example to explain the math, I've adopted a gender binary and not considered intersex children). If you forget the difference and confuse overlapping with mutually exclusive events, you can get crazy results. Imagine the weathercaster predicting a .5 chance of rain on Saturday, on Sunday, and on Monday, and concluding that the chance of rain over the long weekend was 1.5.

The probability of the *complement* of an event, namely A not happening, is 1 minus the probability of it happening. This comes in handy when we have to estimate the probability of "at least one" event. Remember the Browns with their daughter, or perhaps two? Since having at least one daughter is the same as not having all sons, then instead of calculating the disjunction (first child is a girl OR second child is a girl), we could have calculated the complement of a conjunction: 1 minus the chance of having all boys (which is .25), namely .75. In the case of two events it doesn't make much difference which formula we use. But when we have to calculate the probability of at least one A in a large set, the disjunction rule requires the tedium of adding and subtracting a lot of combinations. It's easier to calculate it as the probability of "not all NOT-A," which is simply 1 minus a big product.

Suppose, for example, that every year there's a 10 percent chance of a war breaking out. What are the chances of at least one war breaking out over a decade? (Let's assume that wars are independent, not contagious, which seems to be true.)[46] Instead of adding up the chance that a war will break out in Year 1 plus the chance that it will break out in Year 2 minus the chance that one will break out in both Year 1 and Year 2, and so on for all combinations, we can simply calculate the chance that *no* war will break out over *all* of the years and subtract it from 1. That is simply the chance that war will not break out in a given year, .9, multiplied by itself for each of the other years (.9 × .9 × . . . .9, or $.9^{10}$, which equals .35), which when subtracted from 1 yields .65.

FINALLY WE GET TO a *conditional* probability: the probability of A given B, written as prob(A | B). A conditional probability is conceptually simple: it's just the probability of the THEN in an IF-THEN. It's also arithmetically simple: it's just the probability of A AND B divided by the probability of B. Nonetheless, it's the source of endless confusions, blunders, and paradoxes in reasoning about probability, starting with the hapless fellow in the *XKCD* cartoon on the following page.[47] His error lies in confusing the simple probability or *base rate* of lightning deaths, prob(struck-by-lightning), with the *conditional* probability of a lightning death given that one is outside during an electrical storm, prob(struck-by-lightning | outside-in-storm).

Though the arithmetic of a conditional probability is simple, it's

THE ANNUAL DEATH RATE AMONG PEOPLE
WHO KNOW THAT STATISTIC IS ONE IN SIX.

unintuitive until we make it concrete and visualizable (as always).
Look at the Venn diagrams on the page after the next one, where the
size of a region on the page corresponds to the number of outcomes.
The rectangle, with an area of 1, embraces all the possibilities. A
circle encloses all the As, and the top left figure shows that the probability of A corresponds to its area (dark) as a proportion of the whole
rectangle (pale)—another way of saying the number of occurrences
divided by the number of opportunities. The top right figure shows
the probability of A OR B, which is the total dark area, namely the
area of A plus the area of B without double-counting the wedge in
the middle they share, that is, the probability of A AND B. That
wedge, prob(A AND B), is shown in the lower left diagram.

The bottom right diagram explains the deal with conditional probabilities. It indicates that we should ignore the vast space of everything that can possibly happen, bleached into white, and focus our
attention only on the incidents in which B happens, the shaded circle.

Now we scrutinize how many of *those* incidents are ones in which A also happens: the size of the A AND B wedge as a proportion of the size of the B circle. Of all the interludes in which people walk in an electrical storm (B), what proportion of them result in a lightning strike (A AND B)? That's why we calculate the conditional, prob(A | B), by dividing the conjunction, prob(A AND B), into the base rate, prob(B).

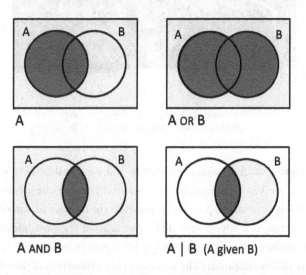

A

A OR B

A AND B

A | B (A given B)

Here's an example. The Grays have two children. The elder is a girl. Knowing this, what is the probability that both are girls? Let's translate the question into a conditional probability, namely the probability that the first is a girl and the second is a girl given that the first is a girl, or, in fancy notation, prob(1st = Girl AND 2nd = Girl | 1st = Girl). The formula tells us to divide the conjunction, which we already calculated to be .25, by the simple probability for the second child, .5, and we get .5. Or, thinking classically and concretely: Girl-Girl (one possibility) divided by Girl-Girl and Girl-Boy (two opportunities) equals one half.

Conditional probabilities add some precision to the concept of statistical independence, which I left hanging in the preceding subsection. The concept may now be defined: A and B are independent if, for all Bs, the probability of A given B is the same as the overall probability of A (and so on for B). Now, remember the illegal multiplication of probabilities for the conjunction of events when they are not independent? What are we supposed to do instead? Easy: the probability of the conjunction of A and B when they are *not* independent is the probability of A times the probability of B given A, to wit, prob(A) × prob(B | A).

Why am I belaboring the concept of conditional probability with all those synonymous representations—English prose, its logical equivalent, the mathematical formula, Venn diagrams, counting up the possibilities? It's because conditional probability is such a source of confusion that you can't have too many explanations.[48]

If you don't believe me, consider the Whites, yet another two-child family. At least one of them is a girl. What is the probability that both are girls, namely, the conditional probability of two girls given at least one girl, or prob(1st = Girl AND 2nd = Girl | 1st = Girl OR 2nd = Girl)? So few people get the answer right that statisticians call it the Boy or Girl paradox. People tend to say .5; the correct answer is .33. In this case concrete thinking can lead to the wrong answer: people visualize an elder girl, realize she could have either a younger sister or a younger brother, and figure that the sister is one possibility out of those two. They forget that there's another way of having at least one girl: she could be the younger of the two. Enumerating the possibilities properly, we get Girl-Girl (one) divided by [Girl-Girl plus Girl-Boy plus Boy-Girl] (three), which equals one third. Or, using the formula, we divide .25 (Girl AND Girl) by .75 (Girl OR Girl).

The Boy or Girl paradox is not just a trick with wording. It comes from a failure of the imagination to enumerate the possibilities, and appears in many guises, including the Monty Hall dilemma. Here's a simpler yet exact equivalent.[49] Some sidewalk card sharks make a living from roping passersby into playing Three Cards in a Hat. The shark shows them a card that is red on both sides, a card that is white on both sides, and a card that is red on one side and white on the other. He mixes them in a hat, draws one, notes that the face is (say) red, and offers the passerby even money that the other side is also red (they pay him a dollar if it's red, he pays them a dollar if it's white). It's a sucker's bet: the odds that it's red are two in three. The rubes mentally count cards instead of *sides* of cards, forgetting that there are two ways that the all-red card, had it been chosen, could have appeared red side up.

And remember the guy who brought his own bomb onto the plane? He calculated the overall probability that a plane would have two bombs on it. But by bringing his own bomb aboard, he had already ruled out most of the possibilities in the denominator. The number he should care about is the conditional probability that a plane will have two bombs *given that* it already has a bomb, namely his own (which has a probability of 1). That conditional is the probability that someone else will have a bomb times 1 (the conjunction of his bomb and the other guy's) divided by 1 (his bomb), which works out, of course, to the probability that someone else will have a bomb, just where he started. The joke was used to good effect in *The World According to Garp*. The Garps are looking over a house when a small plane crashes into it. Garp says, "We'll take the house. The chances of another plane hitting this house are astronomical."[50]

Forgetting to condition a base-rate probability by special circumstances in place—the lightning storm, the bomb you bring aboard—is a common probability blunder. During the 1995 trial of O. J. Simpson, the football star accused of murdering his wife, Nicole, a prosecutor called attention to his history of battering her. A member of Simpson's "Dream Team" of defense attorneys replied that very few batterers go on to kill their wives, perhaps one in 2,500. An English professor, Elaine Scarry, spotted the fallacy. Nicole Simpson was not just any old victim of battering. She was a victim of battering *who had her throat cut*. The relevant statistic is the conditional probability that someone killed his wife *given* that he had battered his wife *and that his wife was murdered by someone*. That probability is eight out of nine.[51]

THE OTHER COMMON ERROR with conditional probability is confusing the probability of A given B with the probability of B given A, the statistical equivalent of affirming the consequent (going from IF P THEN Q to IF Q THEN P).[52] Remember Irwin the hypochondriac, who knew he had liver disease because his symptoms matched the list perfectly, namely no discomfort? Irwin confused the probability of no symptoms given liver disease, which is high, with the probability of liver disease given no symptoms, which is low. That is because the probability of liver disease (its base rate) is low, and the probability of no discomfort is high.

Conditional probabilities cannot be flipped whenever base rates differ. Take a real-life example, the finding that a third of fatal accidents occur in the home, which inspired the headline PRIVATE HOMES ARE DANGEROUS SPOTS. The problem is that the home is where we spend most of our time, so even if homes are not particularly dangerous, a lot of accidents happen to us there because a lot of *everything* happens to us there. The headline writer confused the probability that we were at home given that a fatal accident has occurred—the statistic being reported—with the probability that a fatal accident occurs given that we are at home, which is the propensity that readers are interested in. We can grasp the problem more intuitively by looking at the diagram below, where the base rates are reflected in the relative sizes of the circles (say, with A as days with fatal accidents, B with days at home).

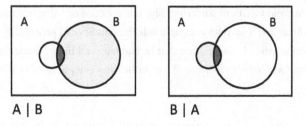

The left diagram shows the probability of A given B (the probability of a fatal accident given that one is at home); it is the area of the dark wedge (A AND B) as a proportion of the big pale circle (B, being at home), which is small. The right diagram shows the probability of B given A (of being at home given there was a fatal accident); it is the area of that same dark wedge but this time as a proportion of the small pale circle, fatal accidents, and is much bigger.

One reason that conditional probabilities are so easy to get backwards is that the English language is ambiguous as to which is intended. "The probability of an accident taking place in the home is .33" could mean "as a proportion of accidents" or "as a proportion of time spent at home." The difference can get lost in translation and spawn bogus estimates of propensities. A majority of bicycle accidents involve boys, so we get the headline BOYS MORE AT RISK ON BICYCLES, implying that boys are more reckless, whereas in fact they may just be more avid bicycle riders. And in what statisticians call the prosecutor's fallacy, the DA announces that the likelihood of the victim's blood type matching that on the defendant's clothing by chance is just 3 percent, and concludes that the probability that the defendant is guilty is 97 percent. He has confused (and hopes the jurors will confuse) the probability of a match given the defendant's innocence with the probability of the defendant's innocence given a match.[53] How to do the arithmetic properly is the topic of the next chapter, Bayesian reasoning.

Ambiguities in conditional probability can be incendiary. In 2019 a pair of social scientists created a furor when they published a study in the prestigious *Proceedings of the National Academy of Sciences* which, citing numbers like the ones I mentioned in an earlier section, claimed that police were likelier to shoot whites than blacks, contrary to the common assumption of racial bias. Critics pointed out that this conclusion pertained to the probability that someone is black given that they were shot, which is indeed lower than the corresponding probability for whites, but only because the country has fewer blacks than whites in the first place, a difference in the base rates. If the police *are* racially biased, that would be a propensity manifesting itself as a higher probability that someone is shot given that they are black, and the data suggest that the probability is indeed higher. Though the original authors noted that the suitable base rate is not obvious—should it be the proportion of blacks in the population, or in encounters with the police?—they realized they had made such a mess in how they had stated the probabilities that they formally retracted the paper.[54]

And the pope from outer space? That's what you get when you confuse the probability of being the pope given that someone is human with the probability that someone is human given that he is the pope.[55]

Prior and Post Hoc Probabilities

A man tries on a custom suit and says to the tailor, "I need this sleeve taken in." The tailor says, "No, just bend your elbow like this. See, it pulls up the sleeve." The customer says, "Well, OK, but when I bend my elbow, the collar goes up the back of my neck." The tailor says, "So? Raise your head up and back. Perfect." The man says, "But now the left shoulder is three inches lower than the right one!" The tailor says, "No problem. Bend at the waist and it evens out." The man leaves the store wearing the suit, his right elbow sticking out, his head craned back, his torso bent to the left, walking with a herky-jerky gait. A pair of pedestrians pass him by. The first says, "Did you see that poor disabled guy? My heart aches for him." The second says, "Yeah, but his tailor is a genius—the suit fits him perfectly!"

The joke illustrates yet another family of probability blunders: confusing prior with post hoc judgments (also called *a priori* and *a posteriori*). The confusion is sometimes called the Texas sharpshooter fallacy, after the marksman who fires a bullet into the side of a barn and then paints a bull's-eye around the hole. In the case of probability, it makes a big difference whether the denominator of the fraction—the number of opportunities for an event to occur—is counted independently of the numerator, the events of interest. Confirmation bias, discussed in chapter 1, sets up the error: once we expect a pattern, we seek out examples and ignore the counterexamples. If you take note of the predictions by a psychic that are borne out by events, but don't divide by the total number of predictions, correct and incorrect, you can get any probability you want. As Francis Bacon noted in 1620, such is the way of all superstitions, whether in astrology, dreams, omens, or divine judgments.

Or financial markets. An unscrupulous investment advisor with a 100,000-person mailing list sends a newsletter to half of the list predicting that the market will rise and a version to the other half predicting that it will fall. At the end of every quarter he discards the names of the people to whom he sent the wrong prediction and repeats the process with the remainder. After two years he signs up the

390 recipients who are amazed at his track record of predicting the market eight quarters in a row.[56]

Though this scam is illegal if carried out knowingly, when it's carried out naïvely it's the lifeblood of the finance industry. Traders are lightning-quick at snapping up bargains, so very few stock-pickers can outperform a mindless basket of securities. One exception was Bill Miller, anointed by *CNN Money* in 2006 as "The Greatest Money Manager of Our Time" for beating the S&P 500 stock market index fifteen years in a row. How impressive is that? One might think that if a manager is equally likely to outperform or underperform the index in any year, the odds of that happening by chance are just 1 in 32,768 (2^{15}). But Miller was singled out *after* his amazing streak had unfolded. As the physicist Len Mlodinow pointed out in *The Drunkard's Walk: How Randomness Rules Our Lives*, the country has more than six thousand fund managers, and modern mutual funds have been around for about forty years. The chance that *some* manager had a fifteen-year winning streak *sometime* over those forty years is not at all unlikely; it's 3 in 4. The *CNN Money* headline could have read EXPECTED 15-YEAR RUN FINALLY OCCURS: BILL MILLER IS THE LUCKY ONE. Sure enough, Miller's luck ran out, and in the following two years the market "handily pulverized him."[57]

On top of confirmation bias, a major contributor to post hoc probability fallacies is our failure to appreciate how many opportunities there are for coincidences to occur. When we are allowed to identify them post hoc, coincidences are not unlikely at all; they're pretty much guaranteed to happen. In one of his *Scientific American* columns, the recreational mathematician Martin Gardner asked, "Would you notice it if the license plate of a car just ahead of you bore digits that, read backward, gave your telephone number? Who except a numerologist or logophile would see the letters U, S, A symmetrically placed in LOUISIANA or at the end of JOHN PHILIP SOUSA, the name of the composer of our greatest patriotic marches? It takes an odd sort of mind to discover that Newton was born the same year that Galileo died, or that Bobby Fischer was born under the sign of Pisces (the Fish)."[58] But these numerologists and odd sorts of minds exist, and their post hoc sharpshooting can be spun into highfalutin theories. The psychoanalyst Carl Jung proposed a mystical force called synchronicity to explain the quintessential thing that needs no explanation, the prevalence of coincidence in the world.

When I was a child, what we now call memes were circulated in comic books and popular magazines. One that made the rounds was a list of the incredible similarities between Abraham Lincoln and John F. Kennedy. Honest Abe and JFK were both elected to Congress in '46 and to the presidency in '60. Both were shot in the head in the presence of their wives on a Friday. Lincoln had a secretary named Kennedy; Kennedy had a secretary named Lincoln. Both were succeeded by Johnsons who were born in '08. Their assassins were both born in '39 and had three names which add up to fifteen letters. John Wilkes Booth ran from a theater and was caught in a warehouse; Lee Harvey Oswald ran from a warehouse and was caught in a theater. What do these eerie parallels tell us? With all due respect to Dr. Jung, absolutely nothing, other than that coincidences happen more often than our statistically untutored minds appreciate. Not to mention the fact that when spooky coincidences are noticed, they tend to get embellished (Lincoln did not have a secretary named Kennedy), while pesky noncoincidences (like their different days, months, and years of birth and death) are ignored.

Scientists are not immune to the Texas sharpshooter fallacy. It's one of the explanations for the replicability crisis that rocked epidemiology, social psychology, human genetics, and other fields in the 2010s.[59] Think of all the foods that are good for you which used to be bad for you, the miracle drug that turns out to work no better than the placebo, the gene for this or that trait which was really noise in the DNA, the cute studies showing that people contribute more to the coffee fund when two eyespots are posted on the wall and that they walk more slowly to the elevator after completing an experiment that presented them with words associated with old age.

It's not that the investigators faked their data. It's that they engaged in what is now known as questionable research practices, the garden of forking paths, and p-hacking (referring to the probability threshold, p, that counts as "statistically significant").[60] Imagine a scientist who runs a laborious experiment and obtains data that are the opposite of "Eureka!" Before cutting his losses, he may be tempted to wonder whether the effect really is there, but only with the men, or only with the women, or if you throw out the freak data from the participants who zoned out, or if you exclude the crazy Trump years, or if you switch to a statistical test which looks at the ranking of the data rather than their values down to the last decimal place. Or you can continue to test participants until the precious asterisk appears in the statistical printout, being sure to quit while you're ahead.

None of these practices is inherently unreasonable if it can be justified before the data are collected. But if they are tried after the fact, some combination is likely to capitalize on chance and cough up a spurious result. The trap is inherent to the nature of probability and has been known for decades; I recall being warned against "data snooping" when I took statistics in 1974. But until recently few scientists intuitively grasped how a smidgen of data snooping could lead to a boatload of error. My professor half-jokingly suggested that scientists be required to write down their hypotheses and methods on a piece of paper before doing an experiment and safekeep it in a lockbox they would open and show to reviewers after the study was done.[61] The only problem, he noted, was that a scientist could secretly keep several lockboxes and then open the one he knew "predicted" the data. With the advent of the web, the problem has been solved, and the state of the art in scientific methodology is to "preregister" the details of a study in a public registry that reviewers and editors can check for post hoc hanky-panky.[62]

ONE KIND OF post hoc probability illusion is so common that it has its own name: the cluster illusion.[63] We're good at spotting tightly packed collections of things or events, because they often are part of a single happening: a barking dog that won't shut up, a weather system that drenches a city for several days, a burglar on a spree who robs several stores in a block. But not all clusters have a root cause—indeed, most of them don't. When there are lots of events, it's inevitable that some will wander into each other's neighborhoods and rub shoulders, unless some nonrandom process tries to keep them apart.

The cluster illusion makes us think that random processes are nonrandom and vice versa. When Tversky and Kahneman showed people (including statisticians) the results of real strings of coin flips, like TTHHTHTTTT, which inevitably have runs of consecutive heads or tails, they thought the coin was rigged. They would say a coin looked fair only if it *was* rigged to prevent the runs, like HTHTTHTHHT, which "looks" random even though it isn't.[64] I witnessed a similar illusion when I worked in an auditory perception lab. The participants had to detect faint tones, which were presented at random times so they couldn't guess when a tone would come. Some said the random-event generator must be broken because the tones came in bursts. They didn't realize that that's what randomness sounds like.

Phantom clusters arise in space as well. The stars making up the ram, lion, crab, virgin, archer, and other constellations are not neighbors in any galaxy but are randomly sprinkled across the night sky from our terrestrial vantage point and only grouped into shapes by our pattern-seeking brains. Spurious clusters also arise in the calendar. People are surprised to learn that if 23 people are in a room, the chances that two will share a birthday are better than even. With 57 in the room, the odds rise to 99 percent. Though it's unlikely that anyone in the room will share *my* birthday, we're not looking for matches with me, or with anyone else singled out a priori. We're counting matches post hoc, and there are 366 ways for a match to occur.

The cluster illusion, like other post hoc fallacies in probability, is the source of many superstitions: that bad things happen in threes, people are born under a bad sign, or an *annus horribilis* means the world is falling apart. When a series of plagues is visited upon us, it does not mean there is a God who is punishing us for our sins or testing our faith. It means there is not a God who is spacing them apart.

EVEN FOR THOSE WHO GRASP the mathematics of chance in all its bedeviling unintuitiveness, a lucky streak can seize the imagination. The underlying odds will determine how long, on average, a streak is expected to last, but the exact moment that luck runs out is an unfathomable mystery. This tension was explored in my favorite essay by the paleontologist, science writer, and baseball fan Stephen Jay Gould.[65]

Gould discussed one of the greatest achievements in sports, Joe DiMaggio's hitting streak of fifty-six games in 1941. He explained that the streak was statistically extraordinary even given DiMaggio's high batting average and the number of opportunities for streaks to have occurred in the history of the sport. The fact that DiMaggio benefited from some lucky breaks along the way does not diminish the achievement but exemplifies it, because no long streak, however pushed along by favorable odds, can ever unfold without them. Gould explains our fascination with runs of luck:

> The statistics of streaks and slumps, properly understood, do teach an important lesson about epistemology, and life in general. The history of a species, or any natural phenomenon that requires unbroken continuity in a world of trouble, works like a batting streak. All are games of a gambler playing with a limited stake against a house with infinite resources. The gambler must eventually go bust. His aim can only be to stick around as long as possible, to have some fun while he's at it, and, if he happens to be a moral agent as well, to worry about staying the course with honor. . . .
>
> DiMaggio's hitting streak is the finest of legitimate legends because it embodies the essence of the battle that truly defines our lives. DiMaggio activated the greatest and most unattainable dream of all humanity, the hope and chimera of all sages and shamans: he cheated death, at least for a while.

BELIEFS AND EVIDENCE

(BAYESIAN REASONING)

Extraordinary claims require extraordinary evidence.
—Carl Sagan

A heartening exception to the disdain for reason in so much of our online discourse is the rise of a "Rationality Community," whose members strive to be "less wrong" by compensating for their cognitive biases and embracing standards of critical thinking and epistemic humility.[1] The introduction to one of their online tutorials can serve as an introduction to the subject of this chapter:[2]

Bayes' rule or Bayes' theorem is the law of probability governing *the strength of evidence*—the rule saying *how much* to revise our probabilities (change our minds) when we learn a new fact or observe new evidence.

You may want to learn about Bayes' rule if you are:

- A professional who uses statistics, such as a scientist or doctor;
- A computer programmer working in machine learning;
- A human being.

Yes, a human being. Many Rationalistas believe that Bayes's rule is among the normative models that are most frequently flouted in everyday reasoning and which, if better appreciated, could add the biggest kick to public rationality. In recent decades Bayesian thinking has skyrocketed in prominence in every scientific field. Though few laypeople can name or explain it, they have felt its influence in the trendy term "priors," which refers to one of the variables in the theorem.

A paradigm case of Bayesian reasoning is medical diagnosis. Suppose that the prevalence of breast cancer in the population of women is 1 percent. Suppose that the sensitivity of a breast cancer test (its true-positive rate) is 90 percent. Suppose that its false-positive rate is 9 percent. A woman tests positive. What is the chance that she has the disease?

The most popular answer from a sample of doctors given these numbers ranged from 80 to 90 percent.[3] Bayes's rule allows you to calculate the correct answer: 9 percent. That's right, the professionals whom we entrust with our lives flub the basic task of interpreting a medical test, and not by a little bit. They think there's almost a 90 percent chance she has cancer, whereas in reality there's a 90 percent chance she doesn't. Imagine your emotional reaction upon hearing one figure or the other, and consider how you would weigh your options in response. That's why you, a human being, want to learn about Bayes's theorem.

Risky decision making requires both assessing the odds (Do I have cancer?) and weighing the consequences of each choice (If I do nothing and have cancer, I could die; if I undergo surgery and don't have cancer, I will suffer needless pain and disfigurement). In chapters 6 and 7 we'll explore how best to make consequential decisions when we know the probabilities, but the starting point must be the probabilities themselves: given the evidence, how likely is it that some state of affairs is true?

For all the scariness of the word "theorem," Bayes's rule is rather simple, and as we will see at the end of the chapter, it can be made gut-level intuitive. The great insight of the Reverend Thomas Bayes (1701–1761) was that the degree of belief in a hypothesis may be quantified as a probability. (This is the subjectivist meaning of "probability" that we met in the last chapter.) Call it prob(Hypothesis), the probability of a hypothesis, that is, our degree of credence that it is true. (In the case of medical diagnosis, the hypothesis is that the patient has the disease.) Clearly our credence in any idea should depend on the evidence. In probability-speak, we can say that our credence should be *conditional* on the evidence. What we are after is the probability of a hypothesis given the data, or prob(Hypothesis | Data). That's called the *posterior* probability, our credence in an idea after we've examined the evidence.

If you've taken this conceptual step, then you are prepared for Bayes's rule, because it's just the formula for conditional probability, which we met in the last chapter, applied to credence and evidence. Remember that the probability of A given B is the probability of A AND B divided by the probability of B. So the probability of a hypothesis given the data (what we are seeking) is the probability of the hypothesis *and* the data (say, the patient has the disease *and* the test result comes out positive) divided by the probability of the data (the total proportion of patients who test positive, healthy and sick). Stated as an equation: prob(Hypothesis | Data) = prob(Hypothesis AND Data) / prob(Data). One more reminder from chapter 4: the probability of A AND B is the probability of A times the probability of B given A. Make that simple substitution and you get Bayes's rule:

$$\text{prob(Hypothesis | Data)} = \frac{\text{prob(Hypothesis)} \times \text{prob(Data | Hypothesis)}}{\text{prob(Data)}}$$

What does this mean? Recall that prob(Hypothesis | Data), the expression on the left-hand side, is the posterior probability: our updated credence in the hypothesis after we've looked at the evidence. This could be our confidence in a disease diagnosis after we've seen the test results.

Prob(Hypothesis) on the right-hand side means the *prior* probability or "priors," our credence in the hypothesis *before* we looked at the data: how plausible or well established it was, what we would be forced to guess if we had no knowledge of the data at hand. In the case of a disease, this could be its prevalence in the population, the base rate.

Prob(Data | Hypothesis) is called the *likelihood*. In the world of Bayes, "likelihood" is not a synonym for "probability," but refers to how likely it is that the data would turn up *if* the hypothesis is true.[4] If someone does have the disease, how likely is it that they would show a given symptom or get a positive test result?

And prob(Data) is the probability of the data turning up across the board, whether the hypothesis is true or false. It's sometimes called the "marginal" probability, not in the sense of "minor" but in the sense of adding up the totals for each row (or each column) along the margin of the table—the probability of getting those data when the hypothesis is true plus the probability of getting those data when the hypothesis is false. A more mnemonic term is the commonness or ordinariness of the data. In the case of medical diagnosis, it refers to the proportion of *all* the patients who have a symptom or get a positive result, healthy and sick.

Substituting the mnemonics for the algebra, Bayes's rule becomes:

$$\text{Posterior probability} = \frac{\text{Prior probability} \times \text{Likelihood of the data}}{\text{Commonness of the data}}$$

Translated into English, it becomes "Our credence in a hypothesis after looking at the evidence should be our prior credence in the hypothesis, multiplied by how likely the evidence would be *if* the hypothesis is true, scaled by how common that evidence is across the board."

And translated into common sense, it works like this. Now that you've seen the evidence, how much should you believe the idea? First, believe it more if the idea was well supported, credible, or plausible to start with—if it has a high prior, the first term in the numerator. As they say to medical students, if you hear hoofbeats outside the window, it's probably a horse, not a zebra. If you see a patient with muscle aches, he's more likely to have the flu than kuru (a rare disease seen among the Fore tribe in New Guinea), even if the symptoms are consistent with both diseases.

Second, believe the idea more if the evidence is especially likely to occur when the idea is true—namely if it has a high likelihood, the second term in the numerator. It's reasonable to take seriously the possibility of methemoglobinemia, also known as blue skin disorder, if a patient shows up with blue skin, or Rocky Mountain spotted fever if a patient from the Rocky Mountains presents with spots and fever.

And third, believe it *less* if the evidence is commonplace—if it has a high marginal probability, the denominator of the fraction. That's why we laugh at Irwin the hypochondriac, convinced of his liver disease because of the characteristic lack of discomfort. True, his symptomlessness has a high likelihood given the disease, edging up the numerator, but it also has a massive marginal probability (since most people have no discomfort most of the time), blowing up the denominator and thus shrinking the posterior, our credence in Irwin's self-diagnosis.

How does this work with numbers? Let's go back to the cancer example. The prevalence of the disease in the population, 1 percent, is how we set our priors: prob(Hypothesis) = .01. The sensitivity of the test is the likelihood of getting a positive result given that the patient has the disease: prob(Data | Hypothesis) = .9. The marginal probability of a positive test result across the board is the sum of the probabilities of a hit for the sick patients (90 percent of the 1 percent, or .009) and of a false alarm for the healthy ones (9 percent of the 99 percent, or .0891), or .0981, which begs to be rounded up to .1. Plug the three numbers into Bayes's rule, and you get .01 times .9 divided by .1, or .09.

So where do the doctors (and, to be fair, most of us) go wrong? Why do we think the patient almost certainly has the disease, when she almost certainly doesn't?

Base-Rate Neglect and the Representativeness Heuristic

Kahneman and Tversky singled out a major ineptitude in our Bayesian reasoning: we neglect the *base rate*, which is usually the best estimate of the prior probability.[5] In the medical diagnosis problem, our heads are turned by the positive test result (the likelihood) and we forget about how rare the disease is in the population (the prior).

The duo went further and suggested we don't engage in Bayesian reasoning at all. Instead we judge the probability that an instance belongs to a category by how *representative* it is: how similar it is to the prototype or stereotype of that category, which we mentally represent as a fuzzy family with its crisscrossing resemblances (chapter 3). A cancer patient, typically, gets a positive diagnosis. How common the cancer is, and how common a positive diagnosis is, never enter our minds. (Horses, zebras, who cares?) Like the availability heuristic from the preceding chapter, the representativeness heuristic is a rule of thumb the brain deploys in lieu of doing the math.[6]

Tversky and Kahneman demonstrated base-rate neglect in the lab by telling people about a hit-and-run accident by a taxi late at night in a city with two cab companies: Green Taxi, which owns 85 percent of the cabs, and Blue Taxi, which owns 15 percent (those are the base rates, and hence the priors). An eyewitness identified the cab as Blue, and tests showed that he correctly identified colors at night 80 percent of the time (that is the likelihood of the data, namely his testimony given the cab's actual color). What is the probability that the cab involved in the accident was Blue? The correct answer, according to Bayes's rule, is .41. The median answer was .80, almost twice as high. Respondents took the likelihood too seriously, pretty much at face value, and downplayed the base rate.[7]

One of the symptoms of base-rate neglect in the world is hypochondria. Who among us hasn't worried we have Alzheimer's after a memory lapse, or an exotic cancer when we have an ache or pain? Another is medical scaremongering. A friend of mine suffered an interlude of panic when a doctor saw her preschool daughter twitch and suggested that the child had Tourette's syndrome. Once she collected herself, she thought it through like a Bayesian, realized that twitches are common and Tourette's rare, and calmed back down (while giving the doctor a piece of her mind about his statistical innumeracy).

Base-rate neglect is also a driver of thinking in stereotypes. Con-

sider Penelope, a college student described by her friends as impractical and sensitive.[8] She has traveled in Europe and speaks French
and Italian fluently. Her career plans are uncertain, but she's a talented calligrapher and wrote a sonnet for her boyfriend as a birthday
present. Which do you think is Penelope's major, psychology or art
history? Art history, of course! Oh, really? Might it be a wee bit relevant that 13 percent of college students major in psychology, but
only 0.08 percent major in art history, an imbalance of 150 to 1? No
matter where she summers or what she gave her boyfriend, Penelope
is unlikely, a priori, to be an art history major. But in our mind's eye
she is *representative* of an art history major, and the stereotype crowds
out the base rates. Kahneman and Tversky confirmed this in experiments in which they asked participants to consider a sample of 70
lawyers and 30 engineers (or vice versa), provided them with a thumb-
nail sketch that matched a stereotype, such as a dull nerd, and asked
them to put a probability on that person's job. People were swayed by
the stereotype; the base rates went in one ear and out the other.[9]
(This is also why people fall for the conjunction fallacy from chapter
1, in which Linda the social justice warrior is likelier to be a feminist
bank teller than a bank teller. She's representative of a feminist, and
people forget about the relative base rates of feminist bank tellers
and bank tellers.)

A blindness to base rates also leads to public demands for the impossible. Why can't we predict who will attempt suicide? Why don't
we have an early-warning system for school shooters? Why can't we
profile terrorists or rampage shooters and detain them preventively?
The answer comes out of Bayes's rule: a less-than-perfect test for a
rare trait will mainly turn out false positives. The heart of the problem is that only a tiny proportion of the population are thieves, sui-
cides, terrorists, or rampage shooters (the base rate). Until the day
that social scientists can predict misbehavior as accurately as astron-

omers predict eclipses, their best tests would mostly finger the innocent and harmless.

Mindfulness of base rates can be a gift of equanimity as we reflect on our lives. Now and again we long for some rare outcome—a job, a prize, admission to an exclusive school, winning the heart of a dreamboat. We ponder our eminent qualifications and may be crushed and resentful when we are not rewarded with our just deserts. But of course other people are in the running, too, and however superior we think we may be, there are more of them. The judges, falling short of omniscience, cannot be guaranteed to appreciate our virtues. Remembering the base rates—the sheer number of competitors—can take some of the sting out of a rejection. However deserving we think we may be, the base rate—one in five? one in ten? one in a hundred?—should ground our expectations, and we can calibrate our hopes to the degree to which our specialness could reasonably be expected to budge the probability upward.

Priors in Science and the Revenge of the Textbooks

Our neglect of base rates is a special case of our neglect of *priors*: the vital, albeit more nebulous, concept of how much credence we should give a hypothesis before we look at the evidence. Now, believing in something before you look at the evidence may seem like the epitome of irrationality. Isn't that what we disdain as prejudice, bias, dogma, orthodoxy, preconceived notions? But prior credence is simply the fallible knowledge accumulated from all our experience in the past. Indeed, the posterior probability from one round of looking at evidence can supply the prior probability for the next round, a cycle called Bayesian updating. It's simply the mindset of someone who wasn't born yesterday. For fallible knowers in a chancy world, justified belief cannot be equated with the last fact you came across. As Francis Crick liked to say, "Any theory that can account for all the facts is wrong, because some of the facts are wrong."[10]

This is why it is reasonable to be skeptical of claims for miracles, astrology, homeopathy, telepathy, and other paranormal phenomena, even when some eyewitness or laboratory study claims to show it. Why isn't that dogmatic and pigheaded? The reasons were laid out by that hero of reason, David Hume. Hume and Bayes were contemporaries, and though neither read the other, word of the other's ideas may have passed between them through a mutual colleague, and Hume's famous argument against miracles is thoroughly Bayesian:[11]

> Nothing is esteemed a miracle, if it ever happen in the common course of nature. It is no miracle that a man, seemingly in good health, should die on a sudden: because such a kind of death, though more unusual than any other, has yet been frequently observed to happen. But it is a miracle, that a dead man should come to life; because that has never been observed in any age or country.[12]

In other words, miracles such as resurrection must be given a low prior probability. Here is the zinger:

> No testimony is sufficient to establish a miracle, unless the testimony be of such a kind, that its falsehood would be more miraculous, than the fact, which it endeavors to establish.[13]

In Bayesian terms, we are interested in the posterior probability that miracles exist, given the testimony. Let's contrast it with the posterior probability that *no* miracles exist given the testimony. (In

Bayesian reasoning, it's often handy to look at the *odds*, that is, the ratio of the credence of a hypothesis to the credence of the alternative, because it spares us the tedium of calculating the marginal probability of the data in the denominator, which is the same for both posteriors and conveniently cancels out.) The "fact which it endeavors to establish" is the miracle, with its low prior, dragging down the posterior. The testimony "of such a kind" is the likelihood of the data given the miracle, and its "falsehood" is the likelihood of the data given *no* miracle: the possibility that the witness lied, misperceived, misremembered, embellished, or passed along a tall tale he heard from someone else. Given everything we know about human behavior, that's far from miraculous! Which is to say, its likelihood is higher than the prior probability of a miracle. That moderately high likelihood boosts the posterior probability of no miracle, and lowers the overall odds of a miracle compared to no miracle. Another way of putting it is this: Which is more likely—that the laws of the universe as we understand them are false, or that some guy got something wrong?

A pithier version of the Bayesian argument against paranormal claims was stated by the astronomer and science popularizer Carl Sagan (1934–1996) in the slogan that serves as this chapter's epigraph: "Extraordinary claims require extraordinary evidence." An extraordinary claim has a low Bayesian prior. For its posterior credence to be higher than the posterior credence in its opposite, the likelihood of the data given that the hypothesis is true must be far higher than the likelihood of the data given that the hypothesis is false. The evidence, in other words, must be extraordinary.

A failure of Bayesian reasoning among scientists themselves is a contributor to the replicability crisis that we met in chapter 4. The issue hit the fan in 2011 when the eminent social psychologist Daryl Bem published the results of nine experiments in the prestigious

Journal of Personality and Social Psychology which claimed to show that participants successfully predicted (at a rate above chance) random events before they took place, such as which of two curtains on a computer screen hid an erotic image before the computer had selected where to place it.[14] Not surprisingly, the effects failed to replicate, but that was a foregone conclusion given the infinitesimal prior probability that a social psychologist had disproven the laws of physics by showing some undergraduates some porn. When I raised this point to a social psychologist colleague, he shot back, "Maybe Pinker doesn't understand the laws of physics!" But actual physicists, like Sean Carroll in his book *The Big Picture*, have explained why the laws of physics really do rule out precognition and other forms of ESP.[15]

The Bem imbroglio raised an uncomfortable question. If a preposterous claim could get published in a prestigious journal by an eminent psychologist using state-of-the-art methods subjected to rigorous peer review, what does that say about our standards of prestige, eminence, rigor, and the state of the art? One answer, we saw, is the peril of post hoc probability: scientists had underestimated the mischief that could build up from data snooping and other questionable research practices. But another is a defiance of Bayesian reasoning.

Most findings in psychology, as it happens, do replicate. Like many psychology professors, every year I run demos of classic experiments in memory, perception, and judgment to students in my intro and lab courses and get the same results year after year. You haven't heard of these replicable findings because they are unsurprising: people remember the items at the end of a list better than those in the middle, or they take longer to mentally rotate an upside-down letter than a sideways one. The notorious replication failures come from studies that attracted attention because their findings were so counterintuitive. Holding a warm mug makes you friendlier. ("Warm"—get it?) Seeing fast-food logos makes you impatient. Holding a pen between your teeth makes cartoons seem funnier, because it forces your lips into a little smile. People who are asked to lie in writing have positive feelings about hand soap; people who are asked to lie aloud have positive feelings about mouthwash.[16] Any reader of popular science knows of other cute findings which turned out to be suitable for the satirical *Journal of Irreproducible Results*.

The reason these studies were sitting ducks for the replicability snipers is that they had low Bayesian priors. Not as low as ESP, to be sure, but it would be an extraordinary discovery if mood and behavior could be easily pushed around by trivial manipulations of the environment. After all, entire industries of persuasion and psychotherapy try to do exactly that at great cost with only modest success.[17] It was the extraordinariness of the findings that earned them a place in the science sections of newspapers and snazzy ideas festivals, and it's why on Bayesian grounds we should demand extraordinary evidence before believing them. Indeed, a bias toward oddball findings can turn science journalism into a high-volume error dispenser. Editors know they can goose up readership with cover headlines like these:

WAS DARWIN WRONG?

WAS EINSTEIN WRONG?

YOUNG UPSTART OVERTURNS SCIENTIFIC APPLECART

A SCIENTIFIC REVOLUTION IN X

EVERYTHING YOU KNOW ABOUT Y IS WRONG

The problem is that "surprising" is a synonym for "low prior probability," assuming that our cumulative scientific understanding is not worthless. This means that even if the quality of evidence is constant, we should have a *lower* credence in claims that are surprising. But the problem is not just with journalists. The physician John Ioannidis scandalized his colleagues and anticipated the replicability crisis with his 2005 article "Why Most Published Research Findings Are False." A big problem is that many of the phenomena that biomedical researchers hunt for are interesting and a priori unlikely to be true, requiring highly sensitive methods to avoid false positives, while many true findings, including successful replication attempts and null results, are considered too boring to publish.

This does not, of course, mean that scientific research is a waste of time. Superstition and folk belief have an even worse track record than less-than-perfect science, and in the long run an understanding emerges from the rough-and-tumble of scientific disputation. As the physicist John Ziman noted in 1978, "The physics of undergraduate text-books is 90% true; the contents of the primary research journals of physics is 90% false."[18] It's a reminder that Bayesian reasoning recommends against the common practice of using "textbook" as an insult and "scientific revolution" as a compliment.

A healthy respect for the boring would also improve the quality of political commentary. In chapter 1 we saw that the track records of many famous forecasters are risible. A big reason is that their careers depend on attracting attention with riveting predictions, which is to say, those with low priors—and hence, assuming they lack the gift of prophecy, low posteriors. Philip Tetlock has studied "superforecasters," who really do have good track records at predicting economic and political outcomes. A common thread is that they are Bayesian: they start with a prior and update it from there. Asked to provide the probability of a terrorist attack within the next year, for example, they'll first estimate a base rate by going to *Wikipedia* and counting the number of attacks in the region in the years preceding—not a practice you're likely to see in the next op-ed you read on what is in store for the world.[19]

Forbidden Base Rates and Bayesian Taboo

Base-rate neglect is not always a symptom of the representativeness heuristic. Sometimes it's actively prosecuted. The "forbidden base rate" is the third of Tetlock's secular taboos (chapter 2), together with the heretical counterfactual and the taboo tradeoff.[20]

The stage for forbidden base rates is set by a law of social science. Measure any socially significant variable: test scores, vocational interests, social trust, income, marriage rates, life habits, rates of different types of violence (street crime, gang crime, domestic violence, organized crime, terrorism). Now break down the results by the standard demographic dividers: age, sex, race, religion, ethnicity. The averages for the different subgroups are never the same, and sometimes the differences are large. Whether the differences arise from nature, culture, discrimination, history, or some combination is beside the point: the differences are there.

This is hardly surprising, but it has a bloodcurdling implication. Say you sought the most accurate possible prediction about the prospects of an individual—how successful the person would be in college or on the job, how good a credit risk, how likely to have committed a crime, or jump bail, or recidivate, or carry out a terrorist attack. If you were a good Bayesian, you'd start with the base rate for that person's age, sex, class, race, ethnicity, and religion, and adjust by the person's particulars. In other words, you'd engage in profiling. You would perpetrate prejudice not out of ignorance, hatred, supremacy, or any of the -isms or -phobias, but from an objective effort to make the most accurate prediction.

Most people are, of course, horrified at the thought. Tetlock had participants think about insurance executives who had to set premiums for different neighborhoods based on their history of fires. They had no problem with that. But when the participants learned that the neighborhoods also varied in their racial composition, they had second thoughts and condemned the executive for merely being a good actuary. And if they themselves had been playing his role and learned the terrible truth about the neighborhood statistics, they tried to morally cleanse themselves by volunteering for an antiracist cause.

Is this yet another example of human irrationality? Are racism, sexism, Islamophobia, anti-Semitism, and the other bigotries "rational"? Of course not! The reasons hark back to the definition of rationality from chapter 2: the use of knowledge to attain a goal. If actuarial prediction were our *only* goal, then perhaps we should use whatever scrap of information could give us the most accurate prior. But of course it's not our only goal.

A higher goal is fairness. It's wicked to treat an individual according to that person's race, sex, or ethnicity—to judge them by the color of their skin or the composition of their chromosomes rather than the content of their character. None of us wants to be prejudged in this way, and by the logic of impartiality (chapter 2) we must extend that right to everyone else.

Moreover, only when a system is *perceived* to be fair—when people know they will be given a fair shake and not prejudged by features of their biology or history beyond their control—can it earn the trust of its citizens. Why play by the rules when the system is going to ding you because of your race, sex, or religion?

Yet another goal is to avoid self-fulfilling prophecies. If an ethnic group or a sex has been disadvantaged by oppression in the past, its members may be saddled with different average traits in the present. If those base rates are fed into predictive formulas that determine their fate going forward, they would lock in those disadvantages forever. The problem is becoming acute now that the formulas are buried in deep learning networks with their indecipherable hidden layers (chapter 3). A society might rationally want to halt this cycle of injustice even if it took a small hit in predictive accuracy at that moment.

Finally, policies are signals. Forbidding the use of ethnic, sexual, racial, or religious base rates is a public commitment to equality and fairness which reverberates beyond the algorithms permitted in a bureaucracy. It proclaims that prejudice for *any* reason is unthinkable, casting even greater opprobrium on prejudice rooted in enmity and ignorance.

Forbidding the use of base rates, then, has a solid foundation in rationality. But a theorem is a theorem, and the sacrifice of actuarial accuracy that we happily make in the treatment of individuals by public institutions may be untenable in other spheres. One of those spheres is insurance. Unless a company carefully estimates the overall risks for different groups, the payouts would exceed the premiums and the insurance would collapse. Liberty Mutual discriminates against teenage boys by factoring their higher base rate for car accidents into their premiums, because if they didn't, adult women would be subsidizing their recklessness. Even here, though, insurance companies are legally prohibited from using certain criteria in calculating rates, particularly race and sometimes gender.

A second sphere in which we cannot rationally forbid base rates is the understanding of social phenomena. If the sex ratio in a professional field is not 50–50, does that prove its gatekeepers are trying to keep women out, or might there be a difference in the base rate of women trying to get in? If mortgage lenders turn down minority applicants at higher rates, are they racist, or might they, like the hypothetical executive in Tetlock's study, be using base rates for defaulting from different neighborhoods that just happen to correlate with race? Social scientists who probe these questions are often rewarded for their trouble with accusations of racism and sexism. But forbidding social scientists and journalists to peek at base rates would cripple the effort to identify ongoing discrimination and distinguish it from historical legacies of economic, cultural, or legal differences between groups.

Race, sex, ethnicity, religion, and sexual orientation have become war zones in intellectual life, even as overt bigotry of all kinds has dwindled.[21] A major reason, I think, is a failure to think clearly about base rates—to lay out when there are good reasons for forbidding them and when there are not.[22] But that's the problem with a taboo. As with the instruction "Don't think about a polar bear," discussing when to apply a taboo is itself taboo.

Bayesian after All

For all our taboos, neglects, and stereotypes, it's a mistake to write off our kind as hopelessly un-Bayesian. (Recall that the San are Bayesians, requiring that spoor be definitive before inferring it was left by a rarer species.) Gigerenzer has argued that sometimes ordinary people are on solid mathematical ground when they appear to be flouting Bayes's rule.[23] Mathematicians themselves complain that social scientists often use statistical formulas mindlessly: they plug in numbers, turn a crank, and assume that the correct answer pops out. In reality, a statistical formula is only as good as the assumptions behind it. Laypeople can be sensitive to those assumptions, and sometimes when they appear to be blowing off Bayes's rule, they may just be exercising the caution that a good mathematician would advise.

For starters, a prior probability is not the same thing as a base rate, even though base rates are often held up as the "correct" prior in the paper-and-pencil tests. The problem is: *which* base rate? Suppose I get a positive result from a prostate-specific antigen test and want to estimate my posterior probability of having prostate cancer. For the prior, should I use the base rate for prostate cancer in the population? Among white Americans? Ashkenazi Jews? Ashkenazi Jews over sixty-five? Ashkenazi Jews over sixty-five who exercise and have no family history? These rates can be very different. The more specific the reference class, of course, the better—but the more specific the reference class, the smaller the sample on which the estimate is based, and the noisier the estimate. The best reference class would consist of people *exactly* like me, namely me—a class of one that is perfectly accurate and perfectly useless. We have no choice but to use human judgment in trading off specificity against reliability when choosing an appropriate prior, rather than accepting the base rate for an entire population stipulated in the wording of a test.

Another problem with using a base rate as the prior is that base rates can change, and sometimes quickly. Forty years ago around a tenth of veterinary students were women; today it's closer to nine tenths.[24] In recent decades, anyone who was given the historical base rate and plugged it into Bayes's rule would have been worse off than if they had neglected the base rate altogether. With many hypotheses that interest us, no record-keeping agency has even compiled base rates. (Do we know what proportion of veterinary students are Jewish? Left-handed? Transgender?) And of course a lack of data on base rates was our plight through most of history and prehistory, when our Bayesian intuitions were shaped.

Because there is no "correct" prior in a Bayesian problem, people's departure from the base rate provided by an experimenter is not necessarily a fallacy. Take the taxicab problem, where the priors were the proportions of Blue and Green taxis in the city. The participants may well have thought that this simple baseline would be swamped by more specific differences, like the companies' accident rates, the number of their cabs driving during the day and at night, and the neighborhoods they serve. If so, then in their ignorance of these crucial data they may have defaulted to indifference, 50 percent. Follow-up studies showed that participants do become better Bayesians when they are given base rates that are more relevant to being in an accident.[25]

Also, a base rate may be treated as a prior only when the examples at hand are *randomly sampled* from that population. If they have been cherry-picked because of an interesting trait—like belonging to a category with a high likelihood of flaunting those data—all bets are off. Take the demos that presented people with a stereotype, like Penelope the sonnet writer, or the nerd in the pool of lawyers and engineers, and asked them to guess their major or profession. Unless the respondents knew that Penelope had been selected from the pool of students by lottery, which would make it a pretty strange question, they could have suspected she was chosen because her traits provided telltale clues, which would make it a natural question. (Indeed, that question was turned into a classic game show, *What's My Line?*, where a panel had to guess the occupation of a mystery guest— selected not at random, of course, but because the guest's job was so distinctive, like bar bouncer, big-game hunter, Harlem Globetrotter, or Colonel Sanders of Kentucky Fried Chicken fame.) When people's noses are rubbed in the randomness of the sampling (such as by seeing the description pulled out of a jar), their estimates are closer to the correct Bayesian posterior.[26]

Finally, people are sensitive to the difference between probability in the sense of credence in a single event and probability in the sense of frequency in the long run. Many Bayesian problems pose the vaguely mystical question of the probability of a single event—whether Irwin has kuru, or Penelope is an art history major, or the taxi in an accident was Blue. Faced with such problems, it is true that people don't readily compute a subjective credence using the numbers they are given. But since even statisticians are divided on how much sense that makes, perhaps they can be forgiven. Gigerenzer, together with Cosmides and Tooby, argues that people don't connect decimal fractions to single events, because that's not the way the human mind encounters statistical information in the world. We experience *events*, not numbers between 0 and 1. We're perfectly capable of Bayesian reasoning with these "natural frequencies," and when a problem is reframed in those terms, our intuitions can be leveraged to solve it.

Let's go back to the medical diagnosis problem from the beginning of the chapter and translate those metaphysical fractions into concrete frequencies. Forget the generic "a woman"; think about a sample of a thousand women. Out of every 1,000 women, 10 have breast cancer (that's the prevalence, or base rate). Of these 10 women who have breast cancer, 9 will test positive (that's the test's sensitivity). Of the 990 women without breast cancer, about 89 will nevertheless test positive (that's the false-positive rate). A woman tests positive. What is the chance that she actually has breast cancer? It's not that hard: 98 of the women test positive in all, 9 of them have cancer; 9 divided by 98 is around 9 percent—there's our answer. When the problem is framed in this way, 87 percent of doctors get it right (compared with about 15 percent for the original wording), as do a majority of ten-year-olds.[27]

How does this magic work? Gigerenzer notes that the concept of a conditional probability pulls us away from countable things in the world. Those decimal fractions—90 percent true positive, 9 percent false positive, 91 percent true negative, 10 percent false negative—don't add up to 100 percent, so to reckon the proportion of true positives among all the positives (the challenge at hand), we would have to work through three multiplications. Natural frequencies, in contrast, allow you to focus on the positives and add them up: 9 true positives plus 89 false positives equals 98 positives in all, of which the 9 trues make up 9 percent. (What one should *do* with this knowledge, given the costs of acting or failing to act on it, will be the topic of the next two chapters.)

Easier still, we can put our primate visual brains to use and turn the numbers into shapes. This can make Bayesian reasoning eye-poppingly intuitive even with textbook puzzles that are far from our everyday experience, like the classic taxicab problem. Visualize the city's taxi fleet as an array of 100 squares, one per taxi (left-hand diagram on the next page). To depict the base rate of 15 percent Blue taxis, we color in 15 squares in the top left corner. To show the likelihoods of the four possible identifications by our eyewitness, who was 80 percent reliable (middle diagram), we lighten 3 of the Blue taxi squares (the 20 percent of the 15 Blue taxis that he would misidentify as "Green"), and darken 17 of the Green ones (the 20 percent of the 85 Green taxis he'd misidentify as "Blue"). We know that the witness said "Blue," so we can throw away all the squares for the "Green" identifications, both true and false, leaving us with the right-hand diagram, which keeps only the "Blue" IDs. Now it's a cinch to eyeball the shape and espy that the darker portion, the taxis that really are Blue, takes up a bit less than half of the overall area. If we want to be exact, we can count: 12 squares out of 29, or 41 percent. The intuitive key to both the natural frequencies and the visual shapes is that they allow you to zoom in on the data at hand (the positive test result; the "Blue" IDs), and sort them into the ones that are true and the ones that are false.

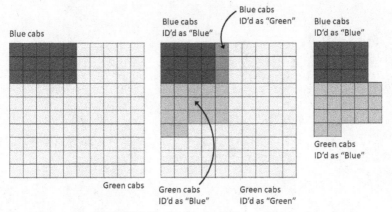

Blue cabs

Blue cabs ID'd as "Blue"

Blue cabs ID'd as "Green"

Blue cabs ID'd as "Blue"

Green cabs

Green cabs ID'd as "Blue"

Green cabs ID'd as "Green"

Green cabs ID'd as "Blue"

Adapted with permission from Presh Talwalkar from Talwalkar 2013

By tapping preexisting intuitions and translating information into mind-friendly formats, it's possible to hone people's statistical reasoning. Hone we must. Risk literacy is essential for doctors, judges, policymakers, and others who hold our lives in their hands. And since we all live in a world in which God plays dice, fluency in Bayesian reasoning and other forms of statistical competence is a public good that should be a priority in education. The principles of cognitive psychology suggest that it's better to work with the rationality people have and enhance it further than to write off the majority of our species as chronically crippled by fallacies and biases.[28] The principles of democracy suggest that, too.

6

RISK AND REWARD

(RATIONAL CHOICE AND
EXPECTED UTILITY)

Everyone complains about his memory, and no one complains about his judgment.

—La Rochefoucauld

S ome theories are unlovable. No one has much affection for the laws of thermodynamics, and generations of hopeful crackpots have sent patent offices their doomed designs for a perpetual motion machine. Ever since Darwin proposed the theory of natural selection, creationists have choked on the implication that humans descended from apes, and communitarians have looked for loopholes in its tenet that evolution is driven by competition.

One of the most hated theories of our time is known in different versions as rational choice, rational actor, expected utility, and *Homo economicus*.[1] This past Christmas season, *CBS This Morning* ran a heartwarming segment on a study that dropped thousands of money-filled wallets in cities across the world and found that most were returned, especially when they contained more money, reminding us that human beings are generous and honest after all. The grinch in the story? "Rationalist approaches to economics," which supposedly predict that people live by the credo "Finders keepers, losers weepers."[2]

What exactly is this mean-spirited theory? It says that when faced with a risky decision, rational actors ought to choose the option that maximizes their "expected utility," namely the sum of its possible rewards weighted by their probabilities. Outside of economics and a few corners of political science, the theory is about as lovable as Ebenezer Scrooge. People interpret it as claiming that humans are, or should be, selfish psychopaths, or that they are uber-rational brainiacs who calculate probabilities and utilities before deciding whether to fall in love. Discoveries from the psychology lab showing that people seem to violate the theory have been touted as undermining the foundations of classical economics, and with it the rationale for market economies.[3]

In its original form, though, rational choice theory is a theorem of mathematics, considered quite beautiful by aficionados, with no direct implications for how members of our species think and choose. Many consider it to have provided the most rigorous characterization of rationality itself, a benchmark against which to measure human judgment. As we shall see, this can be contested—sometimes when people depart from the theory, it's not clear whether the people are being irrational or the supposed standards of rationality are irrational. Either way, the theory shines a light on perplexing conundrums of rationality, and despite its provenance in pure math, it can be a source of profound life lessons.[4]

The theory of rational choice goes back to the dawn of probability theory and the famous argument by Blaise Pascal (1623–1662) on why you should believe in God: if you did and he doesn't exist, you would just have wasted some prayers, whereas if you didn't and he does exist, you would incur his eternal wrath. It was formalized in 1944 by the mathematician John von Neumann and the economist Oskar Morgenstern. Unlike the pope, von Neumann really might have been a space alien—his colleagues wondered about it because of his otherworldly intelligence. He also invented game theory (chapter 8), the digital computer, self-replicating machines, quantum logic, and key components of nuclear weapons, while making dozens of other breakthroughs in math, physics, and computer science.

Rational choice is not a psychological theory of how human beings choose, or a normative theory of what they ought to choose, but a theory of what makes choices *consistent* with the chooser's values and each other. That ties it intimately to the concept of rationality, which is about making choices that are consistent with our goals. Romeo's pursuit of Juliet is rational, and the iron filings' pursuit of the magnet is not, because only Romeo chooses whichever path brings about his goal (chapter 2). At the other end of the scale, we call people "crazy" when they do things that are patently against their interests, like throwing away their money on things they don't want or running naked into the freezing cold.

The beauty of the theory is that it takes off from a few easy-to-swallow axioms: broad requirements that apply to any decision maker we'd be willing to call "rational." It then deduces how the decider would have to make decisions in order to stay true to those requirements. The axioms have been lumped and split in various ways; the version I'll present here was formulated by the mathematician Leonard Savage and codified by the psychologists Reid Hastie and Robyn Dawes.[5]

A Theory of Rational Choice

The first axiom may be called Commensurability: for any options A and B, the decider prefers A, or prefers B, or is indifferent between them.[6] This may sound vacuous—aren't those just the logical possibilities?—but it requires the decider to commit to one of the three, even if it's indifference. The decider, that is, never falls back on the excuse "You can't compare apples and oranges." We can interpret it as the requirement that a rational agent must care about things and prefer some to others. The same cannot be said for nonrational entities like rocks and vegetables.

The second axiom, Transitivity, is more interesting. When you compare options two at a time, if you prefer A to B, and B to C, then you must prefer A to C. It's easy to see why this is a nonnegotiable requirement: anyone who violates it can be turned into a "money pump." Suppose you prefer an Apple iPhone to a Samsung Galaxy but are saddled with a Galaxy. I will now sell you a sleek iPhone for $100 with the trade-in. Suppose you also prefer a Google Pixel to an iPhone. Great! You'd certainly trade in that crummy iPhone for the superior Pixel plus a premium of, say, $100. And suppose you prefer a Galaxy to a Pixel—that's the intransitivity. You can see where this is going. For $100 plus a trade-in, I'll sell you the Galaxy. You'd be right where you started, $300 poorer, and ready for another round of fleecing. Whatever you think rationality consists of, it certainly isn't that.

The third is called Closure. With God playing dice and all that, choices are not always among certainties, like picking an ice cream flavor, but may include a collection of possibilities with different odds, like picking a lottery ticket. The axiom states that as long as the decider can consider A and B, that decider can also consider a lottery ticket that offers A with a certain probability, p, and B with the complement probability, $1 - p$.

Within rational choice theory, although the outcome of a chancy option cannot be predicted, the probabilities are fixed, like in a casino. This is called *risk*, and may be distinguished from *uncertainty*, where the decider doesn't even know the probabilities and all bets are off. In 2002, the US defense secretary Donald Rumsfeld famously explained the distinction: "There are known unknowns; that is to say we know there are some things we do not know. But there are also unknown unknowns—the ones we don't know we don't know." The theory of rational choice is a theory of decision making with known unknowns: with risk, not necessarily uncertainty.

I'll call the fourth axiom Consolidation.[7] Life doesn't just present us with lotteries; it presents us with lotteries whose prizes may themselves be lotteries. A chancy first date, if it goes well, can lead to a second date, which brings a whole new set of risks. This axiom simply says that a decider faced with a series of risky choices works out the overall risk according to the laws of probability explained in chapter 4. If the first lottery ticket has a one-in-ten chance of a payout, with the prize being a second ticket with a one-in-five chance of a payout, then the decider treats it as being exactly as desirable as a ticket with a one-in-fifty chance of the payout. (We'll put aside whatever extra pleasure is taken in a second opportunity to watch the bouncing ping-pong balls or to scratch off the ticket coating.) This criterion for rationality seems obvious enough. As with the speed limit and gravity, so with probability theory: It's not just a good idea. It's the law.

The fifth axiom, Independence, is also interesting. If you prefer A to B, then you also prefer a lottery with A and C as the payouts to a lottery with B and C as the payouts (holding the odds constant). That is, adding a chance at getting C to both options should not change whether one is more desirable than the other. Another way of putting it is that how you *frame* the choices—how you present them in context—should not matter. A rose by any other name should smell just as sweet. A rational decider should focus on the choices themselves and not get sidelined by some distraction that accompanies them both.

Independence from Irrelevant Alternatives, as the generic version of Independence is called, is a requirement that shows up in many theories of rational choice.[8] A simpler version says that if you prefer A to B when choosing between them, you should still prefer A to B when choosing among them and a third alternative, C. According to legend, the logician Sidney Morgenbesser (whom we met in chapter 3) was seated at a restaurant and offered a choice of apple pie or blueberry pie. Shortly after he chose apple, the waitress returned and said they also had cherry pie on the menu that day. As if waiting for the moment all his life, Morgenbesser said, "In that case, I'll have blueberry."[9] If you find this funny, then you appreciate why Independence is a criterion for rationality.

The sixth is Consistency: if you prefer A to B, then you prefer a gamble in which you have some chance at getting A, your first choice, and otherwise get B, to the certainty of settling for B. Half a chance is better than none.

The last may be called Interchangeability: desirability and probability trade off.[10] If the decider prefers A to B, and prefers B to C, there must be some probability that would make her indifferent between getting B for sure, her middle choice, and having a shot at getting either A, her top choice, or settling for C. To get a feel for this, imagine the probability starting high, with a 99 percent chance of getting A and only a 1 percent chance of getting C. Those odds make the gamble sound a lot better than settling for your second choice, B. Now consider the other extreme, a 1 percent chance of getting your first choice and a 99 percent chance of getting your last one. Then it's the other way around: the sure mediocre option beats the near certainty of having to settle for the worst. Now imagine a sequence of probabilities from almost-certainly-A to almost-certainly-C. As the odds gradually shift, do you think you'd stick with the gamble up to a certain point, then be indifferent between gambling and settling for B, then switch to the sure B? If so, you agree that Interchangeability is rational.

Now here is the theorem's payoff. To meet these criteria for rationality, the decider must assess the value of each outcome on a continuous scale of desirability, multiply by its probability, and add them up, yielding the "expected utility" of that option. (In this context, *expected* means "on average, in the long run," not "anticipated," and *utility* means "preferable by the lights of the decider," not "useful" or "practical.") The calculations need not be conscious or with numbers; they can be sensed and combined as analogue feelings. Then the decider should pick the option with the highest expected utility. That is guaranteed to make the decider rational by the seven criteria. A rational chooser is a utility maximizer, and vice versa.

To be concrete, consider a choice between games in a casino. In craps, the probability of rolling a "7" is 1 in 6, in which case you would win $4; otherwise you forfeit the $1 cost of playing. Suppose for now that every dollar is a unit of utility. Then the expected utility of betting on "7" in craps is $(1/6 \times \$4) + (5/6 \times -\$1)$, or $-\$0.17$. Compare that with roulette. In roulette, the probability of landing on "7" is 1 in 38, in which case you would win $35; otherwise you forfeit your $1. Its expected utility is $(1/38 \times \$35) + (37/38 \times -\$1)$, or $-\$0.05$. The expected utility of betting "7" in craps is lower than that in roulette, so no one would call you irrational for preferring roulette. (Of course, someone might call you irrational for gambling in the first place, since the expected value of both bets is negative, owing to the house's take, so the more you play, the more you lose. But if you entered the casino in the first place, presumably you place some positive utility on the glamour of Monte Carlo and the frisson of suspense, which boosts the utility of both options into positive territory and leaves open only the choice of which to play.)

Games of chance make it easy to explain the theory of rational choice, because they provide exact numbers we can multiply and add. But everyday life presents us with countless choices that we intuitively evaluate in terms of their expected utilities. I'm in a convenience store and don't remember whether there's milk in the fridge; should I buy a quart? I suspect I'm out, and if that's the case and I forgo the purchase, I'll be really annoyed at having to eat my cereal dry tomorrow morning. On the other hand if there is milk at home and I do buy more, the worst that can happen is that it will spoil, but that's unlikely, and even if it does, I'll only be out a couple of bucks. So all in all I'm better off buying it. The theory of rational choice simply spells out the rationale behind this kind of reasoning.

How Useful Is Utility?

It's tempting to think that the patterns of preferences identified in the axioms of rationality are about people's subjective feelings of pleasure and desire. But technically speaking, the axioms treat the decider as a black box and consider only her patterns of picking one thing over another. The utility scale that pops out of the theory is a hypothetical entity that is reconstructed from the pattern of preferences and recommended as a way to keep those preferences consistent. The theory protects the decider from being turned into a money pump, a dessert flip-flopper, or some other kind of flibbertigibbet. This means the theory doesn't so much tell us how to act in accord with our values as how to discover our values by observing how we act.

That lays to rest the first misconception of the theory of rational choice: that it portrays people as amoral hedonists or, worse, advises them to become one. Utility is not the same as self-interest; it's whatever scale of value a rational decider consistently maximizes. If people make sacrifices for their children and friends, if they minister to the sick and give alms to the poor, if they return a wallet filled with money, that shows that love and charity and honesty go into their utility scale. The theory just offers advice on how not to squander them.

Of course, in pondering ourselves as decision makers, we don't have to treat ourselves as black boxes. The hypothetical utility scale should correspond to our internal sensations of happiness, greed, lust, warm glow, and other passions. Things become interesting when we explore the relationship, starting with the most obvious object of desire, money. Whether or not money can buy happiness, it can buy utility, since people trade things for money, including charity. But the relationship is not linear; it's concave. In jargon, it shows "diminishing marginal utility."

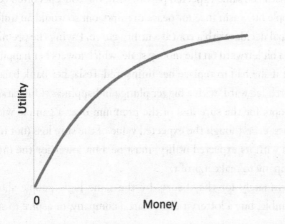

The psychological meaning is obvious: an extra hundred dollars increases the happiness of a poor person more than the happiness of a rich person.[11] (This is the moral argument for redistribution: transferring money from the rich to the poor increases the amount of happiness in the world, all things being equal.) In the theory of rational choice, this curve doesn't actually come from the obvious source, namely asking people with different amounts of money how happy they are, but from looking at people's preferences. Which would you rather have: a thousand dollars for sure, or a 50–50 chance of winning two thousand dollars? Their expected value is the same, but most people opt for the sure thing. This doesn't mean they are flouting rational choice theory; it just means that utility is not the same as value in dollars. The utility of two thousand dollars is less than twice the utility of one thousand dollars. Fortunately for our understanding, people's ratings of their satisfaction and their choice of gambles point to the same bent-over curve relating money and utility.

Economists equate a concave utility curve with being "risk-averse." That's a bit confusing, because the term does not refer to being a nervous Nellie as opposed to a daredevil, only to preferring a sure thing to a bet with the same expected payoff. Still, the concepts often coincide. People buy insurance for peace of mind, but so would an unfeeling rational decider with a concave utility curve. Paying the premium tugs her a bit leftward on the money scale, which lowers her happiness a bit, but if she had to replace her uninsured Tesla, her bank balance would lurch leftward, with a bigger plunge in happiness. Our rational chooser opts for the sure loss of the premium over a gamble with a bigger loss, even though the expected value of the sure loss (not to be confused with its expected utility) must be a bit lower for the insurance company to make a profit.

Unfortunately for the theory, by the same logic people should never gamble, buy a lottery ticket, start a company, or aspire to stardom rather than become a dentist. But of course some people do, a paradox that tied classical economists in knots. The human utility curve can't be both concave, explaining why we avert risk with insurance, and convex, explaining why we seek risk by gambling. Perhaps we gamble for the thrill, just as we buy insurance for the peace of mind, but this appeal to emotions just pushes the paradox up a level: why did we evolve with the contradictory motives to jack ourselves up and calm ourselves down, paying for both privileges? Perhaps we're irrational and that's all there is to it. Perhaps the showgirls, spinning cherries, and other accoutrements of gambling are a form of entertainment that high rollers are willing to spend for. Or perhaps the curve has a second bend and shoots upward at the high end, making the expected utility of a jackpot higher than that of a mere increment in one's bank balance. This could happen if people felt that the prize would vault them into a different social class and lifestyle: the life of a glamorous and carefree millionaire, not just a better-heeled member of the bourgeoisie. Many ads for state lotteries encourage that fantasy.

Though it's easiest to think through the implications of the theory when the utility is reckoned in cash, the logic applies to anything of value we can place along a scale. This includes the public valuation of human life. The saying falsely attributed to Josef Stalin, "One death is a tragedy, a million deaths is a statistic," gets the numbers wrong but captures the way we treat the moral cost of lives lost in a disaster like a war or pandemic. The curve bends over, like the one for the utility of money.[12] On a normal day, a terrorist attack or an incident of food poisoning with a dozen victims can get wall-to-wall coverage. But in the midst of a war or pandemic, a thousand lives lost in a day is taken in stride—even though each of those lives, unlike a diminishing dollar, was a real person, a sentient being who loved and was loved. In *The Better Angels of Our Nature*, I suggested that our morally misguided sense of the diminishing marginal utility of human lives is a reason why small wars can escalate into humanitarian catastrophes.[13]

Violating the Axioms: How Irrational?

You might think that the axioms of rational choice are so obvious that any normal person would respect them. In fact, people frequently cock a snook at them.

Let's begin with Commensurability. It would seem impossible to flout: it's just the requirement that you must prefer A to B, prefer B to A, or be indifferent between them. In chapter 2 we witnessed the act of rebellion, the taboo tradeoff.[14] People treat certain things in life as sacrosanct and find the very thought of comparing them immoral. They feel that anyone who obeys the axiom is like Oscar Wilde's "cynic": a person who knows the price of everything and the value of nothing. How much should we spend to save an endangered species from extinction? To save the life of a little girl who fell down a well? Should we balance the budget by cutting funds for education, seniors, or the environment? A joke from another era begins with a man asking, "Would you sleep with me for a million dollars?"[15] The idiom "Sophie's choice" originated in William Styron's harrowing novel, where it referred to the protagonist having to surrender one of her two children to be gassed in Auschwitz. We saw in chapter 2 how recoiling from the demand to compare sacred entities can be both rational, when it affirms our commitment to a relationship, and irrational, when we look away from painful choices but in fact make them capriciously and inconsistently.

A different family of violations involve a concept introduced by the psychologist Herbert Simon called *bounded rationality*.[16] Theories of rational choice assume an angelic knower with perfect information and unlimited time and memory. For mortal deciders, uncertainty in the odds and payoffs, and the costs of obtaining and processing the information, have to be factored into the decision. It makes no sense to spend twenty minutes figuring out a shortcut that will save you ten minutes in travel time. The costs are by no means trifling. The world is a garden of forking paths, with every decision taking us into a situation in which new decisions confront us, exploding into a profusion of possibilities that could not possibly be tamed by the Consolidation axiom. Simon suggested that a flesh-and-blood decider rarely has the luxury of optimizing but instead must *satisfice*, a portmanteau of "satisfy" and "suffice," namely settle for the first alternative that exceeds some standard that's good enough. Given the costs of information, the perfect can be the enemy of the good.

Unfortunately, a decision rule that makes life simpler can violate the axioms, including Transitivity. Even Transitivity? Could I make a living by finding a human money pump and selling him the same things over and over, like Sylvester McMonkey McBean in Dr. Seuss's *The Sneetches*, who repeatedly charged the Sneetches three dollars to affix a star to their bellies and ten dollars to have it removed? ("Then, when every last cent of their money was spent / The Fix-It-Up Chappie packed up. And he went.") Though intransitivity is the epitome of irrationality, it can easily arise from two features of bounded rationality.

One is that we don't do all the multiplications and additions necessary to melt down the attributes of an item into a glob of utility. Instead we may consider its attributes one by one, whittling down the choices by a process of elimination.[17] In choosing a college, we might first rule out the ones without a lacrosse team, then the ones without a medical school, then the ones too far from home, and so on.

The other shortcut is that we may ignore a small difference in the values of one attribute when others seem more relevant. Savage asks us to consider a tourist who can't decide between visiting Paris and Rome.[18] Suppose instead she was given a choice between visiting Paris and visiting Paris plus receiving a dollar. Paris + $1 is unquestionably more desirable than Paris alone. But that does *not* mean that Paris + $1 is unquestionably more desirable than Rome! We have a kind of intransitivity: the tourist prefers A (Paris + $1) to B (Paris), and is indifferent between B and C (Rome), but does not prefer A to C. Savage's example was rediscovered by a *New Yorker* cartoonist:

"How much would you pay for all the secrets of the universe? Wait, don't answer yet. You also get this six-quart covered combination spaghetti pot and clam steamer. Now how much would you pay?"

A decider who chooses by a process of elimination can fall into full-blown intransitivity.[19] Tversky imagines three job candidates differing in their scores on an aptitude test and years of experience:

	Aptitude	Experience
Archer	200	6
Baker	300	4
Connor	400	2

A human resources manager compares them two at a time with this policy: If one scores more than 100 points higher in aptitude, choose that candidate; otherwise pick the one with more experience. The manager prefers Archer to Baker (more experience), Baker to Connor (more experience), and Connor to Archer (higher aptitude). When experimental participants are put in the manager's shoes, many of them make intransitive sets of choices without realizing it.

So have behavioral economists been able to fund their research by using their participants as money pumps? Mostly not. People catch on, think twice about their choices, and don't necessarily buy something just because they momentarily prefer it.[20] But without these double-takes from System 2, the vulnerability is real. In real life, the process of making decisions by comparing alternatives one aspect at a time can leave a decider open to irrationalities that we all recognize in ourselves. When deciding among more than two choices, we may be swayed by the last pair we looked at, or go around in circles as each alternative seems better than the other two in a different way.[21]

And people really can be turned into money pumps, at least for a while, by preferring A to B but putting a higher price on B.[22] (You would sell them B, trade them A for it, buy back A at the lower price, and repeat.) How could anyone land in this crazy contradiction? It's easy: when faced with two choices with the same expected value, people may prefer the one with the higher probability but pay more for the one with the higher payoff. (Concretely, consider two tickets to play roulette that have the same expected value, $3.85, but from different combinations of odds and payoffs. Ticket A gives you a 35/36 chance of winning $4 and a 1/36 chance of losing $1. Ticket B gives you an 11/36 chance of winning $16 and a 25/36 chance of losing $1.50.[23] Given the choice, people pick A. Asked what they would pay for each, they offer a higher price for B.) It's barmy—when people think about a price, they glom onto the bigger number after the dollar sign and forget the odds—and the experimenter can act as an arbitrageur and pump money out of some of them. The bemused victims say, "I just can't help it," or "I know it's silly and you're taking advantage of me, but I really do prefer that one."[24] After a few rounds, almost everyone wises up. Some of the churning in real financial markets may be stirred by naïve investors being swayed by risks at the expense of rewards or vice versa and arbitrageurs swooping in to exploit the inconsistencies.

WHAT ABOUT THE Independence from Irrelevant Alternatives, with its ditzy dependence on context and framing? The economist Maurice Allais uncovered the following paradox.[25] Which of these two tickets would you prefer?

Supercash:	100% chance of $1 million	Powerball:	10% chance of $2.5 million
			89% chance of $1 million

Though the expected value of the Powerball ticket is larger ($1.14 million), most people go for the sure thing, avoiding the 1 percent chance of ending up with nothing. That doesn't violate the axioms; presumably their utility curve bends over, making them risk-averse. Now which of *these* two would you prefer?

Megabucks:	11% chance of $1 million	LottoUSA:	10% chance of $2.5 million

With this choice, people prefer LottoUSA, which tracks their expected values ($250,000 versus $110,000). Sounds reasonable, right? While pondering the first choice, the homunculus in your head is saying, "The Powerball lottery may have a bigger prize, but if you take it, there's a chance you would walk away with nothing. You'd feel like an idiot, knowing you had blown a million dollars!" When looking at the second choice, it says, "Ten percent, eleven percent, what's the difference? Either way, you have some chance at winning—might as well go for the bigger prize."

Unfortunately for the theory of rational choice, the preferences violate the Independence axiom. To see the paradox, let's carve the probabilities of the two left-hand choices into pieces, keeping everything the same except the way they're presented:

Supercash:	10% chance of $1 million	Powerball:	10% chance of $2.5 million
	1% chance of $1 million		89% chance of $1 million
	89% chance of $1 million		

Megabucks: 10% chance of $1 million	**LottoUSA:** 10% chance of $2.5 million
1% chance of $1 million	

We now see that the choice between Supercash and Powerball is just the choice between Megabucks and LottoUSA with an extra 89 percent chance of winning a million dollars tacked on to each. But that extra chance made you flip your pick. I added cherry pie to each ticket, and you switched from apple to blueberry. If you're getting sick of reading about cash lotteries, Tversky and Kahneman offer a nonmonetary example.[26] Would you prefer a raffle ticket offering a 50 percent chance of a three-week tour of Europe, or a voucher giving you a one-week tour of England for sure? People go for the sure thing. Would you prefer a raffle ticket giving you a 5 percent chance of the three-week tour, or a ticket with a 10 percent chance of the England tour? Now people go for the longer tour.

Psychologically, it's clear what's going on. The difference between a probability of 0 and a probability of 1 percent isn't just any old one-percentage-point gap; it's the distinction between impossibility and possibility. Likewise, the difference between 99 percent and 100 percent is the distinction between possibility and certainty. Neither is commensurable with differences along the rest of the scale, like the difference between 10 percent and 11 percent. Possibility, however small, allows for hope looking forward, and regret looking back. Whether a choice driven by these emotions is "rational" depends on whether you think that emotions are natural responses we should respect, like eating and staying warm, or evolutionary nuisances our rational powers should override.

The emotions triggered by possibility and certainty add an extra ingredient to chance-laden choices like insurance and gambling which cannot be explained by the shapes of the utility curves. Tversky and Kahneman note that no one would buy probabilistic insurance, with premiums at a fraction of the cost but coverage only on certain days of the week, though they happily incur the same overall risk by insuring themselves against some hazards, like fires, but not others, like hurricanes.[27] They buy insurance for peace of mind—to give themselves one less thing to worry about. They would rather banish the fear of one kind of disaster from their anxiety closet than make their lives safer across the board. This may also explain societal decisions such as banning nuclear power, with its tiny risk of a disaster, rather than reducing the use of coal, with its daily drip of many more deaths. The American Superfund law calls for eliminating certain pollutants from the environment completely, though removing the last 10 percent may cost more than the first 90 percent. The US Supreme Court justice Stephen Breyer commented on litigation to force the cleanup of a toxic waste site: "The forty-thousand-page record of this ten-year effort indicated (and all the parties seemed to agree) that, without the extra expenditure, the waste dump was clean enough for children playing on the site to eat small amounts of dirt daily for 70 days each year without significant harm. . . . But there were no dirt-eating children playing in the area, for it was a swamp. . . . To spend $9.3 million to protect non-existent dirt-eating children is what I mean by the problem of 'the last 10 percent.'"[28]

I once asked a family member who bought a lottery ticket every week why he was throwing his money away. He explained to me, as if I were a slow child, "You can't win if you don't play." His answer was not necessarily irrational: there may be a psychological advantage to holding a portfolio of prospects which includes the possibility of a windfall rather than single-mindedly maximizing expected utility, which guarantees it can't happen. The logic is reinforced in a joke. A pious old man beseeches the Almighty. "O Lord, all my life I have obeyed your laws. I have kept the Sabbath. I have recited the prayers. I have been a good father and husband. I make only one request of you. I want to win the lottery." The skies darken, a shaft of light penetrates the clouds, and a deep voice bellows, "I'll see what I can do." The man is heartened. A month passes, six months, a year, but fortune does not find him. In his despair he cries out again, "Lord Almighty, you know I am a pious man. I have beseeched you. Why have you forsaken me?" The skies darken, a shaft of light bursts forth, and a voice booms out, "Meet me halfway. Buy a ticket."

IT'S NOT JUST THE framing of risks that can flip people's choices; it's also the framing of rewards. Suppose you have just been given $1,000. Now you must choose between getting another $500 for sure and flipping a coin that would give you another $1,000 if it lands heads. The expected value of the two options is the same ($500), but by now you have learned that most people are risk-averse and go for the sure thing. Now consider a variation. Suppose you have been given $2,000. You now must choose between giving back $500 and flipping a coin that would require you to give back $1,000 if it lands heads. Now most people flip the coin. But do the arithmetic: in terms of where you would end up, the choices are identical. The only difference is the starting point, which frames the outcomes as a "gain" with the first choice and a "loss" with the second. With this shift in framing, people's risk aversion goes out the window: now they *seek* a risk if it offers the hope of avoiding a loss. Kahneman and Tversky conclude that people are not risk-averse across the board, though they are *loss*-averse: they seek risk if it may avoid a loss.[29]

Once again, it's not just in contrived gambles. Suppose you have been diagnosed with a life-threatening cancer and can have it treated either with surgery, which incurs some risk of dying on the operating table, or with radiation.[30] Experimental participants are told that out of every 100 patients who chose surgery, 90 survived the operation, 68 were alive after a year, and 34 were alive after five years. In contrast, out of every 100 who chose radiation, 100 survived the treatment, 77 were alive after a year, and 22 were alive after five years. Fewer than a fifth of the subjects opt for radiation—they go with the expected utility over the long term.

But now suppose the options are described differently. Out of every 100 patients who chose surgery, 10 died on the operating table, 32 died after a year, and 66 were dead within five years. Out of every 100 who chose radiation, none died during the treatment, 23 died after a year, and 78 died within five years. Now almost half choose radiation. They accept a greater overall chance of dying with the guarantee that they won't be killed by the treatment right away. But the two pairs of options pose the same odds: all that changed was whether they were framed as the number who lived, perceived as a gain, or the number who died, perceived as a loss.

Once again, the violation of the axioms of rationality spills over from private choices into public policy. In an eerie premonition, forty years before Covid-19 Tversky and Kahneman asked people to "imagine that the U.S. is preparing for the outbreak of an unusual Asian disease."[31] I will update their example. The coronavirus, if left untreated, is expected to kill 600,000 people. Four vaccines have been developed, and only one can be distributed on a large scale. If Miraculon is chosen, 200,000 people will be saved. If Wonderine is chosen, there's a ⅓ chance that 600,000 people will be saved and a ⅔ chance that no one will be saved. Most people are risk-averse and recommend Miraculon.

Now consider the other two. If Regenera is chosen, 400,000 people will die. If Preventavir is chosen, there's a ⅓ chance that no one will die and a ⅔ chance that 600,000 people will die. By now you've developed an eye for trick questions in rationality experiments and have surely spotted that the two choices are identical, differing only in whether the effects are framed as gains (lives saved) or losses (deaths). But the flip in wording flips the preference: now a majority of people are risk-*seeking* and favor the Preventavir, which holds out the hope that the loss of life can be avoided entirely. It doesn't take much imagination to see how these framings could be exploited to manipulate people, though they can be avoided with careful presentations of the data, such as always mentioning both the gains *and* the losses, or displaying them as graphs.[32]

Kahneman and Tversky combined our misshapen sense of probability with our squirrelly sense of gains and losses into what they call Prospect theory.[33] It is an alternative to rational choice theory, intended to describe how people do choose rather than prescribe how they ought to choose. The graph below shows how our "decision weights," the subjective sense of probability we apply to a choice, are related to objective probability.[34] The curve is steep near 0 and 1

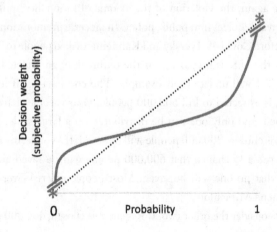

(and with a discontinuity at the boundaries near those special values), more or less objective around .2, and flattish in the middle, where we don't differentiate, say, .10 from .11.

A second graph displays our subjective value.[35] Its horizontal axis is centered at a movable baseline, usually the status quo, rather than at 0. The axis is demarcated not in absolute dollars, lives, or other valued commodities but in relative gains or losses with respect to that baseline. Both gains and losses are concave—each additional unit gained or lost counts for less than the ones already incurred— but the slope is steeper on the downside; a loss is more than twice as painful as the equivalent gain is pleasurable.

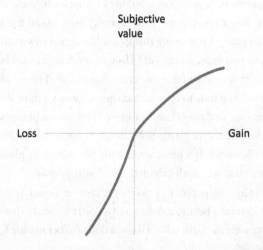

Of course, merely plotting phenomena as curves does not explain them. But we can make sense of these violations of the rational axioms. Certainty and impossibility are epistemologically very different from very high and very low probabilities. That's why, in this book, logic is in a separate chapter from probability theory. ("P OR Q; NOT P; therefore, Q" is not just a statement with a very high probability;

it's a logical truth.) It's why patent officers send back applications for perpetual motion machines unopened rather than taking a chance that some genius has solved our energy problems once and for all. Benjamin Franklin was correct in at least the first half of his statement that nothing is certain but death and taxes. Intermediate probabilities, in contrast, are matters of conjecture, at least outside casinos. They are estimates with margins of error, sometimes large ones. In the real world it's not foolish to treat the difference between a probability of .10 and a probability of .11 with a grain of salt.

The asymmetry between gains and losses, too, becomes more explicable when we descend from mathematics to real life. Our existence depends on a precarious bubble of improbabilities, with pain and death just a misstep away. As Tversky once asked me when we were colleagues, "How many things could happen to you today that could make you much better off? How many things could happen to you today that could make you much *worse* off? The second list is bottomless." It stands to reason that we are more vigilant about what we have to lose, and take chances to avoid precipitous plunges in well-being.[36] And at the negative pole, death is not just something that really, really sucks. It's game over, with no chance to play again, a singularity that makes all calculations of utility moot.

That is also why people can violate yet another axiom, Interchangeability. If I prefer a beer to a dollar, and a dollar to death, that does not mean that, with the right odds, I'd pay a dollar to bet my life for a beer.

Or does it?

Rational Choices after All?

In cognitive science and behavioral economics, showing all the ways in which people flout the axioms of rational choice has become something of a sport. (And not just a sport: five Nobel Prizes have gone to discoverers of the violations.)[37] Part of the fun comes from showing how irrational humans are, the rest from showing what bad psychologists the classical economists and decision theorists are. Gigerenzer loves to tell a true story about a conversation between two decision theorists, one of whom was agonizing over whether to take an enticing job offer at another university.[38] His colleague said, "Why don't you write down the utilities of staying where you are versus taking the job, multiply them by their probabilities, and choose the higher of the two? After all, that's what you advise in your professional work." The first one snapped, "Come on, this is serious!"

But von Neumann and Morgenstern may deserve the last laugh. All those taboos, bounds, intransitivities, flip-flops, regrets, aversions, and framings merely show that people flout the axioms, not that they ought to. To be sure, in some cases, like the sacredness of our relationships and the awesomeness of death, we really may be better off not doing the sums prescribed by the theory. But we do always want to keep our choices consistent with our values. That's all that the theory of expected utility can deliver, and it's a consistency we should not take for granted. We call our decisions foolish when they subvert our values and wise when they affirm them. We have already seen that some breaches of the axioms truly are foolhardy, like avoiding tough societal tradeoffs, chasing zero risk, and being manipulated by a choice of words. I suspect there are countless decisions in life where if we did multiply the risks by the rewards we would choose more wisely.

When you buy a gadget, should you also buy the extended warranty pushed by the salesperson? About a third of Americans do, forking over $40 billion a year. But does it really make sense to take out a health insurance policy on your toaster? The stakes are smaller than insurance on a car or house, where the financial loss would have an impact on your well-being. If consumers thought even crudely about the expected value, they'd notice that an extended warranty can cost almost a quarter of the price of the product, meaning that it would pay off only if the product had more than a 1 in 4 chance of breaking. A glance at *Consumer Reports* would then show that modern appliances are nowhere near that flimsy: fewer than 7 percent of televisions, for example, need any kind of repair.[39] Or consider deductibles on home insurance. Should you pay an extra $100 a year to reduce your out-of-pocket expense in the event of a claim from $1,000 to $500? Many people do it, but it makes sense only if you expect to make a claim every five years. The average claim rate for homeowners insurance is in fact around once every *twenty* years, which means that the people are paying $100 for $25 in expected value (5 percent of $500).[40]

Weighing risks and rewards can, with far greater consequences, also inform medical choices. Doctors and patients alike are apt to think in terms of propensities: cancer screening is good because it can detect cancers, and cancer surgery is good because it can remove them. But thinking about costs and benefits weighted by their probabilities can flip good to bad. For every thousand women who undergo annual ultrasound exams for ovarian cancer, 6 are correctly diagnosed with the disease, compared with 5 in a thousand unscreened women—and the number of deaths in the two groups is the same, 3. So much for the benefits. What about the costs? Out of the thousand who are screened, another 94 get terrifying false alarms, 31 of whom suffer unnecessary removal of their ovaries, of whom 5 have serious complications to boot. The number of false alarms and unnecessary surgeries among women who are not screened, of course, is zero. It doesn't take a lot of math to show that the expected utility of ovarian cancer screening is negative.[41] The same is true for men when it comes to screening for prostate cancer with the prostate-specific antigen test. These are easy cases; we'll take a deeper dive into how to compare the costs and benefits of hits and false alarms in the next chapter.

Even when exact numbers are unavailable, there is wisdom to be had in mentally multiplying probabilities by outcomes. How many people have ruined their lives by taking a gamble with a large chance at a small gain and a small chance at a catastrophic loss—cutting a legal corner for an extra bit of money they didn't need, risking their reputation and tranquility for a meaningless fling? Switching from losses to gains, how many lonely singles forgo the small chance of a lifetime of happiness with a soul mate because they think only of the large chance of a tedious coffee with a bore?

As for betting your life: Have you ever saved a minute on the road by driving over the speed limit, or indulged your impatience by checking your new texts while crossing the street? If you weighed the benefits against the chance of an accident multiplied by the price you put on your life, which way would it go? And if you don't think this way, can you call yourself rational?

7

HITS AND FALSE ALARMS

(SIGNAL DETECTION AND STATISTICAL DECISION THEORY)

The cat that sits down on a hot stove-lid . . . will never sit down on a hot stove-lid again, and that is well; but also she will never sit down on a cold one any more.

—MARK TWAIN[1]

Rationality requires that we distinguish what is true from what we want to be true—that we not bury our heads in the sand, build castles in the air, or decide that the grapes just out of reach are sour. The temptations of wishful and magical thinking are always with us because our fortunes hinge on the state of the world, which we can never know with certainty. To keep up our gumption and safeguard against taking painful measures that may prove unnecessary, we are apt to see what we want to see and disregard the rest. We teeter on the edge of the bathroom scale in a way that minimizes our weight, procrastinate getting a medical test that may return an unwelcome result, and try to believe that human nature is infinitely malleable.

There is a more rational way to reconcile our ignorance with our desires: the tool of reason called Signal Detection Theory or statistical decision theory. It combines the big ideas of the two preceding chapters: estimating the probability that something is true of the world (Bayesian reasoning) and deciding what to do about it by weighing its expected costs and benefits (rational choice).[2]

The signal detection challenge is whether to treat some indicator as a genuine signal from the world or as noise in our imperfect perception of it. It's a recurring dilemma in life. A sentry sees a blip on a radar screen. Are we being attacked by nuclear bombers, or is it a flock of seagulls? A radiologist sees a blob on a scan. Does the patient have cancer, or is it a harmless cyst? A jury hears eyewitness testimony in a trial. Is the defendant guilty, or did the witness misremember? We meet a person who seems vaguely familiar. Have we met her before, or is it a free-floating pang of déjà vu? A group of patients improves after taking a drug. Did the drug do anything, or was it a placebo effect?

The output of statistical decision theory is not a degree of credence but an actionable decision: to have surgery or not, to convict or acquit. In coming down on one side or the other, we are not deciding what to believe about the state of the world. We're committing to an action in expectation of its likely costs and benefits. This cognitive tool clobbers us with the distinction between what is true and what to do. It acknowledges that different states of the world can call for different risky choices, but shows that we need not fool ourselves about reality to play the odds. By sharply distinguishing our assessment of the state of the world from what we decide to do about it, we can rationally act *as if* something is true without necessarily *believing* that it is true. As we shall see, this makes a huge but poorly appreciated difference in understanding the use of statistics in science.

Signals and Noise, Yeses and Nos

How should we think about some erratic indicator of the state of the world? Begin with the concept of a statistical distribution.[3] Suppose we measure something that varies unpredictably (a "random variable"), like scores on a test of introversion from 0 to 100. We sort the scores into bins—0 to 9, 10 to 19, and so on—and count the number of people who fall into each bin. Now we stack them in a *histogram*, a graph that differs from the usual ones we see in that the variable of interest is plotted along the horizontal axis rather than the vertical one. The up-and-down dimension simply piles up the number of people falling into each bin. Here is a histogram of introversion scores from 20 people, one person per square.

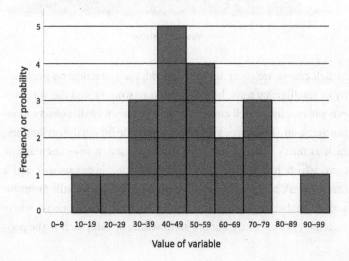

Now imagine that we tested several *million* people, enough so that we no longer have to sort them into bins but can arrange them left to right by their original scores. As we pile up more and more squares and stand farther and farther back, the ziggurat will blur into a smooth mound, the familiar bell-shaped curve below. It has lots of observations heaped up at an average value in the middle, and fewer and fewer as you look at values that are smaller and smaller to the left or larger and larger to the right. The most familiar mathematical model for a bell curve is called the normal or Gaussian distribution.

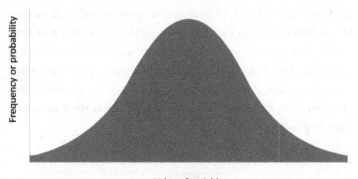

Value of variable

Bell curves are common in the world, such as scores on personality or intelligence tests, heights of men or women, and the speeds of cars on a highway. Bell curves are not the only way that observations can stack up. There are also two-humped or bimodal distributions, such as men's relative degree of sexual attraction to women and to men, which has a large peak at one end for heterosexuals and a smaller peak at the other end for homosexuals, with still fewer bisexuals in between. And there are fat-tailed distributions, where extreme values are rare but not astronomically rare, such as the populations of cities, the incomes of individuals, or the number of visitors to websites. Many of these distributions, such as those generated by "power laws," have a high spine on the left with lots of low values and a long, thick tail on the right with a modicum of extreme ones.[4] But bell curves—unimodal, symmetrical, thin-tailed—are common in the world; they arise whenever a measurement is the sum of a large number of small causes, like many genes together with many environmental influences.[5]

Let's turn to the subject at hand, observations on whether or not something happened in the world. We can't divine it perfectly—we're not God—but only through our measurements, such as blips on a radar screen coming from an aircraft, or the opacity of blobs on a scan from a tumor. Our measurements don't come out precisely the same, on the dot, every time. Instead they tend to be distributed in a bell curve, shown in the diagram below. You can think of it as a plot of the Bayesian likelihood: the probability of an observation given that a signal is present.[6] On average the observation has a certain value (the dashed vertical line), but sometimes it's a bit higher or lower.

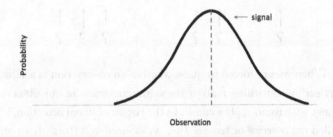

But here's a tragic twist. You might think that when nothing is happening in the world—no bomber, no tumor—we'd get a measurement of zero. Unfortunately, that never happens. Our measurements are always contaminated by noise—radio static, nuisances like flocks of birds, harmless cysts that show up on the scan—and they, too, will vary from measurement to measurement, falling into their own bell curve. More unfortunate still, the upper range of the measurements triggered by noise can overlap with the lower range of the measurements triggered by the thing in the world:

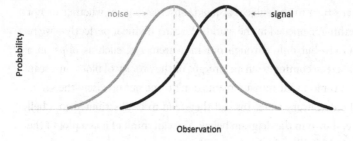

The tragedy is that only God can see the diagram and know whether an observation comes from a signal or from noise. All we mortals see are our observations:

When we are forced to guess whether an observation is a signal (reflecting something real) or noise (the messiness in our observations), we have to apply a cutoff. In the jargon of signal detection, it's called the *criterion* or *response bias*, symbolized as β (beta). If an observation is above the criterion, we say "Yes," acting as if it is a signal (whether or not it is, which we can't know); if it is below, we say "No," acting as if it is noise:

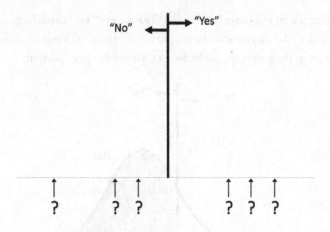

Let's rise back to the God's-eye view and see how well we do, on average, with this cutoff. There are four possibilities. When we say "Yes" and it really is a signal (the bomber or tumor is there), it's called a hit, and the proportion of signals that we correctly identify is shown as the dark shaded portion of the distribution:

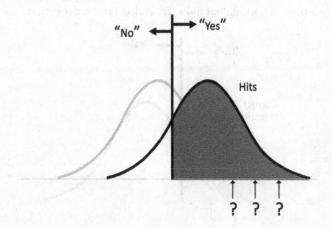

What if it was nothing but noise? When we say "Yes" to nothing, it's called a false alarm, and the proportion of these nothings in which we jump the gun is shown below as the medium-gray portion:

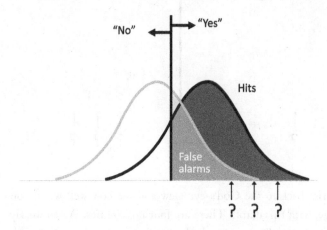

What about the occasions when the observation falls *below* our criterion and we say "No"? Again, there are two possibilities. When there really is something happening in the world, it's called a miss. When there is nothing but noise, it's called a correct rejection.

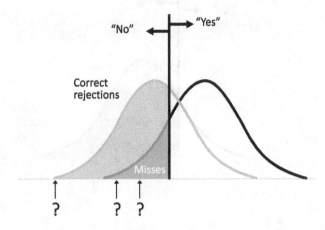

Here is how the four possibilities carve up the space of happenings:

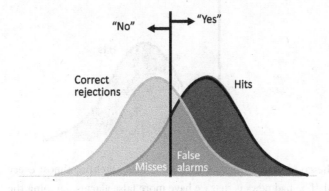

Since we say "Yes" or "No" every time, the proportions of hits and misses when there is a real signal (right heap) must add up to 100 percent. So must the proportions of false alarms and correct rejections when there's nothing but noise (left heap). If we were to lower our criterion leftward, becoming more trigger-happy, or raise it rightward, becoming more gun-shy, then we would be trading hits for misses, or false alarms for correct rejections, as a matter of sheer arithmetic. Less obviously, because the two curves overlap, we would *also* be trading hits for false alarms (when we say "Yes") and misses for correct rejections (when we say "No"). Let's have a closer look at what happens when we relax the response criterion, becoming more trigger-happy or yea-saying:

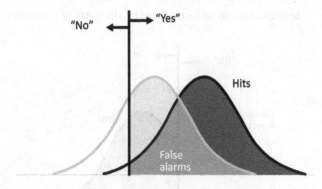

The good news is that we have more hits, catching almost every signal. The bad news is that we have more false alarms, jumping the gun a majority of the time when there's nothing but noise. What if instead we adopt a more stringent response bias, becoming a gun-shy naysayer who demands a high burden of proof?

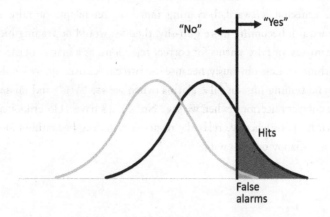

Now the news is reversed: we hardly ever cry wolf with a false alarm, which is good, but we miss most of the signals, which is bad. In the extreme case, if we mindlessly said "Yes" every time, we'd always be right when there was a signal and would always be wrong when there was noise, and vice versa if we said "No" every time.

This sounds obvious, but confusing response bias with accuracy by looking only at the signals or only at the noise is a surprisingly common fallacy. Suppose an examiner separately analyzes performance on the true and false items in a true-false test. He thinks he's seeing whether people are better at detecting truths or rejecting falsehoods, but all he's really seeing is whether they're the kind of person who likes saying "Yes" or "No." I was appalled when a doctor gave me a hearing test that presented a series of beeps increasing in loudness from inaudible to unmissable and asked me to raise a finger when I started hearing them. It wasn't a test of my hearing. It was a test of my impatience and my willingness to go out on a limb when I couldn't honestly say whether I was hearing a tone or tinnitus. The theory of signal detection provides a number of ways to do this right, including penalizing respondents for false alarms, forcing them to say "Yes" a certain percentage of the time, asking them for a confidence rating instead of a thumbs-up/thumbs-down, and making the test multiple-choice instead of true-false.

Costs and Benefits, and Setting a Cutoff

With the tragic tradeoff between hits and false alarms (or misses and correct rejections), what's a rational observer to do? Assuming for the moment that we are stuck with the senses and measuring instruments we have, together with their annoyingly overlapping bell curves, the answer comes right out of expected utility theory (chapter 6): it depends on the benefits of each kind of correct guess and the costs of each kind of error.[7]

Let's go back to the scenario in which Signal Detection Theory arose, detecting incoming bombers from radar blips. The four possibilities are arrayed below, each row representing a state of the world, each column a response of our radar operator, with the outcome listed in each cell:

	"Yes"	**"No"**
Signal (bomber)	Hit (city spared)	Miss (city bombed)
Noise (seagulls)	False alarm (wasted mission, escalated tensions)	Correct rejection (all calm)

In deciding where to set the criterion for responding, our decision maker has to ponder the combined costs (the expected utility) of each column.[8] "Yes" responses will spare the targeted city when it truly is under attack (a hit), which is a massive benefit, while incurring a moderate cost when it is not (a false alarm), including the waste of sending interceptor planes scrambling for no reason, together with fear at home and tensions abroad. "No" responses will expose a city to the attack when there is one (a miss), a massive cost, while keeping the blessed peace and quiet when there is not (a correct rejection). Overall the balance sheet would seem to call for a low or relatively trigger-happy response criterion: a few days when interceptors are needlessly sent scrambling would seem a small price to pay for the day when it would spare a city from being bombed.

The calculation would be different if the costs were different. Suppose the response was not sending planes to intercept the bombers but sending nuclear-tipped ICBMs to destroy the enemy's cities, guaranteeing a thermonuclear World War III. In that case the catastrophic cost of a false alarm would call for being absolutely sure you are being attacked before responding, which means setting the response criterion very, very high.

Also relevant are the base rates of the bombers and seagulls that trigger those blips (the Bayesian priors). If seagulls were common but bombers rare, it would call for a high criterion (not jumping the gun), and vice versa.

As we saw in the previous chapter, we face the same dilemma on a personal scale in deciding whether to have surgery in response to an ambiguous cancer test result:

	"Yes"	"No"
Signal **(cancer)**	Hit (life saved)	Miss (death)
Noise **(benign cyst)**	False alarm (pain, disfigurement, expense)	Correct rejection (life as usual)

So where, exactly, should a rational decision maker—an "ideal observer," in the lingo of the theory—place the criterion? The answer is: at the point that would maximize the observer's expected utility.[9] It's easy to calculate in the lab, where the experimenter controls the number of trials with a beep (the signal) and no beep (the noise), pays the participant for each hit and correct rejection, and fines her for every miss and false alarm. Then a hypothetical participant who wants to make the most money would set her criterion according to this formula, where the values are the payoffs and penalties:

$$\beta = \frac{(\text{value of a correct rejection} - \text{value of a false alarm}) \times \text{prob(noise)}}{(\text{value of a hit} - \text{value of a miss}) \times \text{prob(signal)}}$$

The exact algebra is less important than simply noticing what's on the top and the bottom of the ratio and what's on each side of the minus sign. An ideal observer would set her criterion higher (need better evidence before saying "Yes") to the degree that noise is likelier than a signal (a low Bayesian prior). It's common sense: if signals are rare, you should say "Yes" less often. She should also set a higher bar when the payoffs for hits are lower or for correct rejections are higher, while the penalties for false alarms are higher or for misses are lower. Again, it's common sense: if you're paying big fines for false alarms, you should be more chary of saying "Yes," but if you're getting windfalls for hits, you should be more keen. In laboratory experiments participants gravitate toward the optimum intuitively.

When it comes to decisions involving life and death, pain and disfigurement, or the salvation or destruction of civilization, assigning numbers to the costs is obviously more problematic. Yet the dilemmas are just as agonizing if we don't assign numbers to them, and pondering each of the four boxes, even with a crude sense of which costs are monstrous and which bearable, can make the decisions more consistent and justifiable.

Sensitivity versus Response Bias

Tradeoffs between misses and false alarms are agonizing, and can instill a tragic vision of the human condition. Are we mortals perpetually doomed to choose between the awful cost of mistaken inaction (a city bombed, a cancer left to spread) and the dreadful cost of mistaken action (a ruinous provocation, disfiguring surgery)? Signal Detection Theory says we are, but it also shows us how to mitigate the tragedy. We can bend the tradeoff by increasing the *sensitivity* of our observations. The costs in a signal detection task depend on two parameters: where we set the cutoff (our response bias, criterion, trigger-happiness, or β), and how far apart the signal and noise distributions are, called the "sensitivity," symbolized as *d'*, pronounced "d-prime."[10]

Imagine that we perfected our radar so that it weeds out the seagulls, or at worst registers them as faint snow, while displaying the bombers as big bright spots. That means the bell curves for the noise and signal would be pushed farther apart (lower diagram). This in turn means that, regardless of where you put the response cutoff, you will have both fewer misses *and* fewer false alarms:

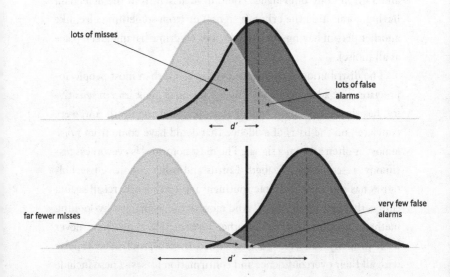

And, by the laws of arithmetic, you will enjoy a greater proportion of hits and correct rejections. While sliding the cutoff back and forth tragically trades off one error for another, pulling the two curves apart—better instruments, more sensitive diagnostics, more reliable forensics—is an unmitigated good, reducing errors of both types. Enhancing sensitivity should always be our aspiration in signal detection challenges, and that brings us to one of its most important applications.

Signal Detection in the Courtroom

An investigation into a wrongdoing is a signal detection task. A judge, jury, or disciplinary panel is faced with evidence about the possible malfeasance of a defendant. The evidence varies in strength, and a given body of evidence could have arisen from the defendant having committed the crime (a signal) or from something else, like another person having done the deed or no crime having taken place at all (noise).

The distributions of evidence overlap more than most people appreciate. The advent of DNA fingerprinting (a giant leap in sensitivity) has shown that many innocent people, some on death row, were convicted on the basis of evidence that could have come from noise almost as often as from a signal. The most notorious is eyewitness testimony: research by Elizabeth Loftus and other cognitive psychologists has shown that people routinely and confidently recall seeing things that never happened.[11] And most of the sciency-techy-looking methods featured in *CSI* and other forensic TV shows have never been properly validated but are shilled by self-proclaimed experts with all their overconfidence and confirmation biases. These include analyses of bullets, bite marks, fibers, hair, shoe prints, tire tracks, tool marks, handwriting, blood spatters, fire accelerants, even fingerprints.[12] DNA is the most reliable forensic technique, but remember the difference between a propensity and a frequency: some percentage of DNA testimony is corrupted by contaminated samples, botched labels, and other human error.

A jury faced with noisy evidence has to apply a criterion and return a yes-or-no verdict. Its decision matrix has costs and benefits that are reckoned in practical and moral currencies: the malefactors who are removed from the streets or allowed to prey on others, the abstract value of justice meted out or miscarried.

	"Convict"	"Acquit"
Signal (guilty)	Hit (justice done; criminal incapacitated)	Miss (justice denied; criminal free to prey on others)
Noise (innocent)	False alarm (miscarriage of justice; an innocent punished)	Correct rejection (justice done; though with costs of a trial)

As we saw in the discussion of forbidden base rates (chapter 5), no one would tolerate a justice system that worked purely on the practical grounds of the costs and benefits to society; we insist on fairness to the individual. But given that juries lack divine omniscience, how should we trade off the incommensurable injustices of a false conviction and a false acquittal? In the language of signal detection, where do we place the response criterion?

The standard presumption has been to assign a high moral cost to false alarms. As the jurist William Blackstone (1723–1780) put it in his eponymous rule, "It is better that ten guilty persons escape than that one innocent suffer." And so juries in criminal trials make a "presumption of innocence," and may convict only if the defendant is "guilty beyond a reasonable doubt" (a high setting for β, the criterion or response bias). They may not convict based on a mere "preponderance of the evidence," also known as "fifty percent plus a feather."

Blackstone's 10:1 ratio is arbitrary, of course, but the lopsidedness is eminently defensible. In a democracy, freedom is the default, and government coercion an onerous exception that must meet a high burden of justification, given the awesome power of the state and its constant temptation to tyrannize. Punishing the innocent, particularly by death, shocks the conscience in a way that failing to punish the guilty does not. A system that does not capriciously target people for ruination marks the difference between a regime of justice and a regime of terror.

As with all settings of a response criterion, the setting based on Blackstone's ratio depends on the valuation of the four outcomes, which may be contested. In the wake of 9/11, the administration of George W. Bush believed that the catastrophic cost of a major terrorist act justified the use of "enhanced interrogation," a euphemism for torture, outweighing the moral cost of false confessions by tortured innocents.[13] In 2011, the US Department of Education set off a firestorm with a new guideline (since rescinded) that colleges must convict students accused of sexual misconduct based on a preponderance of the evidence.[14] Some defenders of such policies acknowledged the tradeoff but argued that sexual infractions are so heinous that the price of convicting a few innocents is worth paying.[15]

There is no "correct" answer to these questions of moral valuation, but we can use signal detection thinking to ascertain whether our practices are consistent with our values. Suppose we believe that no more than one percent of guilty people should be acquitted and no more than one percent of the innocent convicted. Suppose, too, that juries were ideal observers who applied Signal Detection Theory optimally. How strong would the evidence have to be to meet those targets? To be precise, how large does d' have to be, namely the distance between the distributions for the signal (guilty) and the noise (innocent)? The distance may be measured in standard deviations, the most common estimate of variability. (Visually it corresponds to the width of the bell curve, that is, the horizontal distance from the mean to the inflection point, where convex shifts to concave.)

The psychologists Hal Arkes and Barbara Mellers did the math and calculated that to meet these goals the d' for the strength of evidence would have to be 4.7—almost five standard deviations separating the evidence for guilty parties from the evidence for innocent ones.[16] That's an Olympian level of sensitivity which is not met even by our most sophisticated medical technologies. If we were willing to relax our standards and convict up to 5 percent of the innocent and acquit 5 percent of the guilty, d' would "only" have to be 3.3 standard deviations, which is still a princess-and-the-pea level of sensitivity.

Does this mean that our moral aspirations for justice outstrip our probative powers? Almost certainly. Arkes and Mellers probed a sample of students to see what those aspirations really are. The students ventured that a just society should convict no more than 5 percent of the innocent and acquit no more than 8 percent of the guilty. A sample of judges had similar intuitions. (We can't tell whether that's more or less stringent than Blackstone's ratio because we don't know what percentage of defendants really are guilty.) These aspirations demand a d' of 3.0—the evidence left by guilty defendants would have to be three standard deviations stronger than the evidence left by innocent ones.

How realistic is that? Arkes and Mellers dipped into the literature on the sensitivity of various tests and techniques and found that the answer is: not very. When people are asked to distinguish liars from truth-tellers, their d' is approximately 0, which is to say, they can't. Eyewitness testimony is better than that, but not much better, at a modest 0.8. Mechanical lie detectors, that is, polygraph tests, are better still, around 1.5, but they are inadmissible in most courtrooms.[17]

Turning from forensics to other kinds of tests to calibrate our expectations, they found d's of around 0.7 for military personnel screening tests, 0.8–1.7 for weather forecasting, 1.3 for mammograms, and 2.4–2.9 for CT scans of brain lesions (estimated, admittedly, for the technologies of the late twentieth century; all should be higher today).

Suppose that the typical quality of evidence in a jury trial has a d' of 1.0 (that is, one standard deviation higher for guilty than for innocent defendants). If juries adopt a tough response criterion, anchored, say, by a prior belief that a third of defendants are guilty, they will acquit 58 percent of guilty defendants and convict 12 percent of the innocent ones. If they adopt a lax one, corresponding to a prior belief that two thirds of defendants are guilty, they'll acquit 12 percent of guilty defendants and convict 58 percent of innocent ones. The heart-sinking conclusion is that juries acquit far more guilty people, and convict far more innocent ones, than any of us would deem acceptable.

Now, the criminal justice system may strike a better bargain with the devil than that. Most cases don't go to trial but are dismissed because the evidence is so weak, or plea-bargained (ideally) because the evidence is so strong. Still, the signal detection mindset could steer our debates on judicial proceedings toward greater justice. Currently many of the campaigns are naïve to the tradeoff between hits and false alarms and treat the possibility of false convictions as inconceivable, as if the triers of fact were infallible. Many advocates of justice, that is, argue for pulling the decision cutoff downward. Put more criminals behind bars. Believe the woman. Monitor the terrorists and lock them up before they attack. If someone takes a life, they deserve to lose their own. But of mathematical necessity, lowering the response criterion can only trade one kind of injustice for another. The arguments could be restated as: Put more innocent people behind bars. Accuse more blameless men of rape. Lock up harmless youths who shoot off their mouths on social media. Execute more of the guiltless.[18] These paraphrases do not, by themselves, refute the arguments. At a given time a system may indeed be privileging the accused over their possible victims or vice versa and be due for an adjustment. And if less-than-omniscient humans are to have a system of justice at all, they must face up to the grim necessity that some innocents will be punished.

But being mindful of the tragic tradeoffs in distinguishing signals from noise can bring greater justice. It forces us to face the enormity of harsh punishments like the death penalty and long sentences, which are not only cruel to the guilty but inevitably will be visited upon the innocent. And it tells us that the real quest for justice should consist of increasing the sensitivity of the system, not its bias: to seek more accurate forensics, fairer protocols for interrogation and testimony, restraints on prosecutorial zealotry, and other safeguards against miscarriages of both kinds.

Signal Detection and Statistical Significance

The tradeoff between hits and false alarms is inherent to any decision that is based on imperfect evidence, which means that it hangs over every human judgment. I'll mention one more: decisions about whether an empirical finding should license a conclusion about the truth of a hypothesis. In this arena, Signal Detection Theory appears in the guise of statistical decision theory.[19]

Most scientifically informed people have heard of "statistical significance," since it's often reported in news stories on discoveries in medicine, epidemiology, and social science. It's based on pretty much the same mathematics as Signal Detection Theory, pioneered by the statisticians Jerzy Neyman (1894–1981) and Egon Pearson (1895–1980). Seeing the connection will help you avoid a blunder that even a majority of scientists routinely commit. Every statistics student is warned that "statistical significance" is a technical concept that should not be confused with "significance" in the vernacular sense of noteworthy or consequential. But most are misinformed about what it does mean.

Suppose a scientist observes some things in the world and converts her measurements into data that represent the effect she is interested in, like the difference in symptoms between the group that got the drug and the group that got the placebo, or the difference in verbal skills between boys and girls, or the improvement in test scores after students enrolled in an enrichment program. If the number is zero, it means there's no effect; greater than zero, a possible eureka. But human guinea pigs being what they are, the data are noisy, and an average score above zero may mean that there is a real difference in the world, or it could be sampling error, the luck of the draw. Let's go back to the God's-eye view and plot the distribution of scores that the scientist would obtain if there's no difference in reality, called the null hypothesis, and the distribution of scores she would obtain if something is happening, an effect of a given size. The distributions overlap—that's what makes science hard. The diagram should look familiar:

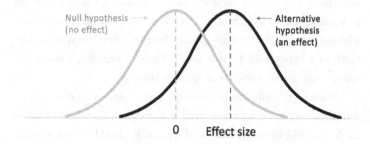

The null hypothesis is the noise; the alternative hypothesis is the signal. The size of the effect is like the sensitivity, and it determines how easy it is to tell signal from noise. The scientist needs to apply some criterion or response bias before breaking out the champagne, called the critical value: below the critical value, she fails to reject the null hypothesis and drowns her sorrows; above the critical value, she rejects it and celebrates—she declares the effect to be "statistically significant."

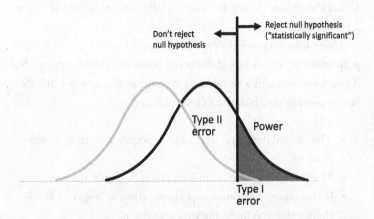

But where should the critical value be placed? The scientist must trade off two kinds of error. She could reject the null hypothesis when it is true, namely a false alarm, or in the argot of statistical decision theory, a Type I error. Or she could fail to reject the null hypothesis when it is false—a miss, or in the patois, a Type II error. Both are bad: a Type I error introduces falsehood into the scientific record; a Type II error represents a waste of effort and money. It happens when the methodology was not designed with sufficient "power" (the hit rate, or 1 minus the Type II error rate) to detect the effect.

Now, deep in the mists of time it was decided—it's not completely clear by whom—that a Type I error (proclaiming an effect when there is none) is especially damaging to the scientific enterprise, which can tolerate only a certain number of them: 5 percent of the studies in which the null hypothesis is true, to be exact. And so the convention arose that scientists should adopt a critical level that ensures that the probability of rejecting the null hypothesis when it is true is less than 5 percent: the coveted "$p < .05$." (Though one might have thought that the costs of a Type II error should also be factored in, as it is in Signal Detection Theory, for some equally obscure historical reason it never was.)

That's what "statistical significance" means: it's a way to keep the proportion of false claims of discoveries beneath an arbitrary cap. So if you have obtained a statistically significant result at $p < .05$, that means you can conclude the following, right?

- The probability that the null hypothesis is true is less than .05.
- The probability that there is an effect is greater than .95.
- If you rejected the null hypothesis, there is less than a .05 chance that you made the wrong decision.
- If you replicated the study, the chance that you would succeed is greater than .95.

Ninety percent of psychology professors, including 80 percent of those who teach statistics, think so.[20] But they're wrong, wrong, wrong, and wrong. If you've followed the discussion in this chapter and in chapter 5, you can see why. "Statistical significance" is a Bayesian *likelihood*: it reflects the probability of obtaining the data (or data even more extreme) given the hypothesis (in this case, the null hypothesis).[21] But each of those statements is a Bayesian *posterior*: the probability of the hypothesis given the data. That's ultimately what we want—it's the whole point of doing a study—but it's not what a significance test delivers. If you remember why Irwin does not have liver disease, why private homes are not necessarily dangerous, and why the pope is not a space alien, you know that these two conditional probabilities must not be switched around. The scientist cannot use a significance test to assess whether the null hypothesis is true or false unless she also considers the *prior*—her best guess of the probability that the null hypothesis is true before doing the experiment. And in the mathematics of null hypothesis significance testing, a Bayesian prior is nowhere to be found.

Most social scientists are so steeped in the ritual of significance testing, starting so early in their careers, that they have forgotten its actual logic. This was brought home to me when I collaborated with a theoretical linguist, Jane Grimshaw, who tutored herself in statistics and said to me, "Let me get this straight. The only thing these tests show is that when some effect doesn't exist, one of every twenty scientists looking for it will falsely claim it does. What makes you so sure it isn't *you*?" The honest answer is: Nothing. Her skepticism anticipated yet another explanation for the replicability imbroglio. Suppose that, like Lewis Carroll's snark hunters, twenty scientists go chasing after a figment. Nineteen file their null results in a drawer, and the one who is lucky (or unlucky) enough to make the Type I error publishes his "finding."[22] In an *XKCD* cartoon, a pair of scientists test for a correlation between jelly beans and acne separately for each of twenty colors, and become famous for linking green jellybeans to acne at $p < .05$.[23] Scientists have finally gotten the joke, are getting into the habit of publishing null results, and have developed techniques to compensate for the file drawer problem when reviewing the literature in a meta-analysis, a study of studies. Null results are conspicuous by their absence, and the analyst can detect the nothing that is not there as well as the nothing that is.[24]

The scandalous misunderstanding of significance testing bespeaks a human yearning. Philosophers since Hume have noted that induction—drawing a generalization from observations—is an inherently uncertain kind of inference.[25] An infinite number of curves can be drawn through any finite set of points; an unlimited number of theories are logically consistent with any body of data. The tools of rationality explained in these chapters offer different ways of coping with this cosmic misfortune. Statistical decision theory cannot ascertain the truth, but it can cap the damage from the two kinds of error. Bayesian reasoning can adjust our credence in the truth, but it must begin with a prior, with all the subjective judgment that goes into it. Neither one provides what everyone longs for: a turnkey algorithm for determining the truth.

SELF AND OTHERS

(GAME THEORY)

Your corn is ripe today; mine will be so tomorrow. 'Tis profitable for us both, that I should labour with you today, and that you should aid me tomorrow. I have no kindness for you, and know you have as little for me. I will not, therefore, take any pains upon your account; and should I labour with you upon my own account, in expectation of a return, I know I should be disappointed, and that I should in vain depend upon your gratitude. Here then I leave you to labour alone: You treat me in the same manner. The seasons change; and both of us lose our harvests for want of mutual confidence and security.

—DAVID HUME[1]

N ot long ago I had a friendly argument with a colleague on the messages our university should be sending on climate change. Professor J argued that we just need to persuade people that it is in their self-interest to reduce their emission of greenhouse gases, since a warmer planet would bring floods, hurricanes, wildfires, and other disasters that would make their lives worse. I replied that it is *not* in their self-interest, because no individual's sacrifice alone can prevent climate change, while the sacrificer would sweat in the summer, shiver in the winter, and wait for buses in the rain while her polluting neighbors stayed comfy and dry. Only if *everyone* eliminated their emissions would *anyone* benefit, and the only way it would be in anyone's interest to do that was if clean energy was cheaper for everyone (via technological advances) and dirty energy more expensive (via carbon pricing). My colleague had a point: in one sense, it is irrational to ruin the planet. But I could not convince Dr. J that in another sense it is, tragically, all too rational.

At that moment I realized that a critical concept was missing from the good doctor's worldview: game theory, the analysis of how to make rational choices when the payoffs depend on someone *else's* rational choices.

Game theory was presented to the world by von Neumann and Morgenstern in the same book in which they explained expected utility and rational choice.[2] But unlike the dilemmas in which we take our chances against a brainless wheel of fortune and the best strategies turn out to be pretty intuitive, game theory deals with dilemmas that pit us against equally cunning deciders, and the outcomes can turn our intuitions upside down and sideways. The games of life sometimes leave rational actors no choice but to do things that make themselves and everyone else worse off; to be random, arbitrary, or out of control; to cultivate sympathies and nurse grievances; to willingly submit to penalties and punishments; and sometimes, to refuse to play at all. Game theory unveils the strange rationality beneath many of the perversities of social and political life, and as we will see in a later chapter, it helps explain the central mystery of this book: how a rational species can be so irrational.

A Zero-Sum Game: Scissors-Paper-Rock

The quintessential game-theoretic dilemma, which lays bare how the payoff of a choice depends on the other guy's choice, is the game of Scissors-Paper-Rock.[3] Two players simultaneously display a hand gesture—two fingers for scissors, flat for paper, clenched for rock—and the winner is determined by the rule "Scissors cuts paper, paper covers rock, rock blunts scissors." The game can be displayed as a matrix in which the possible choices of the first player, Amanda, are shown as rows, the choices of the second, Brad, are shown as columns, and the payoffs are written in each cell, Amanda's in the lower left corner, Brad's in the upper right. Let's give numerical values to the outcomes: 1 for a win, –1 for a loss, 0 for a tie.

Brad's choices

		Scissors	Paper	Rock
Amanda's choices	Scissors	Tie 0 / Tie 0	Lose –1 / Win 1	Win 1 / Lose –1
	Paper	Win 1 / Lose –1	Tie 0 / Tie 0	Lose –1 / Win 1
	Rock	Lose –1 / Win 1	Win 1 / Lose –1	Tie 0 / Tie 0

Amanda's payoffs and Brad's payoffs sum to 0 in every cell, giving us a technical term that has crossed over from game theory into everyday life: the zero-sum game. Amanda's gain is Brad's loss, and vice versa. They're locked in a state of pure conflict, fighting over a fixed pie.

Which move (row) should Amanda choose? The crucial technique in game theory (and indeed in life) is to see the world from the other player's point of view. Amanda must examine Brad's choices, the columns, one at a time. Going from left to right, if Brad picks Scissors, she should pick Rock. If he picks Paper, she should pick Scissors. And if he picks Rock, she should pick Paper. There is no "dominant" choice, one that is superior regardless of what Brad does, and of course she does not know what Brad will do.

But this does not mean that Amanda should pick an arbitrary move, say, Paper, and stick with it. If she did, then Brad would catch on, go with Scissors, and beat her every time. In fact, even if she tilted a bit toward Paper, choosing it, say, 40 percent of the time and the other two strategies 30 percent apiece, Brad could play Scissors and beat her four times out of seven. Amanda's best strategy is to turn herself into a human roulette wheel and play each move at random with the same probability, stifling any skew, tilt, lean, or drift away from a perfect ⅓–⅓–⅓ split.

Since the table is symmetrical along the diagonal, Brad's machinations are identical. As he considers what Amanda might do, row by row, he has no reason to pick one of his moves over the other two, and will arrive at the same "mixed" strategy, playing each option with a probability of ⅓. If Brad were to deviate from this strategy, Amanda would change hers to exploit him, and vice versa. They are locked in a *Nash equilibrium*, named after the mathematician John Nash (the subject of the movie *A Beautiful Mind*). Each is playing the best strategy given the opponent's best strategy; any unilateral change would make them worse off.

The discovery that in some situations a rational agent must be superhumanly random is just one of the conclusions from game theory that seems outlandish until you realize that the situations are not uncommon in life. The equilibrium in Scissors-Paper-Rock is called an outguessing standoff, and examples are common in sports like tennis, baseball, hockey, and soccer. A penalty kicker in soccer can kick right or left, and the goalie can guard right or left; unpredictability is a cardinal virtue. Bluffs in poker and surprise attacks in military strategy are also outguessing standoffs. Even when a move is not literally picked at random (presumably in 1944 the Allies did not roll a die before deciding whether to invade Normandy or Calais), the player must assume a poker face and suppress any tell or leak, making the choice *appear* random to their opponents. The philosophers Liam Clegg and Daniel Dennett have argued that human behavior is inherently unpredictable not just because of random neural noise in the brain but as an adaptation that makes it harder for our rivals to outguess us.[4]

A Non-Zero-Sum Game: The Volunteer's Dilemma

Rational actors can end up in outguessing standoffs not just in games that pit them in zero-sum competition but in ones that partly align them with common interests. An example is the Volunteer's Dilemma, which may be illustrated by the medieval story Belling the Cat. A mouse proposes to his housemates that one of them put a bell around the neck of the cat while she sleeps so they will be alerted to her approach. The problem, of course, is who will bell the cat and incur the risk of awakening her and getting eaten. Parallel dilemmas for humans include which passenger will overcome an aircraft hijacker, which bystander will rescue a person in distress, and which office worker will refill the coffeepot in a communal kitchen.[5] Everyone wants someone to help but prefers that it not be them. If we translate the benefits and costs into numerical units, with 0 as the worst thing that can happen, we get the matrix below. (Technically it should be a hypercube with as many dimensions as there are players, but I've collapsed everyone but the self into a single layer.)

Others' choices

		Help	Shirk
Own choices	Help	50 / 50	100 / 50
	Shirk	50 / 100	0 / 0

Once again there is no dominant strategy that makes the choice easy. If one mouse knew the others would shirk, then he should help, and vice versa. But if each mouse decided whether to bell the cat with a certain probability (one that equated the *other* mice's expected payoffs of belling and shirking), then the mice would fall into an out-guessing standoff, each being willing to bell while hoping another mouse goes first.

Unlike Scissors-Paper-Rock, the Volunteer's Dilemma is not zero-sum: some outcomes are better for everyone than others. (The outcomes are "win–win"—another concept from game theory that has crossed over into everyday parlance.) The mice are collectively worst off if none of them volunteers and best off if one of them does—which does not guarantee that they will arrive at this happy ending, since there's no Head Mouse to draft one of them into possible martyrdom for the good of the horde. Rather, each mouse rolls the die because no mouse would do better by unilaterally switching to a different strategy. Here again they are in a Nash equilibrium, a standoff in which all the players stick with their best choice in response to the others' best choices.

Rendezvous and Other
Coordination Games

A dog-eat-dog contest like Scissor-Paper-Rock and a nervous hypo-critical standoff like the Volunteer's Dilemma involve a degree of competition. But in some games of life everybody wins, if only they can figure out how. These are called coordination games, like Rendezvous. Caitlin and Dan enjoy each other's company and plan to have coffee one afternoon, but Caitlin's phone goes dead before they can settle on Starbucks or Peet's. Each has a slight preference, but they'd both prefer to meet at either place than to forgo the date. The matrix has two equilibria, the top left and bottom right cells, corresponding to their coordinating on the same choice. (Technically their differing preferences introduce a droplet of competition into the scenario, but we can ignore it for now.)

		Dan's choices	
		Peet's	Starbucks
Caitlin's choices	Peet's	95 100	0 0
	Starbucks	0 0	100 95

Caitlin knows that Dan prefers Peet's, and decides to show up there, but Dan knows that Caitlin prefers Starbucks, so he plans to show up *there*. Caitlin, putting herself in Dan's shoes, anticipates his empathy, so she switches her plan to Starbucks, and Dan, equally empathic about *her* empathy, switches to Peet's—until he realizes that she has anticipated his anticipation, and switches back to Starbucks. And so on, ad infinitum, with neither having a reason to settle on something that both of them want.

What they need is *common knowledge*, which in game theory is a technical term referring to something that each one knows that the other knows that they know, ad infinitum.[6] Though it sounds like common knowledge would make someone's head explode, people need not try to stuff an infinite series of "I know that she knows that I know that she knows . . ." into their craniums. They need only have a sense that the knowledge is "self-evident" or "out there" or "on the record." That intuition can be generated by an overt signal that each perceives with the other's awareness, such as a direct chat between them. With many games, a mere promise is "cheap talk" and dismissible. (In a Volunteer's Dilemma, for example, if a mouse were to announce that he refuses to volunteer, hoping it will pressure some other mouse to do it, the other mice could call his bluff and shirk, knowing he might step into the breach.) But in a coordination game, it's in both parties' interests to end up in the same place, so a statement of intent is credible.

In the absence of direct communication (such as when a cell phone goes dead), the parties can instead converge on a *focal point*: a choice that is mutually salient, each party figuring that the other must have noticed it and be aware that they too have noticed it.[7] If the Peet's is nearby, or recently came up in conversation, or is a familiar landmark in town, that might be all that Caitlin and Dan need to break the impasse, regardless of which site boasts better lattes or plusher seats. In coordination games, an arbitrary, superficial, meaningless attention-getter can provide the rational solution to an intractable problem.

Many of our conventions and standards are solutions to coordination games, with nothing to recommend them other than that everyone has settled on the same ones.[8] Driving on the right, taking Sundays off work, accepting paper currency, adopting technological standards (110 volts, Microsoft Word, the QWERTY keyboard) are equilibria in coordination games. There may be higher payoffs with other equilibria, but we remain locked into the ones we have because we can't get there from here. Unless everyone agrees to switch at once, the penalties for discoordination are too high.

Arbitrary focal points can figure in bargaining. Once a buyer and a seller have converged on a range of prices that make the deal more attractive to both of them than walking away, they are in a kind of coordination game. Either of two equilibria (their current offers) is more attractive than failing to coordinate at all, but each is more attractive to one of them. As each party changes the payoffs, hoping to entice the other into the coordination cell that is more advantageous to him or her, they may seek a focal point that, though arbitrary, gives them something to agree upon, such as a round number or an offer that splits the difference. As Thomas Schelling, who first identified focal points in coordination games, put it, "The salesman who works out the arithmetic for his 'rock-bottom' price on the automobile at $35,017.63 is fairly pleading to be relieved of $17.63."[9] Similarly, "If one has been demanding 60 percent and recedes to 50 percent, he can get his heels in; if he recedes to 49 percent, the other will assume that he has hit the skids and will keep sliding."[10]

Chicken and Escalation Games

Though bargaining has elements of a coordination game, the ability of either party to threaten the other by getting up from the table and leaving them both worse off makes it overlap with another famous game, Chicken, which we met in chapter 2.[11] Here is the matrix. (As always, the exact numbers are arbitrary; only the differences are meaningful.)

Buzz's choices

		Swerve	Straight
James's choices	Swerve	Anticlimax 0 Anticlimax 0	Win 1 "Chicken" –1
	Straight	"Chicken" –1 Win 1	Crash –100 Crash –100

The names of the players come from *Rebel Without a Cause*, but Chicken is not just a suicidal teenage pastime. We play it when we drive or walk along a narrow path and face an oncoming traveler, requiring that someone yield, and when we engage in formal and informal bargaining. Public examples include foreclosing or defaulting on a debt, and brinkmanship standoffs in international relations like the Cuban Missile Crisis of 1962. Chicken has a Nash equilibrium in which each player takes some chance at standing his ground and otherwise swerves, though in real life this solution may be moot because the rules of the game may be enriched to include signaling and alterations to the strategy set. In chapter 2 we saw how a paradoxical advantage can go to a player who is visibly crazy or out of control, making his threats credible enough to coerce his opponent into conceding—though with the shadow of mutual destruction hanging over them if both go crazy or lose control simultaneously.[12]

Some games consist not of a one-shot encounter in which the players make a single move simultaneously and then show their hands, but a series of moves in which each responds to the other, with the payouts settled up at the end. One of these games has mind-bogglingly morbid implications. An Escalation Game may be illustrated by a "Dollar Auction" in an eBay from hell.[13] Imagine an auction with the fiendish rule that the loser, not just the winner, has to pay their last bid. Say the item being auctioned is a tchotchke that can be resold for a dollar. Amanda bids 5 cents, hoping for a 95-cent profit. But of course Brad jumps in at 10 cents, and so on, until Amanda's bid has reached 95 cents, which would shave her margin to a still-profitable 5 cents. It may seem silly at that point for Brad to bid a dollar to win a dollar, but breaking even would be better than losing 90 cents, which the perverse rule of the auction would force him to pay if he dropped out. Even more perversely, Amanda is now faced with the choice of losing 95 cents if she folds or losing 5 cents if she raises, so she bids $1.05, which Brad, preferring to lose a dime than a dollar, bests with $1.10, and so on. They get sucked into furiously outbidding each other to throw more and more money away until one of them goes bust and the other enjoys the Pyrrhic victory of losing a bit less.

The rational strategy in the midst of an Escalation Game is to cut your losses and bow out with a certain probability at each move, hoping that the other bidder, being equally rational, might fold first. It's captured in the saying "Don't throw good money after bad" and in the First Law of Holes: "When you're in one, stop digging." One of the most commonly cited human irrationalities is the *sunk-cost* fallacy, in which people continue to invest in a losing venture because of what they have invested so far rather than in anticipation of what they will gain going forward. Holding on to a tanking stock, sitting through a boring movie, finishing a tedious novel, and staying in a bad marriage are familiar examples. It's possible that people fall prey to the sunk-cost fallacy as a spillover from playing Escalation (and Chicken), where a reputation for standing one's ground, no matter how costly, could convince the other player to back down first.

The Escalation Game is not an exotic brainteaser. Real life presents us with quandaries in which we are, as the saying goes, in for a penny, in for a pound. They include long-running labor strikes, dueling lawsuits, and literal wars of attrition, in which each nation feeds men and matériel into the maw of the war machine hoping the other side will exhaust itself first.[14] The common rationale is "We fight so that our boys will not have died in vain," a textbook example of the sunk-cost fallacy but also a tactic in the pathetic quest for a Pyrrhic victory. Many of the bloodiest wars in history were wars of attrition, showing once again how the infuriating logic of game theory may explain some of the tragedies of the human condition.[15] Though persisting with a certain probability may be the least bad option once one is trapped in an Escalation Game, the truly rational strategy is not to play in the first place.

This includes games we may not even realize we're playing. For many people, one of the benefits of winning an auction is the sheer pleasure of winning. Since the thrill of victory and the agony of defeat are independent of the amount of the winning bid and the value of the item, this can turn any auction into an Escalation Game. Auctioneers exploit this psychology by building suspense and showering kudos on the winner. On the other side, eBay user sites advise bidders to decide beforehand how much the item is worth to them and bid no higher. Some sell a form of Odyssean self-control: they robo-bid up to a limit the bidder sets in advance, tying him to the mast for his own good during the frenzy of an Ego Escalation Game.

The Prisoner's Dilemma and the Tragedy of the Commons

Consider a familiar plot from *Law and Order*. A prosecutor detains partners in crime in separate cells, lacks the evidence to convict them, and offers them a deal. If one agrees to testify against the other, he will go free and his partner will go to prison for ten years. If each rats out the other, they both get six years. If they stay true to the partnership and keep mum, she can only convict them on a lesser charge and they will serve six months.

The payoffs are shown below. In discussions of the Prisoner's Dilemma, Cooperate means staying true to one's partner (it doesn't mean cooperating with the prosecutor), and Defect means ratting him out. The payoffs, too, have mnemonic labels, and their relative degree of badness is what defines the Dilemma. For each player, the best outcome is to defect while the other cooperates (the temptation), the worst is to be the victim of such a betrayal (the sucker's payoff), the second-worst is to be a party to a mutual betrayal (the punishment), and the second-best is to stay true to the partnership when the other does (the reward). The best and worst outcomes for the pair considered together fall along the other diagonal: the worst thing that can happen to them collectively is mutual defection, and the best thing mutual cooperation.

Brutus

		Cooperate (mum)	Defect (rat)
Lefty	Cooperate (mum)	6 months (reward) / 6 months (reward)	Go free (temptation) / 10 years (sucker's payoff)
	Defect (rat)	10 years (sucker's payoff) / Go free (temptation)	6 years (punishment) / 6 years (punishment)

Taking in the whole table from our Olympian vantage point, it's obvious where the partners should try to end up. Neither can count on the other taking the fall, so their only sensible goal is the reward of mutual cooperation. Unfortunately for them in their earthly vantage points, they cannot take in the whole table, because their partner's choice is beyond their control. Lefty is staring rightward at his two moves, and Brutus is staring downward at his. Lefty has to think it through this way: "Suppose he stays mum (Cooperates). Then I'd get six months if I stayed mum too, and I'd go free if I sang (Defected). I'd be better off Defecting. Now suppose he rats me out (Defects). Then I'd get ten years if I stayed mum, but only six if I ratted *him* out. Overall, this means that if he Cooperates, I'm better off Defecting, and if he Defects, I'm better off Defecting. It's a no-brainer." Meanwhile, the thought balloon above Brutus's head contains the same soliloquy. Both defect and are sent away for six years rather than six months—the bitter fruit of each acting in his rational self-interest. Not that either had any choice: it's a Nash equilibrium. Defection is a dominant strategy for both of them, one that leaves each one better off regardless of what the other one does. If one of them had been wise or moral or trusting or far-thinking, he'd be at the mercy of the fear and temptation of the other. Even if his partner had assured him he would do the right thing, it could be cheap talk, not worth the paper it's written on.

Prisoner's Dilemmas are common tragedies. A divorcing husband and wife hire legal barracudas, each fearing the other will take them to the cleaners, while the billable hours drain the marital assets. Enemy nations bust their budgets in an arms race, leaving them both poorer but no safer. Bicycle racers dope their blood and corrupt the sport because otherwise they would be left in the dust by rivals who doped theirs.[16] Everyone crowds a luggage carousel, or stands up at a rock concert, craning for a better view, and no one ends up with a better view.

The Prisoner's Dilemma has no solution, but the rules of the game can be changed. One way is for the players to enter enforceable agreements before playing, or to submit to the rule of an authority, which change the payoffs by adding a reward for cooperation or a penalty for defection. Suppose the partners take an oath of *omertà*, enforced by the Godfather, so that if they keep a code of silence they will be promoted to capo, while if they break it they will end up sleeping with the fishes. That changes the payoff matrix to a different game whose equilibrium is mutual cooperation. It's in the partners' interests to take the oath beforehand even though it forecloses their freedom to defect. Rational actors can escape a Prisoner's Dilemma by submitting to binding contracts and the rule of law.

Another game-changer is to play repeatedly, remembering what the partner did on previous rounds. Now a pair can find their way into the blessed Cooperate-Cooperate cell and stay there by playing a strategy called Tit for Tat. It calls for cooperating on the first move and thereafter treating the partner in kind: cooperate if the partner cooperated, defect if he defected (in some versions, giving him a free defection before defecting in case it was a one-time lapse).

Evolutionary biologists have noted that social animals often find themselves in iterated Prisoner's Dilemmas.[17] An example is the mutual reward in grooming one another, with the temptation to be groomed without grooming in turn. Robert Trivers suggested that *Homo sapiens* evolved a suite of moral emotions that implement Tit for Tat and allow us to enjoy the benefits of cooperation.[18] We're impelled by sympathy to cooperate on the first move, by gratitude to repay cooperation with cooperation, by anger to punish defection with defection, by guilt to atone for our own defection before it is punished, and by forgiveness to prevent a partner's one-time defection from condemning them to mutual defection forever. Many of the dramas of human social life—the sagas of sympathy, trust, favor, debt, revenge, gratitude, guilt, shame, treachery, gossip, reputation—may be understood as the playing of strategies in an iterated Prisoner's Dilemma.[19] The chapter epigraph shows that Hume, once again, got there first.

MANY OF THE DRAMAS of political and economic life may be explained as Prisoner's Dilemmas with more than two players, where they are called Public Goods games.[20] Everyone in a community benefits from a public good such as a lighthouse, roads, sewers, police, and schools. But they benefit even more if everyone else pays for them and they are free riders—once a lighthouse is built, anyone can see it. In a poignant environmental version called the Tragedy of the Commons, every shepherd has an incentive to add one more sheep to his flock and graze it on the town commons, but when everyone fattens their flock, the grass is grazed faster than it can regrow, and all the sheep starve. Traffic and pollution work the same way: my decision to drive won't clog the roads or foul the air, just as my decision to take the bus won't spare them, but when everyone chooses to drive, everyone ends up bumper to bumper on a smoggy freeway. Evading taxes, stinting when the hat is passed, milking a resource to depletion, and resisting public health measures like social distancing and mask-wearing during a pandemic, are other examples of defecting in a Public Goods game: they offer a temptation to those who indulge, a sucker's payoff to those who contribute and conserve, and a common punishment when everyone defects.

To get back to the example with which I opened the chapter, here is the Tragedy of the Carbon Commons. The players can be individual citizens, with the burden consisting of the inconvenience of forgoing meat, plane travel, or gas-guzzling SUVs. Or they can be entire countries, in which case the burden is the drag on the economy from forgoing the cheap and portable energy from fossil fuels. The numbers, as always, are arbitrary, and the tragedy is captured in their pattern: we are headed for the lower right cell.

		Emitter 2	
		Conserve	Emit
Emitter 1	Conserve	Burden –10 Burden –10	Benefit +10 Burden + Climate change –110
	Emit	Burden + Climate change –110 Benefit +10	Climate change –100 Climate change –100

Just as an enforceable oath can spare the prisoners in a two-person Dilemma from mutual defection, enforceable laws and contracts can punish people for their own mutual good in a Public Goods game. A pure example is easy to demonstrate in the lab. A group of participants are given a sum of money and offered the chance to chip in to a communal pot (the public good) which the experimenter then doubles and redistributes. The best strategy for everyone is to contribute the maximum, but the best strategy for each individual is to hoard his own sum and let everyone else contribute. Participants catch on to the grim game-theoretic logic and their contributions dwindle to zero—unless they are also given the opportunity to fine the free riders, in which case contributions stay high and everyone wins.

Outside the lab, a commons in a community where everyone knows everyone else can be protected by a multiplayer version of Tit for Tat:

any exploiter of a resource becomes a target of gossip, shaming, veiled threats, and discreet vandalism.[21] In larger and more anonymous communities, changes to the payoffs must be made by enforceable contracts and regulations. And so we pay taxes for roads, schools, and a court system, with evaders sent to jail. Ranchers buy grazing permits, and fishers respect limits on their catch, as long as they know they're being enforced on the other guy, too. Hockey players welcome mandatory helmet rules, which protect their brains without ceding an advantage of comfort and eyesight to their opponents. And economists recommend a carbon tax and investments in clean energy, which reduce the private benefit of emissions and lower the cost of conservation, steering everyone toward the common reward of mutual conservation.

The logic of Prisoner's Dilemmas and Public Goods undermines anarchism and radical libertarianism, despite the eternal appeal of unfettered freedom. The logic makes it rational to say, "There ought to be a law against what I'm doing." As Thomas Hobbes put it, the fundamental principle of society is "that a man be willing, when others are so too . . . to lay down this right to all things; and be contented with so much liberty against other men, as he would allow other men against himself."[22] This social contract does not just embody the moral logic of impartiality. It also removes wicked temptations, sucker's payoffs, and tragedies of mutual defection.

CORRELATION AND CAUSATION

> One of the first things taught in introductory statistics
> textbooks is that correlation is not causation. It is also one
> of the first things forgotten.
>
> —THOMAS SOWELL[1]

Rationality embraces all spheres of life, including the personal, the political, and the scientific. It's not surprising that the Enlightenment-inspired theorists of American democracy were fanboys of science, nor that real and wannabe autocrats latch onto harebrained theories of cause and effect.[2] Mao Zedong forced Chinese farmers to crowd their seedlings together to enhance their socialist solidarity, and a recent American leader suggested that Covid-19 could be treated with injections of bleach.

From 1985 to 2006, Turkmenistan was ruled by President for Life Saparmurat Niyazov. Among his accomplishments were making his autobiography required reading for the nation's driving test and erecting a massive golden statue of himself that rotated to face the sun. In 2004 he issued the following health notice to his adoring public: "I watched young dogs when I was young. They were given bones to gnaw. Those of you whose teeth have fallen out did not gnaw on bones. This is my advice."[3]

Since most of us are in no danger of being sent to prison in Ashgabat, we can identify the flaw in His Excellency's advice. The president made one of the most famous errors in reasoning, confusing correlation with causation. Even if it were true that toothless Turkmens had not chewed bones, the president was not entitled to conclude that gnawing on bones is what strengthens teeth. Perhaps only people with strong teeth can gnaw bones, a case of reverse causation. Or perhaps some third factor, such as being a member of the Communist Party, caused Turkmens both to gnaw bones (to show loyalty to their leader) and to have strong teeth (if dental care was a perquisite of membership), a case of a confound.

The concept of causation, and its contrast with mere correlation, is the lifeblood of science. What causes cancer? Or climate change? Or schizophrenia? It is woven into our everyday language, reasoning, and humor. The semantic contrast between "The ship sank" and "The ship was sunk" is whether the speaker asserts that there was a causal agent behind the event rather than a spontaneous occurrence. We appeal to causality whenever we ponder what to do about a leak, a draft, an ache or a pain. One of my grandfather's favorite jokes was about the man who gorged himself on cholent (the meat and bean stew simmered for twelve hours during the Sabbath blackout on cooking) with a glass of tea, and then lay in pain moaning that the tea had made him sick. Presumably you had to have been born in Poland in 1900 to find this as uproarious as he did, but if you get the joke at all, you can see how the difference between correlation and causation is part of our common sense.

Nonetheless, Niyazovian confusions are common in our public discourse. This chapter probes the nature of correlation, the nature of causation, and the ways to tell the difference.

What Is Correlation?

A correlation is a dependence of the value of one variable on the value of another: if you know one, you can predict the other, at least approximately. ("Predict" here means "guess," not "foretell"; you can predict the height of parents from the heights of their children or vice versa.) A correlation is often depicted in a graph called a *scatterplot*. In this one, every dot represents a country, and the dots are arrayed from left to right by their average income, and up and down by their average self-rated life satisfaction. (The income has been squeezed onto a logarithmic scale to compensate for the diminishing marginal utility of money, for reasons we saw in chapter 6.)[4]

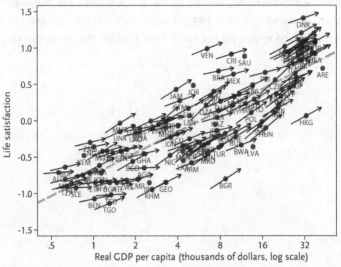

Adapted with permission from Stevenson & Wolfers, 2008

You can immediately spot the correlation: the dots are splayed along a diagonal axis, which is shown as the gray dashed line lurking behind the swarm. Each dot is impaled by an arrow which summarizes a mini-scatterplot for the people *within* the country. The macro- and mini-plots show that happiness is correlated with income, both among the people within a country (each arrow) and across the countries (the dots). And I know that you are resisting the temptation, at least for now, to infer "Being rich makes you happy."

Where do the gray dashed line and the arrows impaling each dot come from? And how might we translate our visual impression that the dots are strung out along the diagonal into something more objective, so that we aren't fooled into imagining a streak in any old pile of pick-up sticks?

This is the mathematical technique called *regression*, the workhorse of epidemiology and social science. Consider the scatterplot below. Imagine that each data point is a tack, and that we connect it to a rigid rod with a rubber band. Imagine that the bands can only

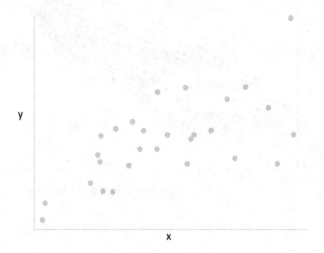

stretch up and down, not diagonally, and that the farther you stretch them, the harder they resist. When all the bands are attached, let go of the rod and let it sproing into place:

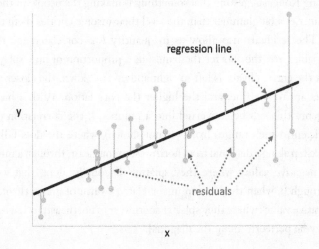

The rod settles into a location and an angle that minimizes the square of the distance between each tack and where it's attached. The rod, thus positioned, is called a regression line, and it captures the linear relationship between the two variables: *y*, corresponding to the vertical axis, and *x*, corresponding to the horizontal one. The length of the rubber band connecting each tack to the line is called the residual, and it captures the idiosyncratic portion of that unit's *y*-value that refuses to be predicted by its *x*. Go back to the happiness–income graph. If income predicted happiness perfectly, every dot would fall exactly along the gray regression line, but with real data, that never happens. Some of the dots float above the line (they have large positive residuals), like Jamaica, Venezuela, Costa Rica, and Denmark. Putting aside measurement error and other sources of noise, the discrepancies show that in 2006 (when the data were gathered) the people of these countries were happier than you would expect based on their income, perhaps because of other traits boasted by the country such as its climate or culture. Other dots hang below the line, like Togo, Bulgaria, and Hong Kong, suggesting that something is making the people in those countries a bit glummer than the level their income entitles them to.

The residuals also allow us to quantify *how* correlated the two variables are: the shorter the bands, as a proportion of how splayed out the entire cluster is left to right and up and down, the closer the dots are to the line, and the higher the correlation. With a bit of algebra this can be converted into a number, r, the correlation co-efficient, which ranges from –1 (not shown), where the dots fall in lockstep along a diagonal from northwest to southeast; through a range of negative values where they splatter diagonally along that axis; through 0, when they are an uncorrelated swarm of gnats; through positive values where they splatter southwest to northeast; to 1, where they lie perfectly along the diagonal.

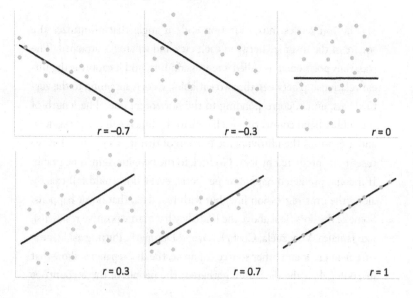

Though the finger-pointing in correlation-versus-causation blunders is usually directed at those who leap from the first to the second, often the problem is more basic: no correlation was established in the first place. Maybe Turkmens who chew more bones don't even *have* stronger teeth ($r = 0$). It's not just presidents of former Soviet republics who fall short of showing correlation, let alone causation. In 2020 Jeff Bezos bragged, "All of my best decisions in business and in life have been made with heart, intuition, guts . . . not analysis," implying that heart and guts lead to better decisions than analysis.[5] But he did not tell us whether all of his *worst* decisions in business and life were *also* made with heart, intuition, and guts, nor whether the good gut decisions and bad analytic ones outnumbered the bad gut decisions and good analytic ones.

Illusory correlation, as this fallacy is called, was first shown in a famous set of experiments by the psychologists Loren and Jean Chapman, who wondered why so many psychotherapists still used the Rorschach inkblot and Draw-a-Person tests even though every study that had ever tried to validate them showed no correlation between responses on the tests and psychological symptoms. The experimenters mischievously paired written descriptions of psychiatric patients with their responses on the Draw-a-Person test, but in fact the descriptions were fake and the pairings were random. They then asked a sample of students to report any patterns they saw across the pairs.[6] The students, guided by their stereotypes, incorrectly estimated that more broad-shouldered men were sketched by hypermasculine patients, more wide-eyed ones came from paranoiacs, and so on—exactly the linkages that professional diagnosticians claim to see in their patients, with as little basis in reality.

Many correlations that have become part of our conventional wisdom, like people pouring into hospital emergency rooms during a full moon, are just as illusory.[7] The danger is particularly acute with correlations that use months or years as their units of analysis (the dots in the scatterplot), because many variables rise and fall in tandem with the changing times. A bored law student, Tyler Vigen, wrote a program that scrapes the web for datasets with meaningless correlations just to show how prevalent they are. The number of murders by steam or hot objects, for example, correlates highly with the age of the reigning Miss America. And the divorce rate in Maine closely tracks national consumption of margarine.[8]

Regression to the Mean

"Regression" has become the standard term for correlational analyses, but the connection is roundabout. The term originally referred to a specific phenomenon that comes along with correlation, regression to the mean. This ubiquitous but counterintuitive phenomenon was discovered by the Victorian polymath Francis Galton (1822–1911), who plotted the heights of children against the average height of their two parents (the "mid-parent" score, halfway between the mother and the father), in both cases adjusting for the average difference between males and females. He found that "when mid-parents are taller than mediocrity, their children tend to be shorter than they. When mid-parents are shorter than mediocrity, their children tend to be taller than they."[9] It's still true, not just of the heights of parents and their children, but of the IQs of parents and their children, and for that matter of any two variables that are not perfectly correlated. An extreme value in one will be tend to be paired with a not-quite-as-extreme value in the other.

This does not mean that tall families are begetting shorter and shorter children and vice versa, so that some day all children will line up against the same mark on the wall and the world will have no jockeys or basketball centers. Nor does it mean that the population is converging on a middlebrow IQ of 100, with geniuses and dullards going extinct. The reason that populations don't collapse into uniform mediocrity, despite regression to the mean, is that the tails of the distribution are constantly being replenished by the occasional very tall child of taller-than-average parents and very short child of shorter-than-average ones.

Regression to the mean is purely a *statistical* phenomenon, a consequence of the fact that in bell-shaped distributions, the more extreme a value, the less likely it is to turn up. That implies that when a value is really extreme, any other variable that is paired with it (such as the child of an outsize couple) is unlikely to live up to its weirdness, or duplicate its winning streak, or get dealt the same lucky hand, or suffer from the same run of bad luck, or weather the same perfect storm, yet again, and will backslide toward ordinariness. In the case of height or IQ, the freakish conspiracy would be whatever unusual combination of genes, experiences, and accidents of biology came together in the parents. Many of the components of that combination will be favored in their children, but the combination itself will not be perfectly reproduced. (And vice versa: because regression is a statistical phenomenon, not a causal one, parents regress to their children's mean, too.)

In a graph, when correlated values from two bell curves are plotted against each other, the scatterplot will usually look like a tilted football. On the following page we have a hypothetical dataset similar to Galton's showing the heights of parents (the average of each couple) and the heights of their adult children (adjusted so that the sons and daughters can be plotted on the same scale).

The gray 45-degree diagonal shows what we would expect on average if children were exactly as exceptional as their parents. The black regression line is what we find in reality. If you zero in on an

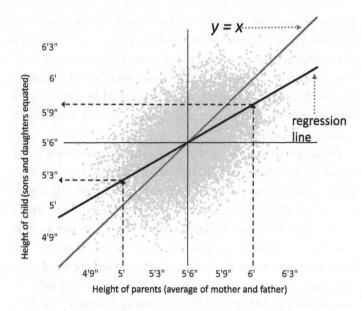

extreme value, say, parents with an average height between them of 6 feet, you'll find that the cluster of points for their children mostly hangs below the 45-degree diagonal, which you can confirm by scanning up along the right dotted arrow to the regression line, turning left, and following the horizontal dotted arrow to the vertical axis, where it points a bit above 5'9", shorter than the parents. If you zero in on the parents with an average height of 5 feet (left dotted arrow), you'll see that their children mostly float above the gray diagonal, and the left turn at the regression line takes you to a value of almost 5'3", taller than the parents.

Regression to the mean happens whenever two variables are imperfectly correlated, which means that we have a lifetime of experience with it. Nonetheless, Tversky and Kahneman have shown that most people are oblivious to the phenomenon (notwithstanding the groaner in the *Frank and Ernest* cartoon).[10]

People's attention gets drawn to an event because it is unusual, and they fail to anticipate that anything associated with that event will probably not be quite as unusual as that event was. Instead, they come up with fallacious causal explanations for what in fact is a statistical inevitability.

A tragic example is the illusion that criticism works better than praise, and punishment better than reward.[11] We criticize students when they perform badly. But whatever bad luck cursed that performance is unlikely to be repeated in the next attempt, so they're bound to improve, tricking us into thinking that punishment works. We praise them when they do well, but lightning doesn't strike twice, so they're unlikely to match that feat the next time, fooling us into thinking that praise is counterproductive.

Unawareness of regression to the mean sets the stage for many other illusions. Sports fans theorize about why a Rookie of the Year is doomed to suffer a sophomore slump, and why the cover subject of a famous magazine will then have to live with the *Sports Illustrated* jinx. (Overconfidence? Impossible expectations? The distractions of fame?) But if an athlete is singled out for an extraordinary week or year, the stars are unlikely to align that way twice in a row, and he or she has nowhere to go but meanward. (Equally meaninglessly, a slumping team will improve after the coach is fired.) After a spree of horrific crimes is splashed across the papers, politicians intervene with SWAT teams, military equipment, Neighborhood Watch signs, and other gimmicks, and sure enough, the following month they congratulate themselves because the crime rate is not as high. Psychotherapists, too, regardless of their flavor of talking cure, can declare unearned victory after treating a patient who comes in with a bout of severe anxiety or depression.

Once again, scientists are not immune. Yet another cause of replication failures is that experimenters don't appreciate a version of regression to the mean called the Winner's Curse. If the results of an experiment seem to show an interesting effect, a lot of things must have gone right, whether the effect is real or not. The gods of chance must have smiled on the experimenters, which they should not count upon a second time, so when they try to replicate the effect, they ought to enlist *more* participants. But most experimenters think that they've already racked up some evidence for the effect, so they can get away with *fewer* participants, not appreciating that this strategy is a one-way path to the *Journal of Irreproducible Results*.[12] A failure to appreciate how regression to the mean applies to striking discoveries led to a muddled 2010 *New Yorker* article called "The Truth Wears Off," which posited a mystical "decline effect," supposedly casting doubt on the scientific method.[13]

The Winner's Curse applies to any unusually successful human venture, and our failure to compensate for singular moments of good fortune may be one of the reasons that life so often brings disappointment.

What Is Causation?

Before we lay out the bridge from correlation to causation, let's spy on the opposite shore, causation itself. It turns out to be a surprisingly elusive concept.[14] Hume, once again, set the terms for centuries of analysis by venturing that causation is merely an expectation that a correlation we experienced in the past will hold in the future.[15] Once we have watched enough billiards, then whenever we see one ball close in on a second, we anticipate that the second will be launched forward, just like all the times before, propped up by our tacit but unprovable assumption that the laws of nature persist over time.

It doesn't take long to see what's wrong with "constant conjunction" as a theory of causality. The rooster always crows just before daybreak, but we don't credit him with causing the sun to rise. Likewise thunder often precedes a forest fire, but we don't say thunder causes fires. These are *epiphenomena*, also known as confounds or nuisance variables: they accompany but do not cause the event. Epiphenomena are the bane of epidemiology. For many years coffee was blamed for heart disease, because coffee drinkers had more heart attacks. It turned out that coffee drinkers also tend to smoke and avoid exercise; the coffee was an epiphenomenon.

Hume anticipated the problem and elaborated on his theory: not only does the cause have to regularly precede its effect, but "if the first object had not been, the second never had existed." The crucial "if it had not been" clause is a *counterfactual*, a "what if." It refers to what would happen in a possible world, an alternate universe, a hypothetical experiment. In a parallel universe in which the cause didn't happen, neither did the effect. This counterfactual definition of causation solves the epiphenomenon problem. The reason we say the rooster doesn't cause the sunrise is that *if* the rooster had become the main ingredient in coq au vin the night before, the sun would still have risen. We say that lightning causes forest fires and thunder doesn't because if there were lightning without thunder, a forest could ignite, but not vice versa.

Causation, then, can be thought of as the difference between outcomes when an event (the cause) takes place and when it does not.[16]

The "fundamental problem of causal inference," as statisticians call it, is that we're stuck in this universe, where either a putative causal event took place or it didn't. We can't peer into that other universe and see what the outcome is over there. We can, to be sure, compare the outcomes in this universe on the various occasions when that kind of event does or does not take place. But that runs smack into a problem pointed out by Heraclitus in the sixth century BCE: You can't step in the same river twice. Between those two occasions, the world may have changed in other ways, and you can't be sure whether one of those other changes was the cause. We also can compare individual things that underwent that kind of event with similar things that did not. But this, too, runs into a problem, pointed out by Dr. Seuss: "Today you are you, that is truer than true. There is no one alive who is youer than you." Every individual is unique, so we can't know whether an outcome experienced by an individual depended on the supposed cause or on that person's myriad idiosyncrasies. To infer causation from these comparisons, we have to assume, as they say less poetically, "temporal stability" and "unit homogeneity." The methods discussed in the next two sections try to make those assumptions reasonable.

Even once we have established that some cause makes a difference to an outcome, neither scientists nor laypeople are content to leave it at that. We connect the cause to its effect with a *mechanism*: the clockwork behind the scenes that pushes things around. People have intuitions that the world is not a video game with patterns of pixels giving way to new patterns. Underneath each happening is a hidden force, power, or oomph. Many of our primitive intuitions of causal powers turn out, in the light of science, to be mistaken, such as the "impetus" that the medievals thought was impressed upon moving objects, and the psi, qi, engrams, energy fields, homeopathic miasms, crystal powers, and other bunkum of alternative medicine. But some intuitive mechanisms, like gravity, survive in scientifically respectable forms. And many new hidden mechanisms have been posited to explain correlations in the world, including genes, pathogens, tectonic plates, and elementary particles. These causal mechanisms are what allow us to predict what would happen in counterfactual scenarios, lifting them from the realm of make-believe: we set up the pretend world and then simulate the mechanisms, which take it from there.

EVEN WITH A GRASP of causation in terms of alternative outcomes and the mechanisms that produce them, any effort to identify "the" cause of an effect raises a thicket of puzzles. One is the elusive difference between a cause and a *condition*. We say that striking a match causes a fire, because without the striking there would be no fire. But without oxygen, without the dryness of the paper, without the stillness of the room, there also would be no fire. So why don't we say "The oxygen caused the fire"?

A second puzzle is *preemption*. Suppose, for the sake of argument, that Lee Harvey Oswald had a co-conspirator perched on the grassy knoll in Dallas in 1963, and they had conspired that whoever got the first clear shot would take it while the other melted into the crowd. In the counterfactual world in which Oswald did not shoot, JFK would still have died—yet it would be wacky to deny that in the world in which he did take the shot before his accomplice, he caused Kennedy's death.

A third is *overdetermination*. A condemned prisoner is shot by a firing squad rather than a single executioner so that no shooter has to live with the dreadful burden of being the one who caused the death: if he had not fired, the prisoner would still have died. But then, by the logic of counterfactuals, *no one* caused his death.

And then there's *probabilistic causation*. Many of us know a nonagenarian who smoked a pack a day all her life. But nowadays few people would say that her ripe old age proves that smoking does not cause cancer, though that was a common "refutation" in the days before the smoking–cancer link became undeniable. Even today, the confusion between less-than-perfect causation and no causation is rampant. A 2020 *New York Times* op-ed argued for abolishing the police because "the current approach hasn't ended [rape]. Most rapists never see the inside of a courtroom."[17] The editorialist did not consider whether, if there were no police, even fewer rapists, or none at all, would see the inside of a courtroom.

We can make sense of these paradoxes of causation only by forgetting the billiard balls and recognizing that no event has a single cause. Events are embedded in a *network* of causes that trigger, enable, inhibit, prevent, and supercharge one another in linked and branching pathways. The four causal puzzlers become less puzzling when we lay out the road maps of causation in each case, shown below.

If you interpret the arrows not as logical implications ("If X

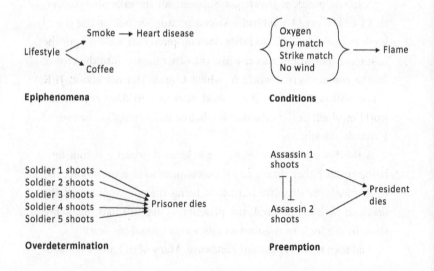

smokes, then X gets heart disease") but as conditional probabilities ("The likelihood of X getting heart disease given that X is a smoker is higher than the likelihood of X getting heart disease given that he is not a smoker"), and the event nodes not as being either on or off but as probabilities, reflecting a base rate or prior, then the diagram is called a causal Bayesian network.[18] One can work out what unfolds over time by applying (naturally) Bayes's rule, node by node through the network. No matter how convoluted the tangle of causes, conditions, and confounds, one can then determine which events are causally dependent on or independent of one another.

The inventor of these networks, the computer scientist Judea Pearl, notes that they are built out of three simple patterns—the chain, the fork, and the collider—each capturing a fundamental (but unintuitive) feature of causation with more than one cause.

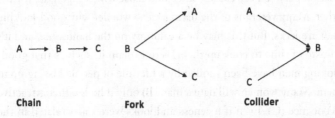

Chain   Fork   Collider

The connections reflect the conditional probabilities. In each case, A and C are not directly connected, which means that the probability of A given B can be specified independently of the probability of C given B. And in each case something distinctive may be said about the relation between them.

In a causal *chain*, the first cause, A, is "screened off" from the ultimate effect, C; its only influence is via B. As far as C is concerned, A might as well not exist. Consider a hotel's fire alarm, set off by the chain "fire → smoke → alarm." It's really not a fire alarm but a smoke alarm, indeed, a haze alarm. The guests may be awakened as readily by someone spray-painting a bookshelf near an intake vent as by an errant crème brûlée torch.

A causal *fork* is already familiar: it depicts a confound or epiphenomenon, with the attendant danger of misidentifying the real cause. Age (B) affects vocabulary (A) and shoe size (C), since older children have bigger feet and know more words. This means that vocabulary is correlated with shoe size. But Head Start would be ill advised to prepare children for school by fitting them with larger sneakers.

Just as dangerous is the *collider*, where unrelated causes converge on a single effect. Actually, it's even more dangerous, because while most people intuitively get the fallacy of a confound (it cracked them up in the shtetl), the "collider stratification selection bias" is almost unknown. The trap in a causal collider is that when you focus on a restricted range of effects, you introduce an artificial negative correlation between the causes, since one cause will compensate for the other. Many veterans of the dating scene wonder why good-looking men are jerks. But this may be a calumny on the handsome, and it's a waste of time to cook up theories to explain it, such as that good-looking men have been spoiled by a lifetime of people kissing up to them. Many women will date a man (B) only if he is either attractive (A) or nice (C). Even if niceness and looks were uncorrelated in the dating pool, the plainer men *had* to be nice or the woman would never have dated them in the first place, while the hunks were not sorted by any such filter. A bogus negative correlation was introduced by her disjunctive choosiness.

The collider fallacy also fools critics of standardized testing into thinking that test scores don't matter, based on the observation that graduate students who were admitted with higher scores are no more likely to complete the program. The problem is that the students who were accepted *despite* their low scores must have boasted *other* assets.[19] If one is unaware of the bias, one could even conclude that maternal smoking is good for babies, since among babies with low birth weights, the ones with mothers who smoked are healthier. That's because low birth weight must be caused by *something*, and the other possible causes, such as alcohol or drug abuse, may be even more harmful to the child.[20] The collider fallacy also explains why Jenny Cavilleri unfairly maintained that rich boys are stupid: to get into Harvard (B), you can be either rich (A) or smart (C).

From Correlation to Causation: Real and Natural Experiments

Now that we've probed the nature of correlation and the nature of causation, it's time to see how to get from one to the other. The problem is not that "correlation does not imply causation." It usually does, because unless the correlation is illusory or a coincidence, *something* must have caused one variable to align with the other. The problem is that when one thing is correlated with another, it does not necessarily mean that the first caused the second. As the mantra goes: When A is correlated with B, it could mean that A causes B, B causes A, or some third factor, C, causes both A and B.

Reverse causation and confounding, the second and third verses of the mantra, are ubiquitous. The world is a huge causal Bayesian network, with arrows pointing every which way, entangling events into knots where everything is correlated with everything else. These gnarls (called multicollinearity and endogeneity) can arise because of the Matthew Effect, pithily explained by Billie Holiday: "Them that's got shall get, them that's not shall lose. So the Bible said, and it still is news."[21] Countries that are richer also tend to be healthier, happier, safer, better educated, less polluted, more peaceful, more democratic, more liberal, more secular, and more gender-egalitarian.[22]

People who are richer also tend to be healthier, better educated, better connected, likelier to exercise and eat well, and likelier to belong to privileged groups.[23]

These snarls mean that almost any causal conclusion you draw from correlations across countries or across people is likely to be wrong, or at best unproven. Does democracy make a country more peaceful, because its leader can't readily turn citizens into cannon fodder? Or do countries facing no threats from their neighbors have the luxury of indulging in democracy? Does going to college equip you with skills that allow you to earn a good living? Or do only smart, disciplined, or privileged people, who can translate their natural assets into financial ones, make it through university?

There is an impeccable way to cut these knots: the randomized experiment, often called a randomized controlled trial or RCT. Take a large sample from the population of interest, randomly divide them into two groups, apply the putative cause to one group and withhold it from the other, and see if the first group changes while the second does not. A randomized experiment is the closest we can come to *creating* the counterfactual world that is the acid test for causation. In a causal network, it consists of surgically severing the putative cause from all its incoming influences, setting it to different values, and seeing whether the probabilities of the putative effects differ.[24]

Randomness is the key: if the patients who were given the drug signed up earlier, or lived closer to the hospital, or had more interesting symptoms, than the patients who were given the placebo, you'll never know whether the drug worked. As one of my graduate school teachers said (alluding to a line from J. M. Barrie's play *What Every Woman Knows*), "Random assignment is like charm. If you have it, you don't need anything else; if you don't have it, it doesn't matter what else you have."[25] It isn't quite true of charm, and it isn't quite true of random assignment either, but it is still with me decades later, and I like it better than the cliché that randomized trials are the "gold standard" for showing causation.

The wisdom of randomized controlled trials is seeping into policy, economics, and education. Increasingly, "randomistas" are urging policymakers to test their nostrums in one set of randomly selected villages, classes, or neighborhoods, and compare the results against a control group which is put on a waitlist or given some meaningless make-work program.[26] The knowledge gained is likely to outperform traditional ways of evaluating policies, like dogma, folklore, charisma, conventional wisdom, and HiPPO (highest-paid person's opinion).

Randomized experiments are no panacea (since nothing is a panacea, which is a good reason to retire that cliché). Laboratory scientists snipe at each other as much as correlational data scientists, because even in an experiment you can't do just one thing. Experimenters may think that they have administered a treatment and only that treatment to the experimental group, but other variables may be confounded with it, a problem called excludability. According to a joke, a sexually unfulfilled couple consults a rabbi with their problem, since it is written in the Talmud that a husband is responsible for his wife's sexual pleasure. The rabbi strokes his beard and comes up with a solution: they should hire a handsome, strapping young man to wave a towel over them the next time they make love, and the fantasies will help the woman achieve climax. They follow the great sage's advice, but it fails to have the desired effect, and they beseech him for guidance once again. He strokes his beard and thinks up a variation. This time, the young man will make love to the woman and the husband will wave the towel. They take his advice, and sure enough, the woman enjoys an ecstatic, earth-moving orgasm. The husband says to the man, "Schmuck! Now *that's* how you wave a towel."

The other problem with experimental manipulations, of course, is that the world is not a laboratory. It's not as if political scientists can flip a coin, impose democracy on some countries and autocracy on others, and wait five years to see which ones go to war. The same practical and ethical problems apply to studies of individuals, as shown in this cartoon.

"The title of my science project is 'My Little Brother: Nature or Nurture.'"

Michael Shaw/The New Yorker Collection/The Cartoon Bank

Though not everything can be studied in an experimental trial, social scientists have mustered their ingenuity to find instances in which the world does the randomization for them. These experiments of nature can sometimes allow one to wring causal conclusions out of a correlational universe. They're a recurring feature in *Freakonomics*, the series of books and other media by the economist Steven Levitt and the journalist Stephen Dubner.[27]

One example is the "regression discontinuity." Say you want to decide whether going to college makes people wealthier or wealth-destined teenagers are likelier to get into college. Though you can't literally randomize a sample of teenagers and force a college to admit one group and reject another, selective colleges effectively do that to students near their cutoff. No one really believes that the student who squeaked in with a test score of 1720 is smarter than the one who fell just short with 1710. The difference is in the noise, and might as well have been random. (The same is true with other qualifications like grades and letters of recommendation.) Suppose one follows both groups for a decade and plots their income against their test scores. If one sees a step or an elbow at the cutoff, with a bigger jump in salary at the reject–admit boundary than for intervals of similar size along the rest of the scale, one may conclude that the magic wand of admission made a difference.

Another gift to causation-hungry social scientists is fortuitous randomization. Does Fox News make people more conservative, or do conservatives gravitate to Fox News? When Fox News debuted in 1996, different cable companies added it to their lineups haphazardly over the next five years. Economists took advantage of the happenstance during the half decade and found that towns with Fox News in their cable lineups voted 0.4 to 0.7 points more Republican than towns that had to watch something else.[28] That's a big enough difference to swing a close election, and the effect could have accumulated in the subsequent decades when Fox News' universal penetration into TV markets made the effect harder to prove but no less potent.

Harder, but not impossible. Another stroke of genius goes by the unhelpful name "instrumental variable regression." Suppose you want to see whether A causes B and are worried about the usual nuisances of reverse causation (B causes A) and confounding (C causes A and B). Now suppose you found some fourth variable, I (the "instrument"), which is correlated with the putative cause, A, but could not possibly be caused by it—say, because it happened earlier in time, the future being incapable of affecting the past. Suppose as well that this pristine variable is also uncorrelated with the confound, C, and that it cannot cause B directly, only through A. Even though A cannot be randomly assigned, we have the next best thing, I. If I, the clean surrogate for A, turns out to be correlated with B, it's an indication that A causes B.

What does this have to do with Fox News? Another gift to social scientists is American laziness. Americans hate getting out of their cars, adding water to a soup mix, and clicking their way up the cable lineup from the single digits. The lower the channel number, the more people watch it. Now, Fox News is assigned different channel numbers by different cable companies pretty much at random (the numbering depended only on when the network struck the deal with each cable company, and was unrelated to the demographics of the viewers). While a low channel number (I) can cause people to watch Fox News (A), and watching Fox News may or may not cause them to vote Republican (B), neither having conservative views (C) nor voting Republican can cause someone's favorite television station to skitter down the cable dial. Sure enough, in a comparison across cable markets, the lower the channel number of Fox News relative to other news networks, the larger the Republican vote.[29]

From Correlation to Causation without Experimentation

When a data scientist finds a regression discontinuity or an instrumental variable, it's a really good day. But more often they have to squeeze what causation they can out of the usual correlational tangle. All is not lost, though, because there are palliatives for each of the ailments that enfeeble causal inference. They are not as good as the charm of random assignment, but often they are the best we can do in a world that was not created for the benefit of scientists.

Reverse causation is the easier of the two to rule out, thanks to the iron law that hems in the writers of science fiction and other time-travel plots like *Back to the Future*: the future cannot affect the past. Suppose you want to test the hypothesis that democracy causes peace, not just vice versa. First, one must avoid the fallacy of all-or-none causation, and get beyond the common but false claim that "democracies never fight each other" (there are plenty of exceptions).[30] The more realistic hypothesis is that countries that are *relatively* more democratic are *less likely* to fall into war.[31] Several research organizations give countries democracy scores from –10 for a full autocracy like North Korea to +10 for a full democracy like Norway. Peace is a bit harder, because (fortunately for humanity, but unfortunately for social scientists) shooting wars are uncommon, so most of the entries in the table would be "0." Instead, one can estimate war-proneness by the number of "militarized disputes" a country was embroiled in over a year: the saber-rattlings, alertings of forces, shots fired across bows, warplanes sent scrambling, bellicose threats, and border skirmishes. One can convert this from a war score to a peace score (so that more-peaceful countries get higher numbers) by subtracting the count from some large number, like the maximum number of disputes ever recorded. Now one can correlate the peace score against the democracy score. By itself, of course, that correlation proves nothing.

But suppose that each variable is recorded *twice*, say, a decade apart. If democracy causes peace, then the democracy score at Time 1 should be correlated with the peace score at Time 2. This, too, proves little, because over a decade the leopard doesn't change its spots: a peaceful democracy then may just be a peaceful democracy now. But as a control one can look to the other diagonal: the correlation between Democracy (the democracy score) at Time 2 and Peace (the peace score) at Time 1. This correlation captures any reverse causation, together with the confounds that have stayed put over the decade. If the first correlation (past cause with present effect) is stronger than the second (past effect with present cause), it's a hint that democracy causes peace rather than vice versa. The technique is called cross-lagged panel correlation, "panel" being argot for a data-set containing measurements at several points in time.

Confounds, too, may be tamed by clever statistics. You may have read in science news articles of researchers "holding constant" or "sta-tistically controlling for" some confounded or nuisance variable. The simplest way to do that is called matching.[32] The democracy–peace relationship is infested with plenty of confounds, such as prosperity, education, trade, and membership in treaty organizations. Let's con-sider one of them, prosperity, measured as GDP per capita. Suppose that for every democracy in our sample we found an autocracy that had the same GDP per capita. If we compare the average peace scores of the democracies with their autocracy doppelgangers, we'd have an estimate of the effects of democracy on peace, holding GDP constant. The logic of matching is straightforward, but it requires a large pool of candidates from which to find good matches, and the number ex-plodes as more confounds have to be held constant. That can work for an epidemiological study with tens of thousands of participants to choose from, but not for a political study in a world with just 195 countries.

The more general technique is called multiple regression, and it capitalizes on the fact that a confound is never *perfectly* correlated with a putative cause. The discrepancies between them turn out to be not bothersome noise but telltale information. Here's how it could work with democracy, peace, and GDP per capita. First, we plot the puta-tive cause, the democracy score, against the nuisance variable (top left graph), one point per country. (The data are fake, made up to illus-trate the logic.) We fit the regression line, and turn our attention to

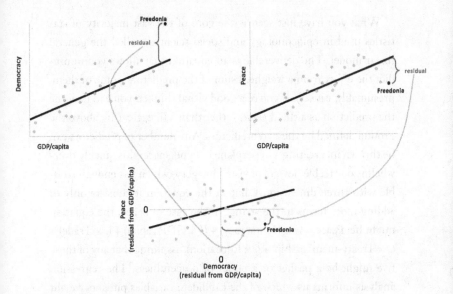

the residuals: the vertical distance between each point and the line, corresponding to the discrepancy between how democratic a country *would be* if income predicted democracy perfectly and how democratic it is in reality. Now we throw away each country's original democracy score and replace it with the residual: the measure of how democratic it is, controlling for its income.

Now do the same with the putative effect, peace. We plot the peace score against the nuisance variable (top right graph), measure the residuals, throw away the original peace data, and replace them with the residuals, namely, how peaceful each country is above and beyond what you would expect from its income. The final step is obvious: correlate the Peace residuals with the Democracy residuals (bottom graph). If the correlation is significantly different from zero, one may venture that democracy causes peacefulness, holding prosperity constant.

What you have just seen is the core of the vast majority of statistics used in epidemiology and social sciences, called the general linear model. The deliverable is an equation that allows you to predict the effect from a weighted sum of the predictors (some of them, presumably, causes). If you're a good visual thinker, you can imagine the prediction as a tilted *plane*, rather than a line, floating above the ground defined by the two predictors. Any number of predictors can be thrown in, creating a hyperplane in hyperspace; this quickly overwhelms our feeble powers of visual imagery (which has enough trouble with three dimensions), but in the equation it consists only of adding more terms to the string. In the case of peace, the equation might be: Peace = (a × Democracy) + (b × GDP/capita) + (c × Trade) + (d × Treaty-membership) + (e × Education), assuming that any of these five might be a pusher or a puller of peacefulness. The regression analysis informs us which of the candidate variables pulls its weight in predicting the outcome, holding each of the others constant. It is not a turnkey machine for proving causation—one still has to interpret the variables and how they are plausibly connected, and watch out for myriad traps—but it is the most commonly used tool for unsnarling multiple causes and confounds.

Multiple Causes, Adding and Interacting

The algebra of a regression equation is less important than the big idea flaunted by its form: events have more than one cause, all of them statistical. The idea seems elementary, but it's regularly flouted in public discourse. All too often, people write as if every outcome had a single, unfailing cause: if A has been shown to affect B, it proves that C cannot affect it. Accomplished people spend ten thousand hours practicing their craft; this is said to show that achievement is a matter of practice, not talent. Men today cry twice as often as their fathers did; this shows that the difference in crying between men and women is social rather than biological. The possibility of multiple causes—nature *and* nurture, talent *and* practice—is inconceivable.

Even more elusive is the idea of *interacting* causes: the possibility that the effect of one cause may depend on another. Perhaps everyone benefits from practice, but talented people benefit more. What we need is a vocabulary for talking and thinking about multiple causes. This is yet another area in which a few simple concepts from statistics can make everyone smarter. The revelatory concepts are *main effect* and *interaction*.

Let me illustrate them with fake data. Suppose we're interested in what makes monkeys fearful: heredity, namely the species they belong to (capuchin or marmoset), or the environment in which they were reared (alone with their mothers or in a large enclosure with plenty of other monkey families). Suppose we have a way of measuring fear—say, how closely the monkey approaches a rubber snake. With two possible causes and one effect, six different things can happen. This sounds complicated, but the possibilities jump off the page as soon as we plot them in graphs. Let's start with the three simplest ones.

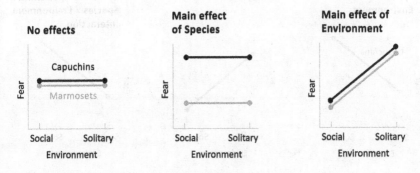

The left graph shows a big fat nothing: a monkey is a monkey. Species doesn't matter (the lines fall on top of each other); Environment doesn't matter either (each line is flat). The middle graph is what we would see if Species mattered (capuchins are more skittish than marmosets, shown by their line floating higher on the graph), while Environment did not (both species are equally fearful whether they are raised alone or with others, shown by each line being flat). In jargon, we say that there is a *main effect* of Species, meaning that the effect is seen across the board, regardless of the environment. The right graph shows the opposite outcome, a main effect of Environment but none of Species. Growing up alone makes a monkey more fearful (seen in the slope of the lines), but it does so for capuchins and marmosets alike (seen in the lines falling on top of each other).

Now let's get even smarter and wrap our minds around multiple causes. Once again we have three possibilities. What would it look like if Species and Environment *both* mattered: if capuchins were innately more fearful than marmosets, *and* if being reared alone makes a monkey more fearful? The leftmost graph shows this situation, namely, two main effects. It takes the form of the two lines having parallel slopes, one hovering above the other.

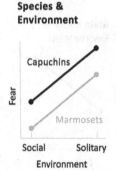

Main effects of Species & Environment

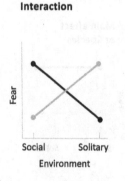

Species × Environment Interaction

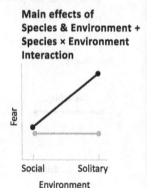

Main effects of Species & Environment + Species × Environment Interaction

Things get really interesting in the middle graph. Here, both factors matter, but each depends on the other. If you're a capuchin, being raised alone makes you bolder; if you're a marmoset, being raised alone makes you meeker. We see an *interaction* between Species and Environment, which visually consists of the lines being nonparallel. In these data, the lines cross into a perfect X, which means that the main effects are canceled out entirely. Across the board, Species doesn't matter: the midpoint of the capuchin line sits on top of the midpoint of the marmoset line. Environment doesn't matter across the board either: the average for Social, corresponding to the point midway between the two leftmost tips, lines up with the average for Solitary, corresponding to the point midway between the rightmost ones. Of course Species and Environment do matter: it's just that *how* each cause matters depends on the other one.

Finally, an interaction can coexist with one or more main effects. In the rightmost graph, being reared alone makes capuchins more fearful, but it has no effect on the always-calm marmosets. Since the effect on the marmosets doesn't perfectly cancel out the effect on the capuchins, we do see a main effect of Species (the capuchin line is higher) and a main effect of Environment (the midpoint of the two left dots is lower than the midpoint of the two right ones). But whenever we interpret a phenomenon with two or more causes, any interaction supersedes the main effects: it provides more insight as to what is going on. An interaction usually implies that the two causes intermingle in a single link in the causal chain, rather than taking place in different links and then just adding up. With these data, the common link might be the amygdala, the part of the brain registering fearful experiences, which may be plastic in capuchins but hardwired in marmosets.

With these cognitive tools, we are now equipped to make sense of multiple causes in the world: we can get beyond "nature versus nurture" and whether geniuses are "born or made." Let's turn to some real data.

What causes major depression, a stressful event or a genetic predisposition? This graph plots the likelihood of suffering a major depressive episode in a sample of women with twin sisters.[33]

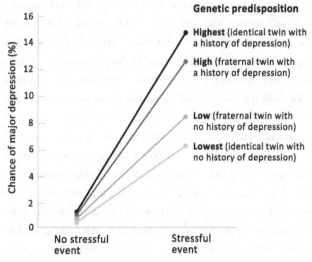

Adapted with permission from Kendler, Kessler, et al., 2010

The sample includes women who had undergone a severe stressor, like a divorce, an assault, or a death of a close relative (the points on the right), and women who had not (the points on the left). Scanning the lines from top to bottom, the first is for women who may be highly predisposed to depression, because their identical twin, with whom they share all their genes, suffered from it. The next line down is for women who are only somewhat predisposed to depression, because a fraternal twin, with whom they share *half* their genes, suffered from it. Below it we have a line for women who are not particularly predisposed, because their fraternal twin did not suffer from depression. At the bottom we find a line for women who are at the lowest risk, because their identical twin did not suffer from it.

The pattern in the graph tells us three things. Experience matters: we see a main effect of Stress in the upward slant of the fan of lines, which shows that undergoing a stressful event ups the odds of getting depressed. Overall, genes matter: the four lines float at different heights, showing that the higher one's genetic predisposition, the greater the chance that one will suffer a depressive episode. But the real takeaway is the *interaction*: the lines are not parallel. (Another way of putting it is that the points fall on top of one another on the left but are spread out on the right.) If you don't suffer a stressful event, your genes barely matter: regardless of your genome, the chance of a depressive episode is less than one percent. But if you do suffer a stressful event, your genes matter a lot: a full dose of genes associated with escaping depression keeps the risk of getting depressed at 6 percent (lowest line); a full dose of genes associated with suffering depression more than doubles the risk to 14 percent (highest line). The interaction tells us not only that both genes and environment are important but that they seem to have their effects on the same link in the causal chain. The genes that these twins share to different degrees are not genes for depression per se; they are genes for vulnerability or resilience to stressful experiences.

Let's turn to whether stars are born or made. The graph on the next page, also from a real study, shows ratings of chess skill in a sample of lifelong players who differ in their measured cognitive ability and in how many games they play per year.[34] It shows that practice makes better, if not perfect: we see a main effect of games played per year, visible in the overall upward slope. Talent will tell: we see a main effect of ability, visible in the gap between the two lines. But the real moral of the story is their *interaction*: the lines are

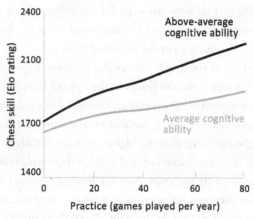

Adapted with permission from Vaci, Edelsbrunner, et al., 2019

not parallel, showing that smarter players gain more with every additional game of practice. An equivalent way of putting it is that without practice, cognitive ability barely matters (the leftmost tips of the lines almost overlap), but with practice, smarter players show off their talent (the rightmost tips are spread apart). Knowing the difference between main effects and interactions not only protects us from falling for false dichotomies but offers us deeper insight into the nature of the underlying causes.

Causal Networks and Human Beings

As a way of understanding the causal richness of the world, a regression equation is pretty simpleminded: it just adds up a bunch of weighted predictors. Interactions can be thrown in as well; they can be represented as additional predictors derived by multiplying together the interacting ones. A regression equation is nowhere near as complex as the deep learning networks we saw in chapter 3, which take in millions of variables and combine them in long, intricate chains of formulas rather than just throwing them into a hopper and adding them up. Yet despite their simplicity, one of the stunning findings of twentieth-century psychology is that a dumb regression equation usually outperforms a human expert. The finding, first noted by the psychologist Paul Meehl, goes by the name "clinical versus actuarial judgment."[35]

Suppose you want to predict some quantifiable outcome—how long a cancer patient will survive; whether a psychiatric patient ends up diagnosed with a mild neurosis or a severe psychosis; whether a criminal defendant will skip bail, blow off parole, or recidivate; how well a student will perform in graduate school; whether a business will succeed or go belly-up; how large a return a stock fund will deliver. You have a set of predictors: a symptom checklist, a set of demographic features, a tally of past behavior, a transcript of under-graduate grades or test scores—anything that might be relevant to the prediction challenge. Now you show the data to an expert—a psychiatrist, a judge, an investment analyst, and so on—and at the same time feed them into a standard regression analysis to get the prediction equation. Who is the more accurate prognosticator, the expert or the equation?

The winner, almost every time, is the equation. In fact, an expert who is given the equation and allowed to use it to supplement his or her judgment often does worse than the equation alone. The reason is that experts are too quick to see extenuating circumstances that they think render the formula inapplicable. It's sometimes called the broken-leg problem, from the idea that a human expert, but not an algorithm, has the sense to know that a guy who has just broken his leg will not go dancing that evening, even if a formula predicts that he does it every week. The problem is that the equation *already* takes into account the likelihood that extenuating circumstances will change the outcome and factors them into the mix with all the other influences, while the human expert is far too impressed with the eye-catching particulars and too quick to throw the base rates out the window. Indeed, some of the predictors that human experts rely on the most, such as face-to-face interviews, are revealed by regression analyses to be perfectly useless.

It's not that humans can be taken out of the loop. A person still is indispensable in supplying predictors that require real comprehension, like understanding language and categorizing behavior. It's just that a human is inept at *combining* them, whereas that is a regression algorithm's stock in trade. As Meehl notes, at a supermarket checkout counter you wouldn't say to the cashier, "It looks to me like the total is around $76; is that OK?" Yet that is what we do when we intuitively combine a set of probabilistic causes.

For all the power of a regression equation, the most humbling discovery about predicting human behavior is how unpredictable it is. It's easy to say that behavior is caused by a combination of heredity and environment. Yet when we look at a predictor that has to be more powerful than the best regression equation—a person's identical twin, who shares her genome, family, neighborhood, schooling, and culture—we see that the correlation between the two twins' traits, while way higher than chance, is way lower than 1, typically around .6.[36] That leaves a *lot* of human differences mysteriously unexplained: despite near-identical causes, the effects are nowhere near identical. One twin may be gay and the other straight, one schizophrenic and the other functioning normally. In the depression graph, we saw that the chance that a woman will suffer depression if she is hit with a stressful event *and* has a proven genetic disposition to depression is not 100 percent but only 14 percent.

A recent lollapalooza of a study reinforces the cussed unpredictability of the human species.[37] One hundred sixty teams of researchers were given a massive dataset on thousands of fragile families, including their income, education, health records, and the results of multiple interviews and in-home assessments. The teams were challenged to predict the families' outcomes, such as the children's grades and the parents' likelihood of being evicted, or employed, or signed up for job training. The competitors were allowed to sic whichever algorithm they wanted on the problem: regression, deep learning, or any other fad or fashion in artificial intelligence. The results? In the understated words of the paper abstract: "The best predictions were not very accurate." Idiosyncratic traits of each family swamped the generic predictors, no matter how cleverly they were combined. It's a reassurance to people who worry that artificial intelligence will soon predict our every move. But it's also a chastening smackdown of our pretensions to fully understand the causal network in which we find ourselves.

And speaking of humility, we have come to the end of seven chapters intended to equip you with what I think are the most important tools of rationality. If I have succeeded, you'll appreciate this final word from *XKCD*.

WHAT'S WRONG WITH PEOPLE?

> Tell people there's an invisible man in the sky who created
> the universe, and the vast majority will believe you. Tell
> them the paint is wet, and they have to touch it to be sure.
>
> —GEORGE CARLIN

This is the chapter most of you have been waiting for. I know this from conversations and correspondence. As soon as I mention the topic of rationality, people ask me why humanity appears to be losing its mind.

At the time of this writing, a glorious milestone in the history of rationality is coming into view: vaccines likely to end a deadly plague are being administered less than a year after the plague emerged. Yet in that same year, the Covid-19 pandemic set off a carnival of cockamamie conspiracy theories: that the disease was a bioweapon engineered in a Chinese lab, a hoax spread by the Democratic Party to sabotage Donald Trump's chances of reelection, a subterfuge by Bill Gates to implant trackable microchips in people's bodies, a plot by a cabal of global elites to control the world economy, a symptom of the rollout of fifth-generation mobile data networks, and a means for

Anthony Fauci (director of the National Institute of Allergy and Infectious Diseases) to earn windfall profits from a vaccine.[1] Shortly before the announcements of the vaccines, a third of Americans said they would reject them, part of an anti-vax movement that opposes the most benevolent invention in the history of our species.[2] Covid quackery has been endorsed by celebrities, politicians, and, disturbingly, the most powerful person on earth at the time of the pandemic, US president Donald Trump.

Trump himself, who was consistently supported by around 40 percent of the American public, raised further doubts throughout his presidency on our collective capacity for reason. He predicted in February 2020 that Covid-19 would disappear "like a miracle," and endorsed quack cures like malaria drugs, bleach injections, and light probes. He disdained basic public health measures like masks and distancing, even after he himself was stricken, inspiring millions of Americans to flout the measures and amplifying the toll of death and financial hardship.[3] It was all part of a larger rejection of the norms of reason and science. Trump told around thirty thousand lies during his term, had a press secretary who touted "alternative facts," claimed that climate change was a Chinese hoax, and suppressed knowledge from scientists in federal agencies overseeing public health and environmental protection.[4] He repeatedly publicized QAnon, the millions-strong conspiracy cult that credits him with combating a cabal of Satan-worshiping pedophiles embedded in the American "deep state." And he refused to acknowledge his defeat in the 2020 election, fighting crackbrained legal battles to overturn the results, led by lawyers who cited yet another conspiracy, this one by Cuba, Venezuela, and several governors and officials of his own party.

Covid quackery, climate denial, and conspiracy theories are symptoms of what some are calling an "epistemological crisis" and a "post-truth era."[5] Another symptom is fake news. In the second decade of

the twenty-first century, social media have become sluices for tor-
rents of tall tales like these:[6]

POPE FRANCIS SHOCKS WORLD, ENDORSES DONALD TRUMP FOR PRESIDENT

YOKO ONO: "I HAD AN AFFAIR WITH HILLARY CLINTON IN THE 1970S"

DEMOCRATS VOTE TO ENHANCE MED CARE FOR ILLEGALS NOW, VOTE DOWN VETS WAITING 10 YEARS FOR SAME SERVICE

TRUMP TO BAN ALL TV SHOWS THAT PROMOTE GAY ACTIVITY

WOMAN SUES SAMSUNG FOR $1.8M AFTER CELL PHONE GETS STUCK IN HER VAGINA

LOTTERY WINNER ARRESTED FOR DUMPING $200,000 OF MANURE ON EX-BOSS'S LAWN

Also rampant are beliefs in ghouls, black magic, and other superstitions. As I mentioned in the first chapter, three quarters of Americans hold at least one paranormal belief. Here are some figures from
the first decade of our century:[7]

Possession by the devil, 42 percent

Extrasensory perception, 41 percent

Ghosts and spirits, 32 percent

Astrology, 25 percent

Witches, 21 percent

Communicating with the dead, 29 percent

Reincarnation, 24 percent

Spiritual energy in mountains, trees, and crystals, 26 percent

Evil eye, curses, spells, 16 percent

Consulted a fortune-teller or psychic, 15 percent

Just as disturbingly to someone like me who likes to plot human progress, these beliefs show few signs of decreasing over the decades, and younger generations are no more skeptical than their elders (with astrology they are more credulous).[8]

Also popular are a miscellany of canards that the historian of science Michael Shermer calls "weird beliefs."[9] Many people endorse conspiracy theories like Holocaust denial, Kennedy assassination plots, and the 9/11 "Truther" theory that the twin towers were felled by a controlled demolition to justify the American invasion of Iraq. Various seers, cults, and ideologies have convinced their followers that the end of the world is nigh; they disagree on when, but are quick to postdate their predictions when they are unpleasantly surprised to find themselves living another day. And a quarter to a third of Americans believe we have been visited by extraterrestrials, either the contemporary ones that mutilate cattle and impregnate women to breed alien–human hybrids, or the ancient ones who built the pyramids and Easter Island statues.

How CAN WE EXPLAIN this pandemic of poppycock? As with Charlie Brown in the *Peanuts* strip, it can make your stomach hurt, especially when Lucy appears to represent a large portion of our compatriots:

Let's begin by setting aside three popular explanations, not because they are wrong but because they are too glib to be satisfying.

The first of these, I must admit, is the inventory of logical and statistical fallacies explained in the preceding chapters. To be sure, many superstitions originate in overinterpreting coincidences, failing to calibrate evidence against priors, overgeneralizing from anecdotes, and leaping from correlation to causation. A prime example is the misconception that vaccines cause autism, reinforced by the observation that autistic symptoms appear, coincidentally, around the age at which children are first inoculated. And all of them represent failures of critical thinking and of the grounding of belief in evidence; that's what entitles us to say they're false in the first place. Yet nothing from the cognitive psychology lab could have predicted QAnon, nor are its adherents likely to be disabused by a tutorial in logic or probability.

A second unpromising lead is to blame today's irrationality on the current scapegoat for everything, social media. Conspiracy theories and viral falsehoods are probably as old as language.[10] What are the accounts of miracles in scriptures, after all, but fake news about paranormal phenomena? For centuries Jews have been accused of conspiring to poison wells, sacrifice Christian children, control the world economy, and foment communist uprisings. At many times in history, other races, minorities, and guilds have also been credited with nefarious plots and targeted with violence.[11] The political scientists Joseph Uscinski and Joseph Parent tracked the popularity of conspiracy theories in letters to the editor of major American newspapers from 1890 to 2010 and found no change over that period; nor did the numbers rise in the subsequent decade.[12] As for fake news, before it was disseminated on Twitter and Facebook, outlandish episodes which happened to a friend of a friend were circulated as urban legends (the Hippie Babysitter, the Kentucky Fried Rat, Halloween Sadists) or emblazoned on the covers of supermarket tabloids (BABY BORN TALKING: DESCRIBES HEAVEN; DICK CHENEY IS A ROBOT; SURGEONS TRANSPLANT

YOUNG BOY'S HEAD ONTO HIS SISTER'S BODY).[13] Social media may indeed be accelerating their spread, but the appetite for florid fantasies lies deep in human nature: people, not algorithms, compose these stories, and it's people they appeal to. And for all the panic that fake news has sown, its political impact is slight: it titillates a faction of partisans rather than swaying a mass of undecideds.[14]

Finally, we must go beyond offhand excuses that just attribute one irrationality to another. It's never a good explanation to say that people embrace some false belief because it gives them comfort or helps them make sense of the world, because that only raises the question of *why* people should get comfort and closure from beliefs that could not possibly do them any good. Reality is a powerful selection pressure. A hominid that soothed itself by believing that a lion was a turtle or that eating sand would nourish its body would be outreproduced by its reality-based rivals.

Nor will it do to write off humans as hopelessly irrational. Just as our foraging ancestors lived by their wits in unforgiving ecosystems, today's conspiracy theorists and miracle-believers pass the demanding tests of their own worlds: they hold down jobs, bring up kids, and keep a roof over their heads and food in the fridge. For that matter, a favorite riposte by Trump's defenders to the charge that he was cognitively impaired was "If he's so dumb, how did he get to be president?" And unless you believe that scientists and philosophers are a superior breed of human, you have to acknowledge that most members of our species have the capacity to discover and accept the canons of rationality. To understand popular delusions and the madness of crowds, we have to examine cognitive faculties that work well in some environments and for some purposes but that go awry when applied at scale, in novel circumstances, or in the service of other goals.

Motivated Reasoning

Rationality is disinterested. It is the same for everyone everywhere, with a direction and momentum of its own. For that reason rationality can be a nuisance, an impediment, an affront. In Rebecca Newberger Goldstein's novel *36 Arguments for the Existence of God: A Work of Fiction*, an eminent literary scholar explains to a graduate student why he abhors deductive thinking:[15]

> It is a form of torture for the imaginatively gifted, the very totalitarianism of thought, one line being made to march strictly in step behind the other, all leading inexorably to a single undeviating conclusion. A proof out of Euclid recalls to my mind nothing so much as the troops goose-stepping before the Supreme Dictator. I have always delighted in my mind's refusal to follow a single line of any mathematical explanation offered to me. Why should these exacting sciences exact anything from me? Or as Dostoevsky's Underground Man shrewdly argues, "Good God, what do I care about the laws of nature and arithmetic if, for one reason or another, I don't like these laws, including the 'two times two is four'?" Dostoevsky spurned the hegemaniacal logic and I can do no less.

The obvious reason that people avoid getting onto a train of reasoning is that they don't like where it takes them. It may terminate in a conclusion that is not in their interest, such as an allocation of money, power, or prestige that is objectively fair but benefits someone else. As Upton Sinclair pointed out, "It is difficult to get a man to understand something, when his salary depends upon his not understanding it."[16]

The time-honored method to head off a line of reasoning before it arrives at an unwanted destination is to derail the reasoner by brute force. But there are less crude methods that exploit the inevitable uncertainties surrounding any issue and steer the argument in a favored direction with sophistry, spin-doctoring, and the other arts of persuasion. Both members of an apartment-hunting couple, for example, may emphasize the reasons why the flat that just happens to be closer to where he or she works is objectively better for the two of them, such as its space or affordability. It's the stuff of everyday arguments.

The mustering of rhetorical resources to drive an argument toward a favored conclusion is called motivated reasoning.[17] The motive may be to end at a congenial conclusion, but it may also be to flaunt the arguer's wisdom, knowledge, or virtue. We all know the barroom blowhard, the debating champ, the legal eagle, the mansplainer, the competitive distance urinator, the intellectual pugilist who would rather *be* right than *get it* right.[18]

Many of the biases that populate the lists of cognitive infirmities are tactics of motivated reasoning. In chapter 1 we saw confirmation bias, such as in the selection task, where people who are asked to turn over the cards that test an "If P then Q" rule choose the P card, which can confirm it, but not the not-Q card, which can falsify it.[19] They turn out to be more logical when they *want* the rule to be false. When the rule says that if someone has their emotional profile, that person is in danger of dying young, they correctly test the rule (and at the same time reassure themselves) by homing in on the people who have their profile and on the people who lived to a ripe old age.[20]

We are also motivated to regulate our information diet. In biased assimilation (or selective exposure), people seek out arguments that ratify their beliefs and shield themselves from those that might disconfirm them.[21] (Who among us doesn't take pleasure in reading editorials that are politically congenial, and get irritated by those from the other side?) Our self-protection continues with the arguments that do reach us. In biased evaluation, we deploy our ingenuity to upvote the arguments that support our position and pick nits in the ones that refute it. And there are the classic informal fallacies we saw in chapter 3: ad hominem, authority, bandwagon, genetic, affective, straw man, and so on. We are even biased about our biases. The psychologist Emily Pronin has found that, as in the mythical town where all the children are above average, a large majority of Americans consider themselves less susceptible to cognitive biases than the average American, and virtually none consider themselves more biased.[22]

So much of our reasoning seems tailored to winning arguments that some cognitive scientists, like Hugo Mercier and Dan Sperber, believe it is the adaptive function of reasoning.[23] We evolved not as intuitive scientists but as intuitive lawyers. While people often try to get away with lame arguments for their own positions, they are quick to spot fallacies in other people's arguments. Fortunately, this hypocrisy can be mobilized to make us more rational collectively than any of us is individually. The wisecrack circulated among veterans of committees that the IQ of a group is equal to the lowest IQ of any member of the group divided by the size of the group turns out to be exactly wrong.[24] When people evaluate an idea in small groups with the right chemistry, which is that they don't agree on everything but have a common interest in finding the truth, they catch on to each other's fallacies and blind spots, and usually the truth wins. When individuals are given the Wason selection task, for example, only one in ten picks the right cards, but when they are put in groups, around seven in ten get it right. All it takes is for one member to see the correct answer, and almost always that person persuades the others.

The Myside Bias

People's desire to get their way or act as know-it-alls can explain only part of our public irrationality. You can appreciate another part by considering this problem in evidence-based policy. Do gun-control measures decrease crime, because fewer criminals can obtain them, or increase it, because law-abiding citizens can no longer protect themselves?

Here are data from a hypothetical study that divided cities into those that adopted a ban on concealed handguns (first row) and those that did not (second row).[25] Laid out in each column are the number of those cities that saw their crime rates improve (left column) or worsen (right column). From these data, would you conclude that gun control is effective at reducing crime?

	Crime rate decreased	Crime rate increased
Gun control	223	75
No gun control	107	21

In fact, the data (which are fake) suggest that gun control *increases* crime. It's easy to get it wrong, because the large number of cities with gun control in which the crime rate declined, 223, pops out. But that could just mean that crime decreased in the whole country, policy or no policy, and that more cities tried gun control than didn't, a trend in political fashions. We need to look at the *ratios*. In cities with gun control, it's around three to one (223 versus 75); in cities without, it is around *five* to one (107 versus 21). On average, the data say, a city was better off without gun control than with it.

As in the Cognitive Reflection Test (chapter 1), getting to the answer requires a bit of numeracy: the ability to set aside first impressions and do the math. People who are so-so in numeracy tend to be distracted by the big number and conclude that gun control works. But the real point of this illustration, devised by the legal scholar Dan Kahan and his collaborators, is what happened with the numerate respondents. The numerate Republicans tended to get the answer right, the numerate Democrats to get it wrong. The reason is that Democrats *start out* believing that gun control is effective and are all too quick to accept data showing they were right all along. Republicans can't stomach the idea and scrutinize the data with a gimlet eye, which, if numerate, spots the real pattern.

Republicans might attribute their success to being more objective than the bleeding-heart libs, but of course the researchers ran a condition in which the knee-jerk wrong answer was congenial to Republicans. They simply flipped the column labels, so that the data now suggested that gun control works: it stanched a fivefold increase in crime, holding it to just a threefold increase. This time the numerate Republicans earned the dunce caps while the Democrats were the Einsteins. In a control condition, the team picked an issue that triggered neither Democrats nor Republicans: whether a skin cream was effective at treating a rash. With neither faction having a dog in the fight, the numerate Republicans and numerate Democrats performed the same. A recent meta-analysis of fifty studies by the psychologist Peter Ditto and his colleagues confirms the pattern. In study after study, liberals and conservatives accept or reject the same scientific conclusion depending on whether or not it supports their talking points, and they endorse or oppose the same policy depending on whether it was proposed by a Democratic or a Republican politician.[26]

Politically motivated numeracy and other forms of biased evaluation show that people reason their way into or out of a conclusion even when it offers them no personal advantage. It's enough that the conclusion enhances the correctness or nobility of their political, religious, ethnic, or cultural tribe. It's called, obviously enough, the myside bias, and it commandeers every kind of reasoning, even logic.[27] Recall that the validity of a syllogism depends on its form, not its content, but that people let their knowledge seep in and judge an argument valid if it ends in a conclusion they know is true or want to be true. The same thing happens when the conclusion is politically congenial:

> If college admissions are fair, then affirmative action laws are
>   no longer necessary.
> College admissions are not fair.
> Therefore, affirmative action laws are necessary.

> If less severe punishments deter people from committing
>   crime, then capital punishment should not be used.
> Less severe punishments do not deter people from
>   committing crime.
> Therefore, capital punishment should be used.

When people are asked to verify the logic of these arguments, both of which commit the formal fallacy of denying the antecedent, liberals mistakenly ratify the first and correctly nix the second; conservatives do the opposite.[28]

In *Duck Soup*, Chico Marx famously asked, "Who ya gonna believe, me or your own eyes?" When people are in the throes of the myside bias, the answer may not be their own eyes. In an update of a classic study showing that football fans always see more infractions by the opposing team, Kahan and collaborators showed a video of a protest in front of a building.[29] When the title labeled it a protest against abortion at a health clinic, conservatives saw a peaceful demonstration, while liberals saw the protesters block the entrance and intimidate the enterers. When it was labeled a protest against the exclusion of gay people at a military recruiting center, it was the conservatives who saw pitchforks and torches and the liberals who saw Mahatma Gandhi.

One magazine reported the gun-control study under the headline THE MOST DEPRESSING DISCOVERY ABOUT THE BRAIN, EVER. Certainly there are reasons to be depressed. One is that opinions that go against the scientific consensus, like creationism and the denial of human-made climate change, may not be symptoms of innumeracy or scientific illiteracy. Kahan has found that most believers and deniers are equally clueless about the scientific facts (many believers in climate change, for example, think that it has something to do with toxic waste dumps and the ozone hole). What predicts their belief is their politics: the farther to the right, the more denial.[30]

Another cause for gloom is that for all the talk of a replicability crisis, the myside bias is only too replicable. In *The Bias That Divides Us*, the psychologist Keith Stanovich finds it in every race, gender, cognitive style, education level, and IQ quantile, even among people who are too clever to fall for other cognitive biases like base-rate neglect and the gambler's fallacy.[31] The myside bias is not an across-the-board personality trait, but presses on whichever trigger or hot button is connected to the reasoner's identity. Stanovich relates it to our political moment. We are not, he suggests, living in a "post-truth" society. The problem is that we are living in a myside society.

The sides are the left and the right, and both sides believe in the truth but have incommensurable ideas of what the truth is. The bias has invaded more and more of our deliberations. The spectacle of face masks during a respiratory pandemic turning into political symbols is just the most recent symptom of the polarization.

WE'VE LONG KNOWN THAT humans are keen to divide themselves into competitive teams, but it's not clear why it's now the left–right split that is pulling each side's rationality in different directions rather than the customary fault lines of religion, race, and class. The right–left axis aligns with several moral and ideological dimensions: hierarchical versus egalitarian, libertarian versus communitarian, throne-and-altar versus Enlightenment, tribal versus cosmopolitan, tragic versus utopian visions, honor versus dignity cultures, binding versus individualizing moralities.[32] But recent flip-flops in which side supports which cause, such as immigration, trade, and sympathy for Russia, suggests that the political sides have become sociocultural tribes rather than coherent ideologies.

In a recent diagnosis, a team of social scientists concluded that the sides are less like literal tribes, which are held together by kinship, than religious sects, which are held together by faith in their moral superiority and contempt for opposing sects.[33] The rise of political sectarianism in the United States is commonly blamed (like everything else) on social media, but its roots lie deeper. They include the fractionation and polarization of broadcast media, with partisan talk radio and cable news displacing national networks; gerrymandering and other geographic distortions of political representation, which incentivize politicians to cater to cliques rather than coalitions; the reliance of politicians and think tanks on ideologically committed donors; the self-segregation of educated liberal professionals into urban enclaves; and the decline of class-crossing civil-society organizations like churches, service clubs, and volunteer groups.[34]

Could the myside bias possibly be rational? There is a Bayesian argument that one *ought* to weigh new evidence against the totality of one's prior beliefs rather than taking every new study at face value. If liberalism has proven itself to be correct, then a study that appears to support a conservative position should not be allowed to overturn one's beliefs. Not surprisingly, this was the response of several liberal academics to Ditto's meta-analysis suggesting that political bias is bipartisan.[35] Nothing guarantees that the favorite positions of the left and right at any historical moment will be aligned with the truth 50–50. Even if both sides interpret reality through their own beliefs, the side whose beliefs are warranted will be acting rationally. Maybe, they continue, the well-documented left-wing lopsidedness of academia is not an irrational bias but an accurate calibration of their Bayesian priors to the fact that the left is always correct.

The response from conservatives is (quoting Hamlet), "Lay not that flattering unction to your soul."[36] Though it may be true that left-wing positions are vindicated more often than right-wing ones (especially if, for whatever reason, the left is more congenial to science than the right), in the absence of disinterested benchmarks neither side is in a position to say. Certainly history has no shortage of examples of both sides getting it wrong, including some real doozies.[37] Stanovich notes that the problem in justifying motivated reasoning with Bayesian priors is that the prior often reflects what the reasoner *wants* to be true rather than what he or she has *grounds for believing* is true.

There is a different and more perverse rationality to the myside bias, coming not from Bayes's rule but from game theory. Kahan calls it expressive rationality: reasoning that is driven by the goal of being valued by one's peer group rather than attaining the most accurate understanding of the world. People express opinions that advertise where their heart lies. As far as the fate of the expresser in a social milieu is concerned, flaunting those loyalty badges is anything but irrational. Voicing a local heresy, such as rejecting gun control in a Democratic social circle or advocating it in a Republican one, can mark you as a traitor, a quisling, someone who "doesn't get it," and condemn you to social death. Indeed, the best identity-signaling beliefs are often the most outlandish ones. Any fair-weather friend can say the world is round, but only a blood brother would say the world is flat, willingly incurring ridicule by outsiders.[38]

Unfortunately, what's rational for each of us seeking acceptance in a clique is not so rational for all of us in a democracy seeking the best understanding of the world. Our problem is that we are trapped in a Tragedy of the Rationality Commons.[39]

Two Kinds of Belief:
Reality and Mythology

The humor in the *Peanuts* strip in which Lucy gets buried in snow while insisting that it rises from the ground exposes a limitation on any explanation of human irrationality that invokes the ulterior motives in motivated reasoning. No matter how effectively a false belief flaunts the believer's mental prowess or loyalty to the tribe, it's still false, and should be punished by the cold, hard facts of the world. As the novelist Philip K. Dick wrote, reality is that which, when you stop believing in it, doesn't go away. Why doesn't reality push back and inhibit people from believing absurdities or from rewarding those who assert and share them?

The answer is that it depends what you mean by "believe." Mercier notes that holders of weird beliefs often don't have the courage of their convictions.[40] Though millions of people endorsed the rumor that Hillary Clinton ran a child sex trafficking ring out of the basement of the Comet Ping Pong pizzeria in Washington (the Pizzagate conspiracy theory, a predecessor of QAnon), virtually none took steps commensurate with such an atrocity, such as calling the police. The righteous response of one of them was to leave a one-star review on Google. ("The pizza was incredibly undercooked. Suspicious professionally dressed men by the bar area that looked like regulars kept staring at my son and other kids in the place.") It's hardly the response most of us would have if we literally thought that children were being raped in the basement. At least Edgar Welch, the man who burst into the pizzeria with his gun blazing in a heroic attempt to rescue the children, took his beliefs seriously. The millions of others must have believed the rumor in a very different sense of "believe."

Mercier also points out that impassioned believers in vast nefarious conspiracies, like the 9/11 Truthers and the chemtrail theorists (who hold that the water-vapor contrails left by jetliners are chemicals dispensed in a secret government program to drug the population), publish their manifestos and hold their meetings in the open, despite their belief in a brutally effective plot by an omnipotent regime to suppress brave truth-tellers like them. It's not the strategy you see from dissidents in undeniably repressive regimes like North Korea or Saudi Arabia. Mercier, invoking a distinction made by Sperber, proposes that conspiracy theories and other weird beliefs are *reflective*, the result of conscious cogitation and theorizing, rather than *intuitive*, the convictions we feel in our bones.[41] It's a powerful distinction, though I draw it a bit differently, closer to the contrast that the social psychologist Robert Abelson (and the comedian George Carlin) drew between *distal* and *testable* beliefs.[42]

People divide their worlds into two zones. One consists of the physical objects around them, the other people they deal with face to face, the memory of their interactions, and the rules and norms that regulate their lives. People have mostly accurate beliefs about this zone, and they reason rationally within it. Within this zone, they believe there's a real world and that beliefs about it are true or false. They have no choice: that's the only way to keep gas in the car, money in the bank, and the kids clothed and fed. Call it the reality mindset.

The other zone is the world beyond immediate experience: the distant past, the unknowable future, faraway peoples and places, remote corridors of power, the microscopic, the cosmic, the counterfactual, the metaphysical. People may entertain notions about what happens in these zones, but they have no way of finding out, and anyway it makes no discernible difference to their lives. Beliefs in these zones are narratives, which may be entertaining or inspiring or morally edifying. Whether they are literally "true" or "false" is the wrong question. The function of these beliefs is to construct a social reality that binds the tribe or sect and gives it a moral purpose. Call it the mythology mindset.

Bertrand Russell famously said, "It is undesirable to believe a proposition when there is no ground whatsoever for supposing it is true." The key to understanding rampant irrationality is to recognize that Russell's statement is not a truism but a revolutionary manifesto. For most of human history and prehistory, there *were* no grounds for supposing that propositions about remote worlds were true. But beliefs about them could be empowering or inspirational, and that made them desirable enough.

Russell's maxim is the luxury of a technologically advanced society with science, history, journalism, and their infrastructure of truth-seeking, including archival records, digital datasets, high-tech instruments, and communities of editing, fact-checking, and peer review. We children of the Enlightenment embrace the radical creed of universal realism: we hold that *all* our beliefs should fall within the reality mindset. We care about whether our creation story, our founding legends, our theories of invisible nutrients and germs and forces, our conceptions of the powerful, our suspicions about our enemies, are true or false. That's because we have the tools to get answers to these questions, or at least to assign them warranted degrees of credence. And we have a technocratic state that should, in theory, put these beliefs into practice.

But as desirable as that creed is, it is not the natural human way of believing. In granting an imperialistic mandate to the reality mindset to conquer the universe of belief and push mythology to the margins, *we* are the weird ones—or, as evolutionary social scientists like to say, the WEIRD ones: Western, Educated, Industrialized, Rich, Democratic.[43] At least, the highly educated among us are, in our best moments. The human mind is adapted to understanding remote spheres of existence through a mythology mindset. It's not because we descended from Pleistocene hunter-gatherers specifically, but because we descended from people who could not or did not sign on to the Enlightenment ideal of universal realism. Submitting all of one's beliefs to the trials of reason and evidence is an unnatural skill, like literacy and numeracy, and must be instilled and cultivated.

And for all the conquests of the reality mindset, the mythology mindset still occupies swaths of territory in the landscape of mainstream belief. The obvious example is religion. More than two billion people believe that if one doesn't accept Jesus as one's savior one will be damned to eternal torment in hell. Fortunately, they don't take the next logical step and try to convert people to Christianity at swordpoint for their own good, or torture heretics who might lure others into damnation. Yet in past centuries, when Christian belief fell into the reality zone, many Crusaders, Inquisitors, conquistadors, and soldiers in the Wars of Religion did exactly that. Like the Comet Ping Pong redeemer, they treated their beliefs as literally true. For that matter, though many people profess to believe in an afterlife, they seem to be in no hurry to leave this vale of tears for eternal bliss in paradise.

Thankfully, Western religious belief is safely parked in the mythology zone, where many people are protective of its sovereignty. In the mid-aughts, the "New Atheists," Sam Harris, Daniel Dennett, Christopher Hitchens, and Richard Dawkins, became targets of vituperation not just from Bible-thumping evangelists but also from mainstream intellectuals. These faitheists (as the biologist Jerry Coyne called them), or believers in belief (Dennett's term), did not counter that God in fact exists.[44] They implied that it is inappropriate, or uncouth, or just not done, to consider God's existence a matter of truth or falsity. Belief in God is an idea that falls outside the sphere of testable reality.

Another zone of mainstream unreality is the national myth. Most countries enshrine a founding narrative as part of their collective consciousness. At one time these were epics of heroes and gods, like the *Iliad*, the *Aeneid*, Arthurian legends, and Wagnerian operas. More recently they have been wars of independence or anticolonial struggles. Common themes include the nation's ancient essence defined by a language, culture, and homeland; an extended slumber and glorious awakening; a long history of victimization and oppression; and a generation of superhuman liberators and founders. Guardians of the mythical heritage don't feel a need to get to the bottom of what actually transpired, and may resent the historians who place it in the reality zone and unearth its shallow history, constructed identity, reciprocal provocations with the neighbors, and founding fathers' feet of clay.

Still another zone of not-quite-true-not-quite-false belief is historical fiction and fictionalized history. It seems pedantic to point out that Henry V did not deliver the stirring words on Saint Crispin's Day that Shakespeare attributed to him. Yet the play purports to be an account of real events rather than a figment of the playwright's imagination, and we would not enjoy it in the same way otherwise. The same is true of fictionalized histories of more recent wars and struggles, which are, in effect, fake news set in the recent past. When the events come too close to the present or the fictionalization rewrites important facts, historians can sound an alarm, as when Oliver Stone brought to life an assassination conspiracy theory in the 1991 movie *JFK*. In 2020, the columnist Simon Jenkins objected to the television series *The Crown*, a dramatized history of Queen Elizabeth and her family which took liberties with many of the depicted events: "When you turn on your television tonight, imagine seeing the news acted rather than read. . . . Afterwards the BBC flashes up a statement saying all this was 'based on true events,' and hoping we enjoyed it."[45] Yet his was a voice crying out in the wilderness. Most critics and viewers had no problem with the sumptuously filmed falsehoods, and Netflix refused to post a warning that some of the scenes were fictitious (though they did post a trigger warning about bulimia).[46]

The boundary between the reality and the mythology zones can vary with the times and the culture. Since the Enlightenment, the tides in the modern West have eroded the mythology zone, a historical shift that the sociologist Max Weber called "the disenchantment of the world." But there are always skirmishes at the borders. The brazen lies and conspiracies of Trumpian post-truth can be seen as an attempt to claim political discourse for the land of mythology rather than the land of reality. Like the plots of legends, scripture, and drama, they are a kind of theater; whether they are provably true or false is beside the point.

The Psychology of Apocrypha

Once we appreciate that humans can hold beliefs they don't treat as factually true, we can begin to make sense of the rationality paradox—how a rational animal can embrace so much claptrap. It's not that the conspiracy theorists, fake-news sharers, and consumers of pseudoscience *always* construe their myths as mythological. Sometimes their beliefs cross the line into reality with tragic results, as in Pizzagate, anti-vaxxers, and the Heaven's Gate cult, whose thirty-nine devotees committed suicide in 1997 in preparation for their souls to be whisked away by a spaceship following the Hale-Bopp comet. But predispositions in human nature can combine with mythological truthiness to make weird beliefs easy to swallow. Let's look at three genres.

Pseudoscience, paranormal woo-woo, and medical quackery engage some of our deepest cognitive intuitions.[47] We are intuitive dualists, sensing that minds can exist apart from bodies.[48] It comes naturally to us, and not just because we can't see the neural networks which underlie the beliefs and desires of ourselves and others. Many of our experiences really do suggest that the mind is not tethered to the body, including dreams, trances, out-of-body experiences, and death. It's not a leap for people to conclude that minds can commune with reality and with each other without needing a physical medium. And so we have telepathy, clairvoyance, souls, ghosts, reincarnation, and messages from the great beyond.

We are also intuitive essentialists, sensing that living things contain invisible substances that give them their form and powers.[49] These intuitions inspire people to probe living things for their seeds, drugs, and poisons. But the mindset also makes people believe in homeopathy, herbal remedies, purging and bloodletting, and a rejection of foreign adulterants such as vaccines and genetically modified foods.

And we are intuitive teleologists.[50] Just as our own plans and artifacts are designed with a purpose, so, we are apt to think, is the complexity of the living and nonliving world. Thus we are receptive to creationism, astrology, synchronicity, and the mystical belief that everything happens for a reason.

A scientific education is supposed to stifle these primitive intuitions, but for several reasons its reach is limited. One is that beliefs that are sacred to a religious or cultural faction, like creationism, the soul, and a divine purpose, are not easily surrendered, and they may be guarded within people's mythology zone. Another is that even among the highly educated, scientific understanding is shallow. Few people can explain why the sky is blue or why the seasons change, let alone population genetics or viral immunology. Instead, educated people trust the university-based scientific establishment: its consensus is good enough for them.[51]

Unfortunately, for many people the boundary between the scientific establishment and the pseudoscientific fringe is obscure. The closest that most people come to science in their own lives is their doctor, and many doctors are more folk healers than experts in randomized clinical trials. Indeed, some of the celebrity doctors who appear on daytime talk shows are charlatans who exuberantly shill new-age flimflam. Mainstream television documentaries and news shows may also blur the lines and credulously dramatize fringe claims like ancient astronauts and crime-fighting psychics.[52]

For that matter, bona fide science communicators must shoulder some of the blame for failing to equip people with the deep understanding that would make pseudoscience incredible on the face of it. Science is often presented in schools and museums as just another form of occult magic, with exotic creatures and colorful chemicals and eye-popping illusions. Foundational principles, such as that the universe has no goals related to human concerns, that all physical interactions are governed by a few fundamental forces, that living bodies are intricate molecular machines, and that the mind is the information-processing activity of the brain, are never articulated, perhaps because they would seem to insult religious and moral sensibilities. We should not be surprised that what people take away from science education is a syncretic mishmash, where gravity and electromagnetism coexist with psi, qi, karma, and crystal healing.

To UNDERSTAND VIRAL HUMBUG such as urban legends, tabloid headlines, and fake news, we have to remember that it is fantastically entertaining. It plays out themes of sex, violence, revenge, danger, fame, magic, and taboo that have always titillated patrons of the arts, high and low. A fake headline like FBI AGENT SUSPECTED IN HILLARY EMAIL LEAKS FOUND DEAD IN APPARENT MURDER-SUICIDE would be an excellent plot in a suspense thriller. A recent quantitative analysis of the content of fake news concluded that "the same features that make urban legends, fiction, and in fact any narrative, culturally attractive also operate for online misinformation."[53]

Often the entertainment spills into genres of comedy, including slapstick, satire, and farce: MORGUE EMPLOYEE CREMATED BY MISTAKE WHILE TAKING A NAP; DONALD TRUMP ENDS SCHOOL SHOOTINGS BY BANNING SCHOOLS; BIGFOOT KEEPS LUMBERJACK AS LOVE SLAVE. QAnon falls into still another genre of entertainment, the multi-platform alternate-reality game.[54] Adherents parse cryptic clues periodically dropped by Q (the hypothetical government whistleblower), crowdsource their hypotheses, and gain internet fame by sharing their discoveries.

It's no surprise that people seek out all manner of entertainment.

What shocks us is that each of these works of art makes a factual claim. Yet our queasiness about blurring fact and fiction is not a universal human reaction, particularly when it pertains to zones that are remote from immediate experience, like faraway places and the lives of the rich and powerful. Just as religious and national myths become entrenched in the mainstream when they are felt to provide moral uplift, fake news may go viral when its spreaders think a higher value is at stake, like reinforcing solidarity within their own side and reminding comrades about the perfidiousness of the other one. Sometimes the moral is not even a coherent political strategy but a sense of moral superiority: the impression that rival social classes, and powerful institutions from which the sharers feel alienated, are decadent and corrupt.

CONSPIRACY THEORIES, for their part, flourish because humans have always been vulnerable to real conspiracies.[55] Foraging people can't be too careful. The deadliest form of warfare among tribal peoples is not the pitched battle but the stealthy ambush and the predawn raid.[56] The anthropologist Napoleon Chagnon writes that the Amazonian Yanomamö have the word *nomohori*, "dastardly trick," for acts of treachery such as inviting neighbors to a feast and then massacring them on cue. Plots by enemy coalitions are unlike other hazards such as predators and lightning bolts because they deploy their ingenuity to penetrate the targets' defenses and cover their own tracks. The only safeguard against this cloak-and-dagger subterfuge is to outthink them preemptively, which can lead to convoluted trains of conjecture and a refusal to take obvious facts at face value. In signal detection terms, the cost of missing a real conspiracy is higher than that of false-alarming to a suspected one. This calls for setting our bias toward the trigger-happy rather than the gun-shy end of the scale, adapting us to try to get wind of possible conspiracies even on tenuous evidence.[57]

Even today, conspiracies small and large really do exist. A group of employees may meet behind the back of an unpopular colleague to recommend that he be let go; a government or insurgency may plan a clandestine coup or invasion or sabotage. Conspiracy theories, like urban legends and fake news, find their way into rumors, and rumors are the stuff of conversation. Studies of rumors show that they tend to convey threats and dangers, and that they confer an aura of expertise on the spreader. And perhaps surprisingly, when they circulate among people with a vested interest in their content, such as within workplaces, they are usually correct.[58]

In everyday life, then, there are incentives for being a sentinel who warns people of hidden threats, or a relay who disseminates their warnings. The problem is that social and mass media allow rumors to spread through networks of people who have no stake in their truth. They consume the rumors for entertainment and affirmation rather than self-protection, and they lack the interest and means to follow them up. For the same reasons, originators and spreaders suffer no reputational damage for being wrong. Without these veracity checks, social media rumors, unlike workplace rumors, are *incorrect* more often than correct. Mercier suggests that the best way to inhibit the spread of dubious news is to pressure the spreaders to act on it: to call the police, rather than leaving a one-star review.

The remaining key to understanding the appeal of weird beliefs is to put the beliefs themselves under the microscope. Evolution works not just on bodies and brains but on ideas. A meme, as Richard Dawkins defined it when he coined the word, is not a captioned photograph circulated on the internet but an idea that has been shaped by generations of sharing to become highly shareable.[59] Examples include earworms that people can't stop humming or stories they feel compelled to pass along. Just as organisms evolve adaptations that protect them from being eaten, ideas may evolve adaptations that protect them from being refuted. The intellectual ecosystem is filled with these invasive ideas.[60] "God works in mysterious ways." "Denial is a defense mechanism of the ego." "Psychic powers are inhibited by skeptical probing." "If you fail to denounce this person as a racist, that shows you are a racist." "Everyone is always selfish, because helping other people feels good." And, of course, "The lack of evidence for this conspiracy shows what a diabolical conspiracy it is." Conspiracy theories, by their very nature, are adapted to be spread.

Reaffirming Rationality

To understand is not to forgive. We can see why humans steer their reasoning toward conclusions that work to the advantage of themselves or their sects, and why they distinguish a reality in which ideas are true or false from a mythology in which ideas are entertaining or inspirational, without conceding that these are good things. They are not good things. Reality is that which, when you apply motivated or myside or mythological reasoning to it, does not go away. False beliefs about vaccines, public health measures, and climate change threaten the well-being of billions. Conspiracy theories incite terrorism, pogroms, wars, and genocide. A corrosion of standards of truth undermines democracy and clears the ground for tyranny.

But for all the vulnerabilities of human reason, our picture of the future need not be a bot tweeting fake news forever. The arc of knowledge is a long one, and it bends toward rationality. We should not lose sight of how much rationality is out there. Few people in developed countries today believe in werewolves, animal sacrifice, bloodletting, miasmas, the divine right of leaders, or omens in eclipses and comets, though all were mainstream in centuries past. None of Trump's thirty thousand falsehoods involved occult or paranormal forces, and each of these forces is rejected by a majority of Americans.[61] Though a few scientific issues become religious or political bloody shirts, most do not: there are factions that distrust vaccines, but not antibiotics; climate change, but not coastal erosion.[62] Despite their partisan biases, most people are pretty good at judging the veracity of headlines, and when they are presented with clear and trustworthy corrections of a false claim, they change their minds, whether it was politically congenial or not.[63]

We also have a beachhead of rationality in the cognitive style called Active Open-Mindedness, especially the subtype called Openness to Evidence.[64] This is Russell's credo that beliefs should be based on good grounds. It is a rejection of motivated reasoning; a commitment to placing all beliefs within the reality zone; an endorsement of the statement attributed to John Maynard Keynes, "When the facts change, I change my mind. What do you do, sir?"[65] The psychologist Gordon Pennycook and his colleagues measured the attitude by having people fill out a questionnaire with items like these, where the parenthesized response ups the openness score:[66]

People should always take into consideration evidence that
    goes against their beliefs. (AGREE)
Certain beliefs are just too important to abandon no matter
    how good a case can be made against them. (DISAGREE)
Beliefs should always be revised in response to new
    information or evidence. (AGREE)
No one can talk me out of something I know is right.
    (DISAGREE)
I believe that loyalty to one's ideals and principles is more
    important than "open-mindedness." (DISAGREE)

In a sample of American internet users, about a fifth of the respondents say they are impervious to evidence, but a majority at least aspire to being open to it. The people who are open to evidence are resistant to weird beliefs. They reject conspiracy theories, witchcraft, astrology, telepathy, omens, and the Loch Ness monster, together with a personal God, creationism, a young earth, a vaccine–autism link, and a denial of anthropogenic climate change.[67] They are more trusting in government and science. And they tend to hold more liberal political positions, such as on abortion, same-sex marriage, capital punishment, and war aversion, generally in the same directions that the world as a whole has been trending.[68] (The authors caution, though, that the correlations with conservatism are complicated.)

Openness to Evidence correlates with cognitive reflection (the ability to think twice and not fall for trick questions, which we met in chapter 1) and with a resistance to many of the cognitive illusions, biases, and fallacies we saw in chapters 3–9.[69] This cluster of good cognitive habits, which Stanovich calls the Rationality Quotient (a play on the intelligence quotient or IQ), correlates with raw intelligence, though imperfectly: smart people can be closed-minded and impulsive, and duller ones open and reflective. Together with resisting weird beliefs, reflective people are better at spotting fake news and at rejecting pseudo-profound bullshit like "Hidden meaning transforms unparalleled abstract beauty."[70]

If we could put something in the drinking water that would make everyone more open and reflective, the irrationality crisis would vanish. Failing that, let's consider a broad set of policies and norms that might strengthen the cognitive immune systems in ourselves and our culture.[71]

Most sweeping would be a valorization of the norm of rationality itself. Now, we can no more impose values from the top down than we can dictate any cultural change that depends on millions of individual choices, like tattooing or slang. But norms can change over time, like the decline of ethnic slurs, littering, and wife jokes, when reflexes of tacit approval and disapproval proliferate through social networks. And so we can each do our part in smiling or frowning on rational and irrational habits. It would be nice to see people earn brownie points for acknowledging uncertainty in their beliefs, questioning the dogmas of their political sect, and changing their minds when the facts change, rather than for being steadfast warriors for the dogmas of their clique. Conversely, it could be a mortifying faux pas to overinterpret anecdotes, confuse correlation with causation, or commit an informal fallacy like guilt by association or the argument from authority. The "Rationality Community" identifies itself by these norms, but they should be the mores of the whole society rather than the hobby of a club of enthusiasts.[72]

Though it's hard to steer the aircraft carrier that constitutes an entire society, particular institutions may have pressure points that savvy leaders and activists could prod. Legislatures are largely populated by lawyers, whose professional goal is victory rather than truth. Recently some scientists have begun to infiltrate the chambers, and they could try to spread the value of evidence-based problem solving among their colleagues. Advocates of any policy would be well advised not to brand it with sectarian symbolism; some climate experts, for example, lamented Al Gore becoming the face of climate change activism in the early 2000s, because that pigeonholed it as a left-wing cause, giving the right an excuse to oppose it.

Among politicians, both of the major American parties indulge in industrial-strength myside bias, but the blame is not symmetrical. Even before the Trumpian takeover, thoughtful Republican stalwarts had disparaged their own organization as "the party of stupid" for its anti-intellectualism and hostility to science.[73] Since then, many others have been horrified by their party's acquiescence to Trump's maniacal lying and trolling: his game plan, in the admiring words of onetime strategist Steve Bannon, to "flood the zone with shit."[74] With Trump's defeat, rational heads on the right should seek to restore American politics to a system with two parties that differ over policy rather than over the existence of facts and truth.

We are not helpless against the onslaught of "post-truth" disinformation. Though lying is as old as language, so are defenses against being lied to; as Mercier points out, without those defenses language could never have evolved.[75] Societies, too, protect themselves against being flooded with shit: barefaced liars are held responsible with legal and reputational sanctions. These safeguards are belatedly being deployed. In a single week in early 2021, the companies that made the voting machines and software named in Trump's conspiracy theory sued members of his legal team for defamation; Trump was banned from Twitter for violating its policy against inciting violence; a mendacious senator who pushed the stolen-election conspiracy theory in Congress lost a major book contract; and the editor of *Forbes* magazine announced, "Let it be known to the business world: Hire any of Trump's fellow fabulists, and *Forbes* will assume that everything your company or firm talks about is a lie."[76]

Since no one can know everything, and most people know almost nothing, rationality consists of outsourcing knowledge to institutions that specialize in creating and sharing it, primarily academia, public and private research units, and the press.[77] That trust is a precious resource which should not be squandered. Though confidence in science has remained steady for decades, confidence in universities is sinking.[78] A major reason for the mistrust is the universities' suffocating left-wing monoculture, with its punishment of students and professors who question dogmas on gender, race, culture, genetics, colonialism, and sexual identity and orientation. Universities have turned themselves into laughingstocks for their assaults on common sense (as when a professor was recently suspended for mentioning the Chinese pause word *ne ga* because it reminded some students of the racial slur).[79] On several occasions correspondents have asked me why they should trust the scientific consensus on climate change, since it comes out of institutions that brook no dissent. That is why universities have a responsibility to secure the credibility of science and scholarship by committing themselves to viewpoint diversity, free inquiry, critical thinking, and active open-mindedness.[80]

The press, perennially tied with Congress as the least trusted American institution, also has a special role to play in the infrastructure of rationality.[81] Like universities, news and opinion sites ought to be paragons of viewpoint diversity and critical thinking. And as I argued in chapter 4, they should also become more numerate and data-savvy, mindful of the statistical illusions instilled by sensationalist anecdote chasing. To their credit, journalists have become more mindful of the way they can be played by disingenuous politicians and contribute to post-truth miasmas, and have begun to implement countermeasures like fact-checking, labeling false claims and not repeating them, stating facts affirmatively rather than negatively, correcting errors openly and swiftly, and avoiding a false balance between experts and cranks.[82]

Educational institutions, from elementary schools to universities, could make statistical and critical thinking a greater part of their curricula. Just as literacy and numeracy are given pride of place in schooling because they are a prerequisite to everything else, the tools of logic, probability, and causal inference run through every kind of human knowledge. Rationality should be the fourth R, together with reading, writing, and arithmetic. To be sure, mere instruction in probability fails to provide lifetime immunity to statistical fallacies. Students forget it as soon as the exam is over and they sell their textbooks, and even when they remember the material, almost no one makes the leap from abstract principles to everyday pitfalls.[83] But well-designed courses and video games—ones that single out cognitive biases (the gambler's fallacy, sunk costs, confirmation bias, and so on), challenge students to spot them in lifelike settings, reframe problems in mind-friendly formats, and provide them with immediate feedback on their errors—really can train them to avoid the fallacies outside the classroom.[84]

RATIONALITY IS A PUBLIC GOOD, and a public good sets the stage for a tragedy of the commons. In the Tragedy of the Rationality Commons, motivated reasoning for the benefit of oneself and one's side produces an opportunity to free ride on our collective understanding.[85] Each of us has a motive to prefer *our* truth, but together we're better off with *the* truth.

Tragedies of the commons can be mitigated with informal norms in which members of a community police the grazing lands or fishing grounds by recognizing good citizens and stigmatizing exploiters.[86] The suggestions I have made so far can, at best, fortify individual reasoners and inculcate the norm that sound reasoning is a virtue. But the commons also must be protected with incentives: payoffs that make it in each reasoner's interests to endorse the ideas with the greatest warrant. Obviously we can't implement a fallacy tax, but particular commons can agree on rules that jigger the incentives toward truth.

I've mentioned that successful institutions of rationality never depend on the brilliance of an individual, since not even the most rational among us is bias-free. Instead they have channels of feedback and knowledge aggregation that make the whole smarter than any of its parts.[87] These include peer review in academia, testability in science, fact-checking and editing in journalism, checks and balances in governance, and adversarial proceedings in the judicial system.

The new media of every era open up a Wild West of apocrypha and intellectual property theft until truth-serving countermeasures are put into place.[88] That's what happened with books and then newspapers in the past, and it's happening with digital media today. The media can become either crucibles of knowledge or cesspools of malarkey, depending on their incentive structure. The dream at the dawn of the internet age that giving everyone a platform would birth a new Enlightenment seems cringeworthy today, now that we are living with bots, trolls, flame wars, fake news, Twitter shaming mobs, and online harassment. As long as the currency in a digital platform consists of likes, shares, clicks, and eyeballs, we have no reason to think it will nurture rationality or truth. *Wikipedia*, in contrast, though not infallible, has become an astonishingly accurate resource despite being free and decentralized. That is because it implements intensive error correction and quality control, supported by "pillars" that are designed to marginalize myside biases.[89] These include verifiability, a neutral point of view, respect and civility, and a mission to provide objective knowledge. As the site proclaims, "Wikipedia is not a soapbox, an advertising platform, a vanity press, [or] an experiment in anarchy or democracy."[90]

At the time of this writing, those gargantuan experiments in anarchy and democracy, the social media platforms, have begun to wake up to the Tragedy of the Rationality Commons, having been roused by two alarms that went off in 2020: misinformation about the Covid pandemic, and threats to the integrity of the American presidential election. The platforms have tuned their algorithms to stop rewarding dangerous falsehoods, inserted warning labels and fact-checking links, and damped down the runaway dynamics that can viralize toxic content and send people down extremist rabbit holes. It's too early to say which will work and which will not.[91] Clearly, these efforts should be redoubled, with an eye to revamping the perverse incentive structure which rewards notoriety while providing no recompense to truth.

But just as social media probably get too much blame for partisan irrationality, their algorithmic tweaks won't be enough to repair it. We should be creative in changing the rules in other arenas so that disinterested truth is given an edge over myside bias. In opinion journalism, pundits could be judged by the accuracy of their forecasts rather than their ability to sow fear and loathing or to fire up a faction.[92] In policy, medicine, policing, and other specialties, evidence-based evaluation should be a mainstream, not a niche, practice.[93] And in governance, elections, which can bring out the worst in reasoning, could be supplemented with deliberative democracy, such as panels of citizens tasked with recommending a policy.[94] This mechanism puts to use the discovery that in groups of cooperative but intellectually diverse reasoners, the truth usually wins.[95]

Human reasoning has its fallacies, biases, and indulgence in mythology. But the ultimate explanation for the paradox of how our species could be both so rational and so irrational is not some bug in our cognitive software. It lies in the duality of self and other: our powers of reason are guided by our motives and limited by our points of view. We saw in chapter 2 that the core of morality is impartiality: the reconciliation of our own selfish interests with others'. So, too, is impartiality the core of rationality: a reconciliation of our biased and incomplete notions into an understanding of reality that transcends any one of us. Rationality, then, is not just a cognitive virtue but a moral one.

11

WHY RATIONALITY MATTERS

Beginning to reason is like stepping onto an escalator that
leads upward and out of sight. Once we take the first step,
the distance to be traveled is independent of our will and
we cannot know in advance where we shall end.

—Peter Singer[1]

Offering reasons why rationality matters is a bit like blowing into your sails or lifting yourself by your bootstraps: it
cannot work unless you first accept the ground rule that
rationality is the way to decide what matters. Fortunately, as we saw
in chapter 2, we all do accept the primacy of reason, at least tacitly,
as soon as we discuss this issue, or any issue, rather than coercing
assent by force. It's now time to raise the stakes and ask whether the
conscious application of reason actually improves our lives and makes
the world a better place. It ought to, given that reality is governed by
logic and physical law rather than deviltry and magic. But do people
really suffer harm from their fallacies, and would their lives go better if they recognized and thought their way out of them? Or is gut

feeling a better guide to life decisions than cogitation, with its risk of overthinking and rationalization?

One can ask the same questions about the welfare of the world. Is progress a story of problem solving, driven by philosophers who diagnose ills and scientists and policymakers who find remedies? Or is progress a story of struggle, with the downtrodden rising up and overcoming their oppressors?[2] In earlier chapters we learned to distrust false dichotomies and single-cause explanations, so the answers to these questions will not be just one or the other. I will, though, explain why I believe that exercising our godlike reason rather than allowing it to "fust in us unus'd" can lead to a better life and a better world.

Rationality in Our Lives

Are the fallacies and illusions showcased in the preceding chapters just wrong answers to hard math problems? Are they brainteasers, gotchas, trick questions, laboratory curiosities? Or can poor reasoning lead to real harm, with the implication that critical thinking could protect people from their own worst cognitive instincts?

Certainly many of the biases we have explored would seem to be punished by reality, with all its indifference to our irrational beliefs.[3] We discount the future myopically, but it always arrives, minus the large rewards we sacrificed for the quick high. We try to recoup sunk costs, and so stay too long in bad investments, bad movies, and bad relationships. We assess danger by availability, and so avoid safe planes for dangerous cars, which we drive while texting. We misunderstand regression to the mean, and so pursue illusory explanations for successes and failures.

In dealing with money, our blind spot for exponential growth makes us save too little for retirement and borrow too much with our credit cards. Our failure to discount post hoc sharpshooting, and our misplaced trust in experts over actuarial formulas, lead us to invest in expensively managed funds which underperform simple indexes. Our difficulty with expected utility tempts us with insurance and gambles that leave us worse off in the long run.

In dealing with our health, our difficulty with Bayesian thinking can terrify us into overinterpreting a positive test for an uncommon disease. We can be persuaded or dissuaded from surgery depending on the choice of words in which the risks are framed rather than the balance of risks and benefits. Our intuitions about essences lead us to reject lifesaving vaccines and embrace dangerous quackery. Illusory correlations, and a confusion of correlation with causation, lead us to accept worthless diagnoses and treatments from physicians and psychotherapists. A failure to weigh risks and rewards lulls us into taking foolish risks with our safety and happiness.

In the legal arena, probability blindness can lure judges and juries into miscarriages of justice by vivid conjectures and post hoc probabilities. A failure to appreciate the tradeoff between hits and false alarms leads them to punish many innocents in order to convict a few more of the guilty.

In many of these cases the professionals are as vulnerable to folly as their patients and clients, showing that intelligence and expertise provide no immunity to cognitive infections. The classic illusions have been shown in medical personnel, lawyers, investors, brokers, sportswriters, economists, and meteorologists, all dealing with figures in their own specialties.[4]

These are some of the reasons to believe that failures of rationality have consequences in the world. Can the damage be quantified? The critical-thinking activist Tim Farley tried to do that on his website and Twitter feed named after the frequently asked question "What's the Harm?"[5] Farley had no way to answer it precisely, of course, but he tried to awaken people to the enormity of the damage wreaked by failures of critical thinking by listing every authenticated case he could find. From 1970 through 2009, but mostly in the last decade in that range, he documented 368,379 people killed, more than 300,000 injured, and $2.8 billion in economic damages from blunders in critical thinking. They include people killing themselves or their children by rejecting conventional medical treatments or using herbal, homeopathic, holistic, and other quack cures; mass suicides by members of apocalyptic cults; murders of witches, sorcerers, and the people they cursed; guileless victims bilked out of their savings by psychics, astrologers, and other charlatans; scofflaws and vigilantes arrested for acting on conspiratorial delusions; and economic panics from superstitions and false rumors. Here are a few tweets from 2018–19:

> What's the harm in conspiracy theories? FBI identifies "conspiracy-driven domestic extremists" as a new domestic terror threat.

> What's the harm in getting health advice from an #herbalist? A 13-year-old died after being told not to take insulin. Now the herbalist is headed to jail.

> What's the harm in a #faithhealing church? Ginnifer fought for her life for 4 hours. Travis Mitchell, her father, "laid on hands" and the family took turns praying as she struggled to breathe and changed colors. "I knew she was dead when she didn't cry out anymore," Mitchell said.

> What's the harm in believing in supernatural beings? Sumatran villagers killed an endangered tiger because they thought it was a shape-shifting "siluman."

> What's the harm in seeing a #psychic? Maryland "psychic" convicted of scamming clients out of $340K.

As Farley would be the first to note, not even thousands of anecdotes can prove that surrendering to irrational biases leads to more harm than overcoming them. At the very least we need a comparison group, namely the effects of reason-informed institutions such as medicine, science, and democratic government. That is the topic of the next section.

We do have one study of the effects of rational decision making on life outcomes. The psychologists Wändi Bruine de Bruin, Andrew Parker, and Baruch Fischhoff developed a measure of competence in reasoning and decision making (like Keith Stanovich's Rationality Quotient) by collecting tests for some of the fallacies and biases discussed in the preceding chapters.[6] These included overconfidence, sunk costs, inconsistencies in estimating risks, and framing effects (being affected by whether an outcome is described as a gain or a loss). Not surprisingly, people's skill in avoiding fallacies was correlated with their intelligence, though only partly. It was also correlated with their decision-making style—the degree to which they said they approached problems reflectively and constructively rather than impulsively and fatalistically.

To measure life outcomes, the trio developed a kind of schlimazel scale, a measure of people's susceptibility to mishaps large and small. Participants were asked, for example, whether in the past decade they had ruined clothes by not following the washing instructions on the tag, locked their keys in their car, taken the wrong train or bus, broken a bone, crashed a car, driven drunk, lost money in stocks, gotten into a fight, been suspended from school, quit a job after a week, or accidentally gotten pregnant or gotten someone pregnant. They found that people's reasoning skills did indeed predict their life outcomes: the fewer fallacies in reasoning, the fewer debacles in life.

Correlation, of course, is not causation. Reasoning competence is correlated with raw intelligence, and we know that higher intelligence protects people from bad outcomes in life such as illness, accidents, and job failure, holding socioeconomic status constant.[7] But intelligence is not the same thing as rationality, since being good at computing something is no guarantee that a person will try to compute the right things. Rationality also requires reflectiveness, open-mindedness, and mastery of cognitive tools like formal logic and mathematical probability. Bruine de Bruin and her colleagues did the multiple regression analyses (the method explained in chapter 9) and found that even when they held intelligence constant, better reasoners suffered fewer bad outcomes.[8]

Socioeconomic status, too, confounds one's fortunes in life. Poverty is an obstacle course, confronting people with the risks of unemployment, substance abuse, and other hardships. But here, too, the regression analyses showed that better reasoners had better life outcomes, holding socioeconomic status constant. All this still falls short of proving causation. But we do have some of the needed links: a high prior plausibility, two major confounds statistically controlled, and reverse causation unlikely (getting into a car crash shouldn't make you commit cognitive fallacies). This entitles us to vest some credence in the causal conclusion that competence in reasoning can protect a person from misfortunes in life.

Rationality and Material Progress

Though the availability bias hides it from us, human progress is an empirical fact. When we look beyond the headlines to the trend lines, we find that humanity overall is healthier, richer, longer-lived, better fed, better educated, and safer from war, murder, and accidents than in decades and centuries past.[9]

Having documented these changes in two books, I'm often asked whether I "believe in progress." The answer is no. Like the humorist Fran Lebowitz, I don't believe in anything you have to believe in. Though many measures of human well-being, when plotted over time, show a gratifying increase (though not always or everywhere), it's not because of some force or dialectic or evolutionary law that lifts us ever upward. On the contrary, nature has no regard for our well-being, and often, as with pandemics and natural disasters, it looks as if it's trying to grind us down. "Progress" is shorthand for a set of pushbacks and victories wrung out of an unforgiving universe, and is a phenomenon that needs to be explained.

The explanation is rationality. When humans set themselves the goal of improving the welfare of their fellows (as opposed to other dubious pursuits like glory or redemption), and they apply their ingenuity in institutions that pool it with others', they occasionally succeed. When they retain the successes and take note of the failures, the benefits can accumulate, and we call the big picture progress.

We can start with the most precious thing of all, life. Beginning in the second half of the nineteenth century, life expectancy at birth rose from its historic resting place at around 30 years and is now 72.4 years worldwide, 83 years in the most fortunate countries.[10] This gift of life was not dropped onto our doorsteps. It was the hard-won dividend of advances in public health (motto: "Saving lives, millions at a time"), particularly after the germ theory of disease displaced other causal theories like miasmas, spirits, conspiracies, and divine retribution. The lifesavers included chlorination and other means of safeguarding drinking water, the lowly toilet and sewer, the control of disease vectors like mosquitoes and fleas, programs for large-scale vaccination, the promotion of hand-washing, and basic prenatal and perinatal care such as nursing and body contact. When disease and injuries do strike, advances in medicine keep them from killing as many people as they did in the era of folk healers and barber-surgeons, including antibiotics, antisepsis, anesthesia, transfusions, drugs, and oral rehydration therapy (a salt and sugar solution that stops fatal diarrhea).

Humanity has always struggled to grow enough calories and protein to feed itself, with famine just one bad harvest away. But hunger today has been decimated in most of the world: undernourishment and stunting are in decline, and famines now afflict only the most remote and war-ravaged regions, a problem not of too little food but of barriers to getting it to the hungry.[11] The calories did not come in heavenly manna or from a cornucopia held by Abundantia, the Roman goddess of plenty, but from advances in agronomy. These included crop rotation to replenish depleted soils; technologies for high-throughput planting and harvesting such as seed drills, plows, tractors, and combine harvesters; synthetic fertilizer (credited with saving 2.7 billion lives); a transportation and storage network to bring food from farm to table, including railroads, canals, trucks, granaries, and refrigeration; national and international markets that allow a surplus in one area to fill a shortage in another; and the Green Revolution of the 1960s, which spread productive and vigorous hybrid crops.

Poverty needs no explanation; it is the natural state of humankind. What needs an explanation is wealth. For most of human history, around 90 percent of humanity lived in what we today call extreme poverty. In 2020, less than 9 percent do; still too high, but targeted for elimination in the next decade.[12] The great material enrichment of humanity began with the Industrial Revolution of the nineteenth century. It was literally powered by the capture of energy from coal, oil, wind, and falling water, and later from the sun, the earth, and nuclear fission. The energy was fed into machines that turn heat into work, factories with mass production, and conveyances like railroads, canals, highways, and container ships. Material technologies depended on financial ones, particularly banking, finance, and insurance. And neither of these could have been parlayed into widespread prosperity without governments to enforce contracts, minimize force and fraud, smooth out financial lurches with central banks and reliable money, and invest in wealth-generating public goods such as infrastructure, basic research, and universal education.

The world has not yet put an end to war, as the folk singers of the 1960s dreamed, but it has dramatically reduced their number and lethality, from a toll of 21.9 battle deaths per 100,000 people in 1950 to just 0.7 in 2019.[13] Peter, Paul, and Mary deserve only some of the credit. More goes to institutions that were designed to reduce the incentives of nations to go to war, beginning with Immanuel Kant's plan for "perpetual peace" in 1795. One of them is democracy, which, as we saw in the chapter on correlation and causation, really does reduce the chance of war, presumably because a country's cannon fodder is less keen on the pastime than its kings and generals. Another is international trade and investment, which make it cheaper to buy things than to steal them, and make it unwise for countries to kill their customers and debtors. (The European Union, awarded the Nobel Peace Prize in 2012, grew out of a trade organization, the European Coal and Steel Community.) Yet another is a network of international organizations, particularly the United Nations, which knits countries into a community, mobilizes peacekeeping forces, immortalizes states, grandfathers in borders, and outlaws and stigmatizes war while providing alternative means of resolving disputes.

Brainchildren of human ingenuity have also underwritten other historical boosts in well-being, such as safety, leisure, travel, and access to art and entertainment. Though many of the gadgets and bureaucracies grew organically and were perfected through trial and error, none was an accident. People at the time advocated for them with arguments driven by logic and evidence, costs and benefits, cause and effect, and tradeoffs between individual advantage and the common good. Our ingenuity will have to be redoubled to deal with the trials we face today, particularly the Tragedy of the Carbon Commons (chapter 8). Brainpower will have to be applied to technologies that make clean energy cheap, pricing that makes dirty energy expensive, policies that prevent factions from becoming spoilers, and treaties to make the sacrifices global and equitable.[14]

Rationality and Moral Progress

Progress consists of more than gains in safety and material well-being. It consists also of gains in how we treat each other: in equality, benevolence, and rights. Many cruel and unjust practices have declined over the course of history. They include human sacrifice, slavery, despotism, blood sports, eunuchism, harems, foot-binding, sadistic corporal and capital punishments, the persecution of heretics and dissidents, and the oppression of women and of religious, racial, ethnic, and sexual minorities.[15] None has been extirpated from the face of the earth, but when we chart the historical changes, in every case we see descents and, in some cases, plunges.

How did we come to enjoy this progress? Theodore Parker, and a century later Martin Luther King Jr., divined a moral arc bending toward justice. But the nature of the arc and its power to pull the levers of human behavior are mysterious. One can imagine more prosaic pathways: changing fashions; shaming campaigns; appeals to the heart; popular protest movements; religious and moralistic crusades. A popular view is that moral progress is advanced through struggle: the powerful never hand over their privileges, which must be wrested from them by the might of people acting in solidarity.[16]

My greatest surprise in making sense of moral progress is how many times in history the first domino was a reasoned argument.[17] A philosopher wrote a brief which laid out arguments on why some practice was indefensible, or irrational, or inconsistent with values that everyone claimed to hold. The pamphlet or manifesto went viral, was translated into other languages, was debated in pubs and salons and coffeehouses, and then influenced leaders, legislators, and popular opinion. Eventually the conclusion was absorbed into the conventional wisdom and common decency of a society, erasing the tracks of the arguments that brought it there. Few people today feel the need, or could muster the ability, to formulate a coherent argument on why slavery is wrong, or public disembowelment, or the beating of children; it's just obvious. Yet exactly those debates took place centuries ago.

And the arguments that prevailed, when they are brought to our attention today, continue to ring true. They appeal to a sense of reason that transcends the centuries, because they conform to principles of conceptual consistency that are part of reality itself. Now, as we saw in chapter 2, no logical argument can establish a moral claim. But an argument *can* establish that a claim under debate is inconsistent with another claim a person holds dear, or with values like life and happiness that most people claim for themselves and would agree are legitimate desires of everyone else. As we saw in chapter 3, inconsistency is fatal to reasoning: a set of beliefs that includes a contradiction can be deployed to deduce anything and is perfectly useless.

Wary as I must be of inferring causation from correlation, and of singling out just one cause in a crisscrossing historical mesh, I cannot claim that good arguments are the cause of moral progress. We cannot do a randomized controlled trial on history, with half of a sample of societies exposed to a compelling moral treatise and the other half given a placebo filled with high-minded mumbo jumbo. Nor do we have a large enough dataset of moral triumphs to extract a causal conclusion from the network of correlations. (The closest I can think of are cross-national studies showing that education and access to information in one era, which are indicators of a readiness to exchange ideas, predict democracy and liberal values in a later one, holding socioeconomic confounds constant.)[18] For now I can only give examples of precocious arguments that historians tell us were influential in their day and that remain unimpeachable in ours.

LET'S BEGIN WITH RELIGIOUS PERSECUTION. Did people really need an intellectual argument to understand why something might be a wee bit wrong with burning heretics at the stake? In fact they did. In 1553 the French theologian Sebastian Castellio (1515–1563) composed an argument against religious intolerance, noting the absence of reasoning behind John Calvin's orthodoxies and the "logical outcome" of his practices:

> Calvin says that he is certain, and [other sects] say they are; Calvin says that they are wrong and wishes to judge them, and so do they. Who shall be judge? Who made Calvin the arbiter of all the sects, that he alone should kill? He has the Word of God and so have they. If the matter is certain, to whom is it so? To Calvin? But then why does he write so many books about manifest truth? ... In view of all the uncertainty we must define the heretic simply as one with whom we disagree. And if then we are going to kill heretics, the logical outcome will be a war of extermination, since each is

sure of himself. Calvin would have to invade France and all other nations, wipe out cities, put all the inhabitants to the sword, sparing neither sex nor age, not even babies and the beasts.[19]

The sixteenth century saw another precocious argument against a barbaric practice. Today it seems obvious that war is not healthy for children and other living things. But for most of history, war was seen as noble, holy, thrilling, manly, glorious.[20] Though it was only after the cataclysms of the twentieth century that war ceased to be venerated, the seeds of pacifism had been planted by one of the "fathers of modernity," the philosopher Desiderius Erasmus (1466–1536), in his 1517 essay "The Plea of Reason, Religion, and Humanity against War." After giving a poignant account of the blessings of peace and the horrors of war, Erasmus turned to a rational-choice analysis of war, explaining its zero-sum payoffs and negative expected utility:

> To these considerations add that the advantages derived from peace diffuse themselves far and wide, and reach *great numbers*; while in *war*, if any thing turns out happily . . . the advantage redounds only to a *few*, and those unworthy of reaping it. One man's safety is owing to the destruction of another; one man's prize is derived from the plunder of another. The cause of rejoicings made by one side is to the other a cause of mourning. Whatever is unfortunate in war is severely so indeed, and whatever, on the contrary, is called good fortune, is a savage and a cruel good fortune, an ungenerous happiness, deriving its existence from another's woe. Indeed, at the conclusion, it commonly happens that both sides, the victorious and the vanquished, have cause to deplore. I know not whether any war ever succeeded so

fortunately in all its events but that the conqueror, if he had a heart to feel or an understanding to judge, as he ought to do, repented that he ever engaged in it at all. . . .

If we were to calculate the matter fairly, and form a just computation of the cost attending war and that of procuring peace, we should find that peace might be purchased at a tenth part of the cares, labors, troubles, dangers, expenses, and blood that it costs to carry on a war. . . .

But the object is to do all possible injury to an enemy. A most inhuman object . . . And consider whether you can hurt him essentially without hurting, at the same time, and by the same means, your own people. It surely is to act like a madman to take to yourself so large a portion of certain evil when it must ever be uncertain how the die of war may fall in the ultimate issue.[21]

The eighteenth-century Enlightenment was a font of arguments against other kinds of cruelty and oppression. As with religious persecution, we are left almost speechless when asked what's wrong with the use of sadistic torture for criminal punishment, such as drawing and quartering, breaking on the wheel, burning at the stake, or sawing someone in half from the crotch up. But in a 1764 pamphlet the economist and utilitarian philosopher Cesare Beccaria (1738–1794) laid out arguments against those barbarities by identifying the costs and benefits of criminal punishment. The legitimate goal of punishment, Beccaria argued, is to incentivize people not to exploit others, and the expected utility of wrongdoing should be the metric by which we assess punitive practices.

As punishments become more cruel, the minds of men, which like fluids always adjust to the level of their surroundings, become hardened, and the ever lively power of the emotions brings it about that after a hundred years of cruel tortures, the wheel causes no more fear than prison previously did. For a punishment to serve its purpose, it is only necessary that the harm that it inflicts outweighs the benefit that the criminal can derive from the crime, and into the calculation of this balance, we must add the certainty of punishment and the loss of the good produced by the crime. Anything more than this is superfluous, and therefore tyrannical.[22]

Beccaria's argument, and those by fellow philosophes Voltaire and Montesquieu, influenced the prohibition of "cruel and unusual punishments" by the Eighth Amendment to the US Constitution. In recent years the amendment continues to be invoked to chip away at the range of executions in America, and many legal observers believe it's only a matter of time before the entire practice is ruled unconstitutional.[23]

Other forms of barbarism were also targeted during the Enlightenment by arguments that remain pungent to this day. The other great eighteenth-century utilitarian, Jeremy Bentham (1748–1832), composed the first systematic argument against the criminalization of homosexuality:

As to any primary mischief, it is evident that it produces no pain in anyone. On the contrary it produces pleasure. . . . The partners are both willing. If either of them be unwilling, the act is not that which we have here in view: it is an offence totally different in its nature of effects: it is a personal injury; it is a kind of rape. . . . As to any danger exclusive of pain, the danger, if any, must consist in the tendency of the example. But what is the tendency of this example? To dispose others to engage in the same practises: but this practise for anything that has yet appeared produces not pain of any kind to any one.[24]

Bentham also stated the argument against cruelty to animals in a way that continues to guide the animal protection movement today:

The day *may* come, when the rest of the animal creation may acquire those rights which never could have been withholden from them but by the hand of tyranny. The French have already discovered that the blackness of the skin is no reason why a human being should be abandoned without redress to the caprice of a tormentor. It may come one day to be recognized, that the number of the legs, the villosity [hairiness] of the skin, or the termination of the *os sacrum* [tailbone] are reasons equally insufficient for abandoning a sensitive being to the same fate. What else is it that should trace the insuperable line? Is it the faculty of reason, or, perhaps, the faculty of discourse? But a full-grown horse or dog is beyond comparison a more rational, as well as a more conversable animal, than an infant of a day, or a week, or even a month old. But suppose the case were otherwise, what would it avail? The question is not, Can they *reason*? nor, Can they *talk*? but, Can they *suffer*?[25]

Bentham's juxtaposition of the morally irrelevant differences in skin color among humans with the differences in physical and cognitive traits among species is no mere simile. It is a goad to question our instinctive response to the superficial features of the entities we are being asked to consider (the reaction from System 1, if you will) and to reason our way to coherent beliefs on who is deserving of rights and protections.

The prodding of cognitive reflection by analogizing a protected group with a vulnerable one is a common means by which moral persuaders have awakened people to their biases and bigotries. The philosopher Peter Singer, an intellectual descendant of Bentham and today's foremost proponent of animal rights, calls the process "the expanding circle."[26]

Slavery was a common frame of reference. The Enlightenment hosted a vigorous abolitionist movement, initiated by arguments from Jean Bodin (1530–1596), John Locke (1632–1704), and Montesquieu (1689–1755).[27] With the latter two, their case against slavery also underlay their criticism of absolute monarchy and their insistence that governments are legitimately empowered only by the consent of the governed. The starting point was to undermine the assumption of a natural hierarchy: any ranking of aristocrat and commoner, lord and vassal, owner and slave. "We are born free," Locke wrote, "as we are born rational."[28] Humans are inherently thinking, sentient, volitional beings, none possessing a natural right to dominate any other. In his chapter on slavery in *Two Treatises of Government*, Locke elaborates:

> Freedom of men under government is, to have a standing rule to live by, common to every one of that society, and made by the legislative power erected in it; a liberty to follow my own will in all things, where the rule prescribes not; and not to be subject to the inconstant, uncertain, unknown, arbitrary will of another man: as freedom of nature is, to be under no other restraint but the law of nature.[29]

The keystone idea that equality is the default relationship among people was co-opted by Thomas Jefferson (1743–1826) as the justification for democratic government: "We hold these truths to be self-evident, that all men are created equal, that they are endowed by their Creator with certain unalienable Rights, that among these are Life, Liberty and the pursuit of Happiness. That to secure these rights, Governments are instituted among Men, deriving their just powers from the consent of the governed."

While Locke may have anticipated that his writings would inspire one of the great developments in human history, the rise of democracy, he may not have anticipated another one it would inspire. In her 1730 preface to *Some Reflections upon Marriage*, the philosopher Mary Astell (1666–1731) wrote:

> If absolute Sovereignty be not necessary in a State how comes it to be so in a Family? or if in a Family why not in a State? Since no reason can be alleg'd for the one that will not hold more strongly for the other. . . . If all Men are born free, how is it that all Women are born slaves? As they must be if the being subjected to the inconstant, uncertain, unknown, arbitrary Will of Men, be the perfect Condition of Slavery?[30]

Sound familiar? Astell shrewdly appropriated Locke's argument (including his phrase "the perfect condition of slavery") to undermine the oppression of women, making her the first English feminist. Long before it became an organized movement, feminism began as an argument, picked up after Astell by the philosopher Mary Wollstonecraft (1759–1797). In *A Vindication of the Rights of Woman* (1792), Wollstonecraft not only extended the argument that it was logically inconsistent to deny women the rights granted to men, but

argued that any assumption that women were inherently less intellectual or authoritative than men was spurious because of a confound
between nature and nurture: women were raised without the education and opportunities afforded to men. She began her book with an
open letter to Talleyrand, a major figure in the French Revolution,
who had argued that, *égalité shmégalité*, girls don't need a formal education:

> Consider, I address you as a legislator, whether, when men
> contend for their freedom, and to be allowed to judge for
> themselves, respecting their own happiness, it be not in-
> consistent and unjust to subjugate women, even though you
> firmly believe that you are acting in the manner best calcu-
> lated to promote their happiness? Who made man the exclu-
> sive judge, if woman partake with him the gift of reason?
>
> In this style, argue tyrants of every denomination from
> the weak king to the weak father of a family; they are all eager
> to crush reason; yet always assert that they usurp its throne
> only to be useful. Do you not act a similar part, when you
> force all women, by denying them civil and political rights, to
> remain immured in their families groping in the dark? For
> surely, sir, you will not assert, that a duty can be binding
> which is not founded on reason? If, indeed, this be their des-
> tination, arguments may be drawn from reason; and thus au-
> gustly supported, the more understanding women acquire,
> the more they will be attached to their duty, comprehending
> it, for unless they comprehend it, unless their morals be fixed
> on the same immutable principle as those of man, no author-
> ity can make them discharge it in a virtuous manner. They
> may be convenient slaves, but slavery will have its constant
> effect, degrading the master and the abject dependent.[31]

And speaking of slavery itself, the truly commanding arguments against the abominable institution came from the writer, editor, and statesman Frederick Douglass (1818–1895). Himself born into slavery, Douglass could searingly engage his audiences' empathy for the suffering of the enslaved, and as one of history's greatest orators, he could stir them with the music and imagery of his speech. Yet Douglass deployed these gifts in the service of rigorous moral argumentation. In his most famous speech, "What to the Slave Is the Fourth of July?" (1852), Douglass apophatically rejected any need to provide arguments against slavery using "the rules of logic" because, he said, they were obvious, before proceeding to do exactly that. For example:

> There are seventy-two crimes in the State of Virginia, which, if committed by a black man, (no matter how ignorant he be), subject him to the punishment of death; while only two of the same crimes will subject a white man to the like punishment. What is this but the acknowledgement that the slave is a moral, intellectual and responsible being? The manhood of the slave is conceded. It is admitted in the fact that Southern statute books are covered with enactments forbidding, under severe fines and penalties, the teaching of the slave to read or to write. When you can point to any such laws, in reference to the beasts of the field, then I may consent to argue the manhood of the slave.[32]

Douglass continued, "At a time like this, scorching irony, not convincing argument, is needed," and he then confronted his audience with a copious inventory of inconsistencies in their belief systems:

> You hurl your anathemas at the crowned headed tyrants of Russia and Austria, and pride yourselves on your Demo-

cratic institutions, while you yourselves consent to be the mere tools and body-guards of the tyrants of Virginia and Carolina. You invite to your shores fugitives of oppression from abroad, honor them with banquets, greet them with ovations, cheer them, toast them, salute them, protect them, and pour out your money to them like water; but the fugitives from your own land you advertise, hunt, arrest, shoot and kill. . . .

You can bare your bosom to the storm of British artillery to throw off a threepenny tax on tea; and yet wring the last hard-earned farthing from the grasp of the black laborers of your country.

And foreshadowing Martin Luther King more than a century later, he held the nation to its founding declaration:

You declare, before the world, and are understood by the world to declare, that you "hold these truths to be self evident, that all men are created equal; and are endowed by their Creator with certain inalienable rights; and that, among these are, life, liberty, and the pursuit of happiness;" and yet, you hold securely, in a bondage which, according to your own Thomas Jefferson, "is worse than ages of that which your fathers rose in rebellion to oppose," a seventh part of the inhabitants of your country.

That Douglass and King could approvingly quote Jefferson, himself a hypocritical and in some ways dishonorable man, does not compromise the rationality of their arguments but reinforces it. We should care about people's virtue when considering them as friends, but not when considering the ideas they voice. Ideas are true or false, consistent or contradictory, conducive to human welfare or not, regardless of who thinks them. The equality of sentient beings, grounded in the logical irrelevance of the distinction between "me" and "you," is an idea that people through the ages rediscover, pass along, and extend to new living things, expanding the circle of sympathy like moral dark energy.

Sound arguments, enforcing a consistency of our practices with our principles and with the goal of human flourishing, cannot improve the world by themselves. But they have guided, and should guide, movements for change. They make the difference between moral force and brute force, between marches for justice and lynch mobs, between human progress and breaking things. And it will be sound arguments, both to reveal moral blights and discover feasible remedies, that we will need to ensure that moral progress will continue, that the abominable practices of today will become as incredible to our descendants as heretic burnings and slave auctions are to us.

The power of rationality to guide moral progress is of a piece with its power to guide material progress and wise choices in our lives. Our ability to eke increments of well-being out of a pitiless cosmos and to be good to others despite our flawed nature depends on grasping impartial principles that transcend our parochial experience. We are a species that has been endowed with an elementary faculty of reason and that has discovered formulas and institutions that magnify its scope. They awaken us to ideas and expose us to realities that confound our intuitions but are true for all that.

NOTES

CHAPTER 1: HOW RATIONAL AN ANIMAL?

1. Russell 1950/2009.
2. Spinoza 1677/2000, *Ethics*, III, preface.
3. Data on human progress: Pinker 2018.
4. Kalahari San: Lee & Daly 1999. The San, previously known as the Bushmen, comprise the Ju/ 'hoan (formerly !Kung), Tuu, Gana, /Gwi, and Khoi peoples, variously spelled.
5. Hunter-gatherers: Marlowe 2010.
6. Liebenberg works with the !Xõ, /Gwi, Khomani, and Ju/ 'hoan (formerly !Kung) San. Examples here are from the !Xõ. Liebenberg's experiences with the San, and his theory that scientific thinking evolved from tracking, are presented in *The Origin of Science* (2013/2021), *The Art of Tracking* (1990), and Liebenberg, //Ao, et al. 2021. Additional examples are from Liebenberg 2020. For other descriptions of hunter-gatherer rationality, see Chagnon 1997; Kingdon 1993; Marlowe 2010.
7. A video of a pursuit hunt, narrated by David Attenborough, may be seen here: https://youtu.be/826HMLoiE_o.
8. Liebenberg 2013/2021, p. 57.
9. Personal communication from Louis Liebenberg, Aug. 11, 2020.
10. Liebenberg 2013/2021, p. 104.
11. Liebenberg 2020 and personal communication, May 27, 2020.
12. Moore 2005. See also Pew Forum on Religion and Public Life 2009, and note 8 to chapter 10 below.
13. Vosoughi, Roy, & Aral 2018.
14. Pinker 2010; Tooby & DeVore 1987.

15. Amos Tversky (1937–1996) and Daniel Kahneman (1934–) pioneered the study of cognitive illusions and biases; see Tversky & Kahneman 1974, Kahneman, Slovic, & Tversky 1982, Hastie & Dawes 2010, and Kahneman's best-seller, *Thinking, Fast and Slow* (2011). Their lives and collaboration are described in Michael Lewis's *The Undoing Project* (2016) and Kahneman's autobiographical statement for his 2002 Nobel Prize (Kahneman 2002).

16. Frederick 2005.

17. The psychologists Philip Maymin and Ellen Langer have shown that simply asking people to be mindful of their visual surroundings reduced reasoning errors in 19 of 22 classic problems from the cognitive psychology literature.

18. Frederick 2005.

19. Frederick 2005, p. 28. Actually, "A banana and a bagel cost 37 cents. The banana costs 13 cents more than the bagel. How much does the bagel cost?"

20. Wagenaar & Sagaria 1975; Wagenaar & Timmers 1979.

21. Goda, Levy, et al. 2015; Stango & Zinman 2009.

22. Citations omitted to spare embarrassment to two friends.

23. US deaths (7-day rolling average): Roser, Ritchie, et al. 2020, accessed Aug. 23, 2020. American lethal hazards: Ritchie 2018, accessed Aug. 23, 2020; data are from 2017.

24. Wason 1966; see also Cosmides 1989; Fiddick, Cosmides, & Tooby 2000; Mercier & Sperber 2011; Nickerson 1996; Sperber, Cara, & Girotto 1995.

25. van Benthem 2008, p. 77.

26. Since, logically speaking, the P choice could disconfirm the rule as easily as the not-Q choice, the explanation in terms of confirmation bias is a bit subtler: participants deploy reasoning to justify their initial, intuitive choice, whatever it is; see Nickerson 1998 and Mercier & Sperber 2011. Winning arguments: Dawson, Gilovich, & Regan 2002; Mercier & Sperber 2011.

27. Quoted in Grayling 2007, p. 102.

28. From *Novum Organum*, Bacon 1620/2017.

29. Popper 1983. Wason task vs. scientific hypothesis-testing: Nickerson 1996.

30. Peculiarity of the selection task: Nickerson 1996; Sperber, Cara, & Girotto 1995.

31. Cheng & Holyoak 1985; Cosmides 1989; Fiddick, Cosmides, & Tooby 2000; Stanovich & West 1998. A different take: Sperber, Cara, & Girotto 1995.

32. Ecological rationality: Gigerenzer 1998; Tooby & Cosmides 1993; see Pinker 1997/2009, pp. 302–6.

33. The problem was originated by the recreational mathematician Martin Gardner (1959), who called it the Three Prisoners problem; it was named after Monty Hall by the statistician Steven Selvin (1975).

34. Granberg & Brown 1995; Saenen, Heyvaert, et al. 2018.

35. Crockett 2015; Granberg & Brown 1995; Tierney 1991; vos Savant 1990.

36. Crockett 2015.

37. Vazsonyi 1999. My Erdös number is 3, thanks to Michel, Shen, Aiden, Veres, Gray, The Google Books Team, Pickett, Hoiberg, Clancy, Norvig, Orwant,

Pinker, Nowak, & Lieberman-Aiden 2011. The computer scientist Peter Norvig has coauthored a report with fellow computer scientist (and Erdös coauthor) Maria Klawe.

38. To be fair, normative analyses of the Monty Hall dilemma have inspired voluminous commentary and disagreement; see https://en.wikipedia.org/wiki /Monty_Hall_problem.

39. Try it: Math Warehouse, "Monty Hall Simulation Online," https://www .mathwarehouse.com/monty-hall-simulation-online/.

40. Such as *Late Night with David Letterman*: https://www.youtube.com/watch ?v=EsGc3jC9yas.

41. Vazsonyi 1999.

42. Suggested by Granberg & Brown 1995.

43. Rules of conversation: Grice 1975; Pinker 2007, chap. 8.

44. History and concepts of probability: Gigerenzer, Swijtink, et al. 1989.

45. vos Savant 1990.

46. Thanks to Julian De Freitas for running and analyzing the study. The design was similar to one summarized informally in Tversky & Kahneman 1983, pp. 307–8. The items here were chosen from a larger set pretested in a pilot study. The differences were found in comparisons of the ratings participants gave either for the conjunction or for the single conjunct before they had seen the other one (that is, in a between-participants comparison). When we compared the ratings of both items by the same participant (a within-participant comparison), the conjunction fallacy was seen only with the Russia and Venezuela items. Still, 86 percent of the participants committed at least one conjunction error, and with every item, a majority of participants rated the probability of the conjunction as greater than or equal to the probability of the conjunct.

47. Donaldson, Doubleday, et al. 2011; Tetlock & Gardner 2015.

48. Kaplan 1994.

49. Declines in war, crime, poverty, and disease: Pinker 2011; Pinker 2018.

50. Tversky & Kahneman 1983.

51. Gould 1988.

52. Quoted by Tversky & Kahneman 1983, p. 308.

53. Tversky & Kahneman 1983, p. 313.

54. Quoted in Hertwig & Gigerenzer 1999.

55. Hertwig & Gigerenzer 1999.

56. Hertwig & Gigerenzer 1999; Tversky & Kahneman 1983.

57. Kahneman & Tversky 1996.

58. Mellers, Hertwig, & Kahneman 2001.

59. Purves & Lotto 2003.

60. AI fails: Marcus & Davis 2019.

61. Pinker 1997/2009, chaps. 1, 4.

62. Pinker 2015.

63. Federal Aviation Administration 2016, chap. 17.

CHAPTER 2: RATIONALITY AND IRRATIONALITY

1. Justified true belief, and counterexamples showing that it is necessary but not sufficient for knowledge: Gettier 1963; Ichikawa & Steup 2018.
2. James 1890/1950.
3. Carroll 1895.
4. Just do it: Fodor 1968; Pinker 1997/2009, chap. 2.
5. Nagel 1997.
6. Myers 2008.
7. For many examples, see the sources in note 79 to chapter 10 below.
8. Stoppard 1972, p. 30.
9. Hume 1739/2000, book II, part III, section III, "Of the influencing motives of the will."
10. Cohon 2018.
11. Though that's not what he literally believed about taste in art and wine, as expressed in "Of the standard of taste" (Gracyk 2020). His point here was only that goals are inherently subjective.
12. Bob Dylan, "Mr. Tambourine Man."
13. Pinker 1997/2009; Scott-Phillips, Dickins, & West 2011.
14. Ainslie 2001; Schelling 1984.
15. Mischel & Baker 1975.
16. Ainslie 2001; Laibson 1997; Schelling 1984. See also Pinker 2011, chap. 9, "Self-Control."
17. Frederick 2005.
18. Jeszeck, Collins, et al. 2015.
19. Dasgupta 2007; Nordhaus 2007; Varian 2006; Venkataraman 2019.
20. MacAskill 2015; Todd 2017.
21. Venkataraman 2019.
22. Ainslie 2001; Laibson 1997.
23. McClure, Laibson, et al. 2004.
24. Homer 700 BCE/2018, translation by Emily Wilson.
25. Baumeister & Tierney 2012.
26. Nudges and other behavioral insights: Hallsworth & Kirkman 2020; Thaler & Sunstein 2008. Nudge skeptics: Gigerenzer 2015; Kahan 2013.
27. Rational ignorance: Gigerenzer 2004; Gigerenzer & Garcia-Retamero 2017; Hertwig & Engel 2016; Williams 2020; see also Pinker 2007, pp. 422–25.
28. Schelling 1960.
29. Chicken: J. S. Goldstein 2010. The game played in the movie is a bit different: the teenagers drive their cars toward a cliff, each trying to jump out second.
30. Hotheadedness as a paradoxical tactic: Frank 1988; see also Pinker 1997/2009, chap. 6.
31. Sagan & Suri 2003.
32. Crazy love as a paradoxical tactic: Frank 1988; Pinker 1997/2009, chap. 6, "Fools for Love."
33. Novel by Dashiell Hammett; screenplay by John Huston.

34. Tetlock 2003; Tetlock, Kristel, et al. 2000.

35. Satel 2008.

36. For example, Block 1976/2018.

37. Reframing taboo tradeoffs: Tetlock 2003; Tetlock, Kristel, et al. 2000; Zelizer 2005.

38. Hume 1739/2000, book II, part III, section III, "Of the influencing motives of the will." Hume's moral philosophy: Cohon 2018.

39. Rachels & Rachels 2010.

40. Stoppard 1972, p. 39.

41. Gould 1999.

42. Plato 399-390 BCE/2002. Plato's moral philosophy brought to life: R. Goldstein 2013.

43. God commands child murder: Pinker 2011, chap. 1.

44. "'Tis as little contrary to reason to prefer even my own acknowledg'd lesser good to my greater, and have a more ardent affection for the former than the latter."

45. Morality as impartiality: de Lazari-Radek & Singer 2012; R. Goldstein 2006; Greene 2013; Nagel 1970; Railton 1986; Singer 1981/2011.

46. Terry 2008.

47. Self-interest, sociality, and rationality as sufficient conditions for morality: Pinker 2018, pp. 412-15. Morality as a strategy in positive-sum games: Pinker 2011, pp. 689-92.

48. Chomsky 1972/2006; Pinker 1994/2007, chap. 4.

CHAPTER 3: LOGIC AND CRITICAL THINKING

1. Eliot 1883/2017, pp. 257-58.

2. Leibniz 1679/1989.

3. Accessible introductions to logic: McCawley 1993; Priest 2017; Warburton 2007.

4. Based on Carroll 1896/1977, book II, chap. III, §2, example (4), p. 72.

5. Donaldson, Doubleday, et al. 2011.

6. Logical words in logic versus conversation: Grice 1975; Pinker 2007, chaps. 2, 8.

7. Emerson 1841/1993.

8. Liberman 2004.

9. McCawley 1993.

10. From the *Yang 2020* website, retrieved Feb. 6, 2020: Yang 2020.

11. Curtis 2020; Richardson, Smith, et al. 2020; Warburton 2007; see also the *Wikipedia* article "List of fallacies," https://en.wikipedia.org/wiki/List_of_fallacies.

12. Mercier & Sperber 2011; see Norman 2016, for a critique.

13. Friedersdorf 2018.

14. Shackel 2014.

15. Russell 1969.

16. Basterfield, Lilienfeld, et al. 2020.

17. A common saying loosely based on a passage from Henrik Ibsen's *Enemy of the People*: "The majority never has right on its side. . . . The majority has might on its side—unfortunately; but right it has not."

18. Proctor 2000.

19. For discussion of one example, see Paresky, Haidt, Strossen & Pinker 2020.

20. Haidt 2016.

21. The story is found in many textbooks, usually attributed to Francis Bacon in 1592, but its real source, even as a parody, is obscure, and probably from the early twentieth century; see Simanek 1999.

22. Ecological rationality: Gigerenzer 1998; Pinker 1997/2009, pp. 302–6; Tooby & Cosmides 1993.

23. Cosmides 1989; Fiddick, Cosmides, & Tooby 2000.

24. Weber 1922/2019.

25. Cole, Gay, et al. 1971, pp. 187–88; see also Scribner & Cole 1973.

26. Norenzayan, Smith, et al. 2002.

27. Wittgenstein 1953.

28. Not all philosophers agree: Bernard Suits (1978/2014) defines a game as "the voluntary attempt to overcome unnecessary obstacles." See also McGinn 2012, chap. 2.

29. Pinker 1997/2009, pp. 306–13; Pinker 1999/2011, chap. 10; Pinker & Prince 2013; Rosch 1978.

30. Armstrong, Gleitman, & Gleitman 1983; Pinker 1999/2011, chap 10; Pinker & Prince 2013.

31. Goodfellow, Bengio, & Courville 2016; Rumelhart, McClelland, & PDP Research Group 1986; Aggarwal 2018. For critical views, see Marcus & Davis 2019; Pearl & Mackenzie 2018; Pinker 1999/2011; Pinker & Mehler 1988.

32. Rumelhart, Hinton, & Williams 1986; Aggarwal 2018; Goodfellow, Bengio, & Courville 2016.

33. Lewis-Kraus 2016.

34. The word "algorithm" was originally reserved for such formulas, and they were contrasted with "heuristics" or rules of thumb. But in common parlance today, the word is used for all AI systems, including ones based on neural networks.

35. Marcus & Davis 2019.

36. Kissinger 2018.

37. Lake, Ullman, et al. 2017; Marcus 2018; Marcus & Davis 2019; Pearl & Mackenzie 2018.

38. Ashby, Alfonso-Reese, et al. 1998; Evans 2012; Kahneman 2011; Marcus 2000; Pinker 1999/2011; Pinker & Prince 2013; Sloman 1996.

39. Pinker 1999/2011, chap. 10; Pinker & Prince 2013.

CHAPTER 4: PROBABILITY AND RANDOMNESS

1. Letter to Miss Sophia Thrale, 24 July 1783, in Johnson 1963.

2. *Bartlett's Familiar Quotations*. The citation does not lead to a primary source, but it was probably a letter to Max Born in 1926. A variant occurs in a letter

to Cornelius Lanczos, quoted in Einstein 1981, and three more may be found in Einstein's *Wikiquote* entry, https://en.wikiquote.org/wiki/Albert_Einstein.

3. Eagle 2019; randomness as incompressibility, usually called Kolmogorov complexity, is discussed in section 2.2.1.

4. Millenson 1965.

5. Gravity poster: http://www.mooneyart.com/gravity/historyof_01.html.

6. Gigerenzer, Hertwig, et al. 2005.

7. Quoted in Bell 1947.

8. Interpretations of probability: Gigerenzer 2008a; Gigerenzer, Swijtink, et al. 1989; Hájek 2019; Savage 1954.

9. Quoted in Gigerenzer 1991, p. 92.

10. Gigerenzer 2008a.

11. Tversky & Kahneman 1973.

12. Gigerenzer 2008a.

13. Combs & Slovic 1979; Ropeik 2010; Slovic 1987.

14. McCarthy 2019.

15. Duffy 2018; see also Ropeik 2010; Slovic 1987.

16. Figures from 2014–15, referenced in Pinker 2018, table 13-1, p. 192. See also Ritchie 2018; Roth, Abate, et al. 2018.

17. Savage 2013, table 2. The figure is for commercial aviation in the United States.

18. Gigerenzer 2006.

19. "Mack the Knife," lyrics by Bertolt Brecht, from *The Threepenny Opera*.

20. Cape Cod sharks: Sherman 2019. Cape Cod traffic deaths: Nolan, Bremer, et al. 2019.

21. Caldeira, Emanuel, et al. 2013. See also Goldstein & Qvist 2019; Goldstein, Qvist, & Pinker 2019.

22. Nuclear vs coal: Goldstein & Qvist 2019; Goldstein, Qvist, & Pinker 2019. Coal kills: Lockwood, Welker-Hood, et al. 2009. Nuclear replaced by coal: Jarvis, Deschenes, & Jha 2019. Even if we accept recent claims that authorities covered up thousands of Chernobyl deaths, the death toll from sixty years of nuclear power would still equal about one month of coal-related deaths.

23. Ropeik 2010; Slovic 1987.

24. Pinker 2018, table 13-1, p. 192; Mueller 2006.

25. Walker, Petulla, et al. 2019.

26. Averages are for 2015–19. Number of police shootings: Tate, Jenkins, et al. 2020. Number of homicides: Federal Bureau of Investigation 2019, and previous years.

27. Schelling 1960, p. 90; see also Tooby, Cosmides, & Price 2006. Pearl Harbor and 9/11 as public outrages: Mueller 2006.

28. Chwe 2001; De Freitas, Thomas, et al. 2019; Schelling 1960.

29. Baumeister, Stillwell, & Wotman 1990.

30. Hostility to data on public outrages: Pearl Harbor and 9/11, Mueller 2006; George Floyd killing, Blackwell 2020.

31. Made popular by the Obama chief of staff Rahm Emanuel, but first used by the anthropologist Luther Gerlach. Thanks to Fred Shapiro, editor of *The Yale Book of Quotations*.

32. For an extended argument of this kind regarding terrorism, see Mueller 2006.

33. https://twitter.com/MaxCRoser/status/919921745464905728?s=20.

34. McCarthy 2015.

35. Rosling 2019.

36. Crisis-driven media and political cynicism: Bornstein & Rosenberg 2016.

37. Lankford & Madfis 2018.

38. https://ourworldindata.org/.

39. From Paulos 1988.

40. Edwards 1996.

41. Many books explain probability and its pitfalls, including Paulos 1988; Hastie & Dawes 2010; Mlodinow 2009; Schneps & Colmez 2013.

42. Batt 2004; Schneps & Colmez 2013.

43. *Texas v. Pennsylvania* 2020. Motion: https://www.texasattorneygeneral.gov/sites/default/files/images/admin/2020/Press/SCOTUSFiling.pdf. Docket: https://www.supremecourt.gov/docket/docketfiles/html/public/22O155.html. Analysis: Bump 2020.

44. Gilovich, Vallone, & Tversky 1985.

45. Miller & Sanjurjo 2018; Gigerenzer 2018a.

46. Pinker 2011, pp. 202–7.

47. https://xkcd.com/795/.

48. Krämer & Gigerenzer 2005.

49. Krämer & Gigerenzer 2005; Miller & Sanjurjo 2018; Miller & Sanjurjo 2019.

50. https://www.youtube.com/watch?v=DBSAeqdcZAM.

51. Scarry's criticism is described in Rosen 1996; see also Good 1996.

52. Krämer & Gigerenzer 2005.

53. Krämer & Gigerenzer 2005; Schneps & Colmez 2013.

54. Paper: Johnson, Tress, et al. 2019. Critique: Knox & Mummolo 2020. Reply: Johnson & Cesario 2020. Retraction: Cesario & Johnson 2020.

55. Edwards 1996.

56. Mlodinow 2009; Paulos 1988.

57. Fabrikant 2008; Mlodinow 2009; Serwer 2006.

58. Gardner 1972.

59. Open Science Collaboration 2015; Gigerenzer 2018b; Ioannidis 2005; Pashler & Wagenmakers 2012.

60. Ioannidis 2005; Simmons, Nelson, & Simonsohn 2011. "The garden of forking paths" was coined by the statistician Andrew Gelman (Gelman & Loken 2014).

61. The cognitive psychologist Michael Corballis.

62. For example, the Center for Open Science's OSF Registries, https://osf.io/prereg/.

63. Feller 1968; see Pinker 2011, pp. 202–7.
64. Kahneman & Tversky 1972. Originally shown by William Feller (1968).
65. Gould 1988.

CHAPTER 5: BELIEFS AND EVIDENCE (BAYESIAN REASONING)

1. Rationality Community: Caplan 2017; Chivers 2019; Raemon 2017. Prominent members include Julia Galef of *Rationally Speaking* (https://juliagalef .com/), Scott Alexander of *Slate Star Codex* (https://slatestarcodex.com/), Scott Aaronson of *Shtetl-Optimized* (https://www.scottaaronson.com/blog/), Robin Hanson of *Overcoming Bias* (https://www.overcomingbias.com/), and Eliezer Yudkowsky, who started *Less Wrong* (https://www.lesswrong.com/).
2. Arbital 2020.
3. Gigerenzer 2011.
4. More accurately, prob(Data | Hypothesis) is *proportional* to the likelihood. The term "likelihood" has slightly different technical meanings in different statistical subcommunities; this is the one commonly used in discussions of Bayesian reasoning.
5. Kahneman & Tversky 1972; Tversky & Kahneman 1974.
6. "In his evaluation of evidence, man is apparently not a conservative Bayesian: he is not Bayesian at all." Kahneman & Tversky 1972, p. 450.
7. Tversky & Kahneman 1982.
8. Hastie & Dawes 2010.
9. Tversky & Kahneman 1974.
10. Overheard; there's no print version I can find.
11. Hume, Bayes, and miracles: Earman 2002.
12. Hume 1748/1999, section X, "Of miracles," part 1, 90.
13. Hume 1748/1999, section X, "Of miracles," part 1, 91.
14. French 2012.
15. Carroll 2016. See also Stenger 1990.
16. Open Science Collaboration 2015; Pashler & Wagenmakers 2012.
17. Ineffectiveness of persuasion industries: Mercier 2020.
18. Ziman 1978, p. 40.
19. Tetlock & Gardner 2015.
20. Tetlock 2003; Tetlock, Kristel, et al. 2000.
21. Decline of bigotry: Pinker 2018, pp. 215–19; Charlesworth & Banaji 2019.
22. Politics of base rates in social science: Tetlock 1994.
23. Gigerenzer 1991, 2018a; Gigerenzer, Swijtink, et al. 1989; see also Cosmides & Tooby 1996.
24. Burns 2010; Maines 2007.
25. Bar-Hillel 1980; Tversky & Kahneman 1982; Gigerenzer 1991.
26. Gigerenzer 1991, 1996; Kahneman & Tversky 1996.
27. Cosmides & Tooby 1996; Gigerenzer 1991; Hoffrage, Lindsey, et al. 2000; Tversky & Kahneman 1983. Kahneman and Tversky point out that frequency

formats reduce, but don't always eliminate, base-rate neglect, as we saw in chapter 1 with Kahneman's adversarial collaboration with Gigerenzer's collaborator Ralph Hertwig on whether frequency formats eliminate the conjunction fallacy: Kahneman & Tversky 1996; Mellers, Hertwig, & Kahneman 2001.
28. Gigerenzer 2015; Kahan 2013.

CHAPTER 6: RISK AND REWARD
(RATIONAL CHOICE AND EXPECTED UTILITY)

1. The model of the human as a rational actor is explained in any introductory economics or political science textbook. The theory that relates rational choice to expected utility was developed by von Neumann & Morgenstern 1953/2007 and refined by Savage 1954. I will use "rational choice" and "expected utility" interchangeably for the theory that equates them. See Luce & Raiffa 1957 and Hastie & Dawes 2010 for accessible explanations.
2. Cohn, Maréchal, et al. 2019.
3. Glaeser 2004.
4. Contesting the axioms of rational choice: Arkes, Gigerenzer, & Hertwig 2016; Slovic & Tversky 1974.
5. Hastie & Dawes 2010; Savage 1954.
6. More commonly, it is called Completeness or Comparability.
7. Also known as Distribution of Probabilities across Alternatives, Algebra of Combining, and Reduction of Compound Lotteries.
8. Variants of the Independence axiom include Chernoff's condition, Sen's property, Arrow's Independence of Irrelevant Alternatives (IIA), and Luce's choice axiom.
9. Liberman 2004.
10. More commonly, Continuity or Solvability.
11. Stevenson & Wolfers 2008.
12. Richardson 1960, p. 11; Slovic 2007; Wan & Shammas 2020.
13. Pinker 2011, pp. 219–20.
14. Tetlock 2003; Tetlock, Kristel, et al. 2000.
15. "Gee, a million dollars . . . maybe." "Would you sleep with me for a hundred dollars?" "What kind of woman do you think I am?" "We've already established that; we're just haggling over price."
16. Simon 1956.
17. Tversky 1972.
18. Savage 1954, cited in Tversky 1972, pp. 283–84.
19. Tversky 1969.
20. Arkes, Gigerenzer, & Hertwig 2016.
21. Tversky 1972, p. 298; Hastie & Dawes 2010, p. 251.
22. Called preference reversals: Lichtenstein & Slovic 1971.
23. Rounding results in a difference of a cent or two, but the differences cancel out over the bets used in the study and don't affect the results.

24. No intransitive money pumps: Arkes, Gigerenzer, & Hertwig 2016, p. 23. Preference-reversing money pumps: Hastie & Dawes 2010, p. 76. Wise up: Arkes, Gigerenzer, & Hertwig 2016, pp. 23–24.

25. Allais 1953.

26. Kahneman & Tversky 1979, p. 267.

27. Kahneman & Tversky 1979.

28. Breyer 1993, p. 12.

29. Kahneman & Tversky 1979.

30. McNeil, Pauker, et al. 1982.

31. Tversky & Kahneman 1981.

32. Hastie & Dawes 2010, pp. 282–88.

33. Kahneman & Tversky 1979.

34. The decision weight graph differs from fig. 4 in Kahneman & Tversky 1979 and is instead based on fig. 12.2 in Hastie & Dawes 2010, which I believe is a better visualization of the theory.

35. Based on Kahneman & Tversky 1979.

36. This pervasive asymmetry is called the Negativity bias; Tierney & Baumeister 2019.

37. Maurice Allais, Herbert Simon, Daniel Kahneman, Richard Thaler, George Akerlof.

38. Gigerenzer 2008b, p. 20.

39. Abito & Salant 2018; Braverman 2018.

40. Sydnor 2010.

41. Gigerenzer & Kolpatzik 2017; see also Gigerenzer 2014, for a similar argument on breast cancer screening.

CHAPTER 7: HITS AND FALSE ALARMS
(SIGNAL DETECTION AND STATISTICAL DECISION THEORY)

1. Twain 1897/1989.

2. Signal Detection Theory and expected utility theory: Lynn, Wormwood, et al. 2015.

3. Statistical distributions are explained in any introduction to statistics or psychology. Signal Detection Theory: Green & Swets 1966; Lynn, Wormwood, et al. 2015; Swets, Dawes, & Monahan 2000; Wolfe, Kluender, et al. 2020, chap. 1. For the histories of Signal Detection Theory and statistical decision theory and their connection, see Gigerenzer, Krauss, & Vitouch 2004; Gigerenzer, Swijtink, et al. 1989.

4. Pinker 2011, pp. 210–20.

5. This is called the Central Limit Theorem.

6. "Likelihood" here is being used in the narrow sense common in discussions of Bayes's rule.

7. Lynn, Wormwood, et al. 2015.

8. Lynn, Wormwood, et al. 2015.

9. Lynn, Wormwood, et al. 2015.

10. Confusingly, "sensitivity" is used in medical contexts to refer to the hit rate, namely the likelihood of a positive finding given that a condition is present. It is contrasted with "specificity," the correct rejection rate, the likelihood of a negative finding given that the condition is absent.

11. Loftus, Doyle, et al. 2019.

12. National Research Council 2009; President's Council of Advisors on Science and Technology 2016.

13. Contesting enhanced interrogation: Bankoff 2014.

14. Ali 2011.

15. Contesting sexual misconduct: Soave 2014; Young 2014a. Two surveys of false rape accusations have found rates between 5 and 10 percent: De Zutter, Horselenberg, & van Koppen 2017; Rumney 2006. See also Bazelon & Larimore 2009; Young 2014b.

16. Arkes & Mellers 2002.

17. Arkes and Mellers cite a 1981 study which reported a range of 0.6–0.9, and a set of flawed studies with d's closer to 2.7. My estimate comes from a meta-analysis in National Research Council 2003, p. 122, which reports a median of 0.86 for a related measure of sensitivity, area under the ROC curve. That figure may be converted, under the assumption of equal-variance normal distributions, to a d' of 1.53 by multiplying the corresponding z-score by $\sqrt{2}$.

18. False accusations, convictions, and executions: National Research Council 2009; President's Council of Advisors on Science and Technology 2016. For rape in particular: Bazelon & Larimore 2009; De Zutter, Horselenberg, & van Koppen 2017; Rumney 2006; Young 2014b. For terrorism: Mueller 2006.

19. Statistical decision theory, in particular, null hypothesis significance testing, is explained in every statistics and psychology textbook. For its history and its relation to Signal Detection Theory, see Gigerenzer, Krauss, & Vitouch 2004; Gigerenzer, Swijtink, et al. 1989.

20. Gigerenzer, Krauss, & Vitouch 2004.

21. As with note 6 above, "likelihood" is used in the narrow sense common in discussions of Bayes's rule, namely the probability of the data given a hypothesis.

22. Gigerenzer 2018b; Open Science Collaboration 2015; Ioannidis 2005; Pashler & Wagenmakers 2012.

23. https://xkcd.com/882/.

24. *Nature* editors 2020b. "Nothing that is not there and the nothing that is" is from Wallace Stevens's "The Snow Man."

25. Henderson 2020; Hume 1748/1999.

CHAPTER 8: SELF AND OTHERS (GAME THEORY)

1. Hume 1739/2000, book III, part II, section V, "Of the obligation of promises."

2. von Neumann & Morgenstern 1953/2007. Semitechnical introductions: Binmore 1991; Luce & Raiffa 1957. Mostly nontechnical: Binmore 2007; Rosenthal 2011. Completely nontechnical: Poundstone 1992.

3. Each game presented in this chapter is discussed in most of the sources in note 2 above.

4. Clegg 2012; Dennett 2013, chap. 8.

5. Thomas, De Freitas, et al. 2016.

6. Chwe 2001; De Freitas, Thomas, et al. 2019; Schelling 1960; Thomas, DeScioli, et al. 2014.

7. Pinker 2007, chap. 8; Schelling 1960.

8. Lewis 1969. Skepticism that conventions require common knowledge: Binmore 2008.

9. The example has been adjusted for inflation.

10. Schelling 1960, pp. 67, 71.

11. J. Goldstein 2010.

12. Frank 1988; Schelling 1960; see also Pinker 1997/2009, chap. 6.

13. Dollar auction: Poundstone 1992; Shubik 1971.

14. Dawkins 1976/2016; Maynard Smith 1982.

15. Pinker 2011, pp. 217–20.

16. Shermer 2008.

17. Dawkins 1976/2016; Maynard Smith 1982.

18. Trivers 1971.

19. Pinker 1997/2009, chap. 7; Pinker 2002/2016, chap. 14; Pinker 2011, chap. 8; Trivers 1971.

20. Ridley 1997.

21. Ellickson 1991; Ridley 1997.

22. Hobbes 1651/1957, chap. 14, p. 190.

CHAPTER 9: CORRELATION AND CAUSATION

1. Sowell 1995.

2. Cohen 1997.

3. BBC News 2004.

4. Stevenson & Wolfers 2008, adapted with permission of the authors.

5. Hamilton 2018.

6. Chapman & Chapman 1967, 1969.

7. Thompson & Adams 1996.

8. *Spurious correlations*, https://www.tylervigen.com/spurious-correlations.

9. Galton 1886.

10. Tversky & Kahneman 1974.

11. Tversky & Kahneman 1974.

12. Tversky & Kahneman 1971, 1974.

13. The author, Jonah Lehrer (2010), quoted scientists who explained regression to the mean and questionable research practices to him, but he still maintained that something was happening but they didn't know what it was.

14. Pinker 2007, pp. 208–33.

15. Hume 1739/2000.

16. Holland 1986; King, Keohane, & Verba 1994, chap. 3.

17. Kaba 2020. For accessible reviews of studies that do show a causal effect of policing on crime (using methods explained in this chapter), see Yglesias 2020a, 2020b.
18. Pearl 2000.
19. Weissman 2020.
20. VanderWeele 2014.
21. Lyric from the 1941 recording. So the Bible says: Matthew 25:29, "For unto every one that hath shall be given, and he shall have abundance: but from him that hath not shall be taken away even that which he hath."
22. Social Progress Imperative 2020; Welzel 2013.
23. Deary 2001; Temple 2015; Ritchie 2015.
24. Pearl & Mackenzie 2018.
25. The cognitive psychologist Reid Hastie.
26. Baron 2012; Bornstein 2012; Hallsworth & Kirkman 2020.
27. Levitt & Dubner 2009; https://freakonomics.com/.
28. DellaVigna & Kaplan 2007.
29. Martin & Yurukoglu 2017.
30. See Pinker 2011, pp. 278–84.
31. The example here is adapted from Russett & Oneal 2001, and discussed in Pinker 2011, pp. 278–84.
32. Stuart 2010.
33. Kendler, Kessler, et al. 2010.
34. Vaci, Edelsbrunner, et al. 2019.
35. Dawes, Faust, & Meehl 1989; Meehl 1954/2013. See also Tetlock 2009 regarding political and economic predictions.
36. Polderman, Benyamin, et al. 2015; see Pinker 2002/2016, pp. 395–98, 450–51.
37. Salganik, Lundberg, et al. 2020.

CHAPTER 10: WHAT'S WRONG WITH PEOPLE?

1. Shermer 2020a.
2. O'Keefe 2020.
3. Wolfe & Dale 2020.
4. Kessler, Rizzo, & Kelly 2020; *Nature* editors 2020a; Tollefson 2020.
5. Rauch 2021.
6. Gilbert 2019; Pennycook & Rand 2020a.
7. The first five figures are from a Gallup survey, Moore 2005; the second five from Pew Forum on Religion and Public Life 2009.
8. According to repeated surveys between 1990 and 2005 or 2009, there were slight upward trends for belief in spiritual healing, haunted houses, ghosts, communicating with the dead, and witches, and slight downward trends for belief in possession by the devil, ESP, telepathy, and reincarnation. Consultations with a psychic or fortune-teller, belief in aliens visiting Earth, and channeling were steady (Moore 2005; Pew Forum on Religion and Public

Life 2009). According to reports from the National Science Foundation, from 1979 to 2018 the percentage believing that astrology is "very" or "sort of" scientific declined very slightly, from the low 40s to the high 30s, and in 2018 included 58 percent of 18- to 24-year-olds and 49 percent of 25- to 34-year-olds (National Science Board 2014, 2020). All paranormal beliefs are more popular in younger than in older respondents (Pew Forum on Religion and Public Life 2009). For astrology, the age gradient is stable over the decades, suggesting that the credulity is an effect of youth itself, which many people grow out of, not of being a Gen Z, Millennial, or any other cohort.

9. Shermer 1997, 2011, 2020b.

10. Mercier 2020; Shermer 2020c; Sunstein & Vermeule 2008; Uscinski & Parent 2014; van Prooijen & van Vugt 2018.

11. Horowitz 2001; Sunstein & Vermeule 2008.

12. Statista Research Department 2019; Uscinski & Parent 2014.

13. Brunvand 2014; the tabloid headlines are from my personal collection.

14. Nyhan 2018.

15. R. Goldstein 2010.

16. https://quoteinvestigator.com/2017/11/30/salary/.

17. Kunda 1990.

18. Thanks to the linguist Ann Farmer for her credo "It isn't about being right. It's about getting it right."

19. Though see note 26 to chapter 1 above.

20. Dawson, Gilovich, & Regan 2002.

21. Kahan, Peters, et al. 2017; Lord, Ross, & Lepper 1979; Taber & Lodge 2006; Dawson, Gilovich, & Regan 2002.

22. Pronin, Lin, & Ross 2002.

23. Mercier & Sperber 2011, 2017; Tetlock 2002. But see also Norman 2016.

24. Mercier & Sperber 2011, p. 63; Mercier, Trouche, et al. 2015.

25. Kahan, Peters, et al. 2017.

26. Ditto, Liu, et al. 2019. For replies, see Baron & Jost 2019; Ditto, Clark, et al. 2019.

27. Stanovich 2020, 2021.

28. Gampa, Wojcik, et al. 2019.

29. Kahan, Hoffman, et al. 2012.

30. Kahan, Peters, et al. 2012.

31. Stanovich 2020, 2021.

32. Hierarchical vs. egalitarian and libertarian vs. communitarian: Kahan 2013 and other references in note 39 below. Throne-and-altar vs. Enlightenment, tribal vs. cosmopolitan: Pinker 2018, chaps. 21, 23. Tragic vs. utopian: Pinker 2002/2016, chap. 16; Sowell 1987. Honor vs. dignity: Pinker 2011, chap. 3; Campbell & Manning 2018; Pinker 2012. Binding vs. individualizing: Haidt 2012.

33. Finkel, Bail, et al. 2020.

34. Finkel, Bail, et al. 2020; Wilkinson 2019.

35. Baron & Jost 2019.
36. The epigraph to Sowell 1995.
37. Ditto, Clark, et al. 2019. Doozies from each side: Pinker 2018, pp. 363–66.
38. Mercier 2020, pp. 191–97.
39. Kahan 2013; Kahan, Peters, et al. 2017; Kahan, Wittlin, et al. 2011.
40. Mercier 2020, chap. 10. Mercier quoted the Google review in a guest lecture in my class on rationality, Mar. 5, 2020.
41. Mercier 2020; Sperber 1997.
42. Abelson 1986.
43. Henrich, Heine, & Norenzayan 2010.
44. Coyne 2015; Dawkins 2006; Dennett 2006; Harris 2005. See R. Goldstein 2010 for a fictionalized debate.
45. Jenkins 2020.
46. BBC News 2020.
47. Baumard & Boyer 2013; Hood 2009; Pinker 1997/2009, chaps. 5, 8; Shermer 1997, 2011.
48. Bloom 2004.
49. Gelman 2005; Hood 2009.
50. Kelemen & Rosset 2009.
51. Rauch 2021; Shtulman 2017; Sloman & Fernbach 2017.
52. See the magazines *Skeptical Inquirer* (http://www.csicop.org/si) and *Skeptic* (http://www.skeptic.com/), and the Center for Inquiry (https://centerforin quiry.org/) for regular updates on pseudoscience in mainstream media.
53. Acerbi 2019.
54. Thompson 2020.
55. Mercier 2020; Shermer 2020c; van Prooijen & van Vugt 2018.
56. Pinker 2011, chap. 2; Chagnon 1997.
57. van Prooijen & van Vugt 2018.
58. Mercier 2020, chap. 10.
59. Dawkins 1976/2016.
60. Friesen, Campbell, & Kay 2015.
61. Moore 2005; Pew Forum on Religion and Public Life 2009.
62. Kahan 2015; Kahan, Wittlin, et al. 2011.
63. Nyhan & Reifler 2019; Pennycook & Rand 2020a; Wood & Porter 2019.
64. Baron 2019; Pennycook, Cheyne, et al. 2020; Sá, West, & Stanovich 1999; Tetlock & Gardner 2015.
65. Like most pithy quotes, apocryphally; credit probably should go to fellow economist Paul Samuelson: https://quoteinvestigator.com/2011/07/22/keynes -change-mind/.
66. Pennycook, Cheyne, et al. 2020. The first three items were added to the Active Open-Mindedness test by Sá, West, & Stanovich 1999.
67. Pennycook, Cheyne, et al. 2020. For similar findings, see Erceg, Galić, & Bubić 2019; Stanovich 2012. Pennycook, Cheyne, et al. 2020, Stanovich, West, & Toplak 2016, and Stanovich & Toplak 2019 point out that some of these

correlations may be inflated by the term "belief" in the openness questionnaire, which respondents may have interpreted as "religious belief." When the
word "opinion" is used, the correlations are lower, but still significant.

68. Global trends in political and social beliefs: Welzel 2013; Pinker 2018, chap. 15.

69. Pennycook, Cheyne, et al. 2012; Stanovich 2012; Stanovich, West, & Toplak
2016. Cognitive Reflection Test: Frederick 2005. See also Maymin & Langer
2021, in which it is connected to mindfulness.

70. Pennycook, Cheyne, et al. 2012; Pennycook & Rand 2020b.

71. Cognitive immune system: Norman 2021.

72. Caplan 2017; Chivers 2019; Raemon 2017.

73. "Party of stupid" has been attributed to the former Republican governor of
Louisiana Bobby Jindal, though he himself said "stupid party." Critiques
from within the conservative movement, pre-Trump: M. K. Lewis 2016;
Mann & Ornstein 2012/2016; Sykes 2017. Post-Trump: Saldin & Teles 2020;
see also The Lincoln Project, https://lincolnproject.us/.

74. Quoted in Rauch 2018.

75. Mercier 2020.

76. Lane 2021.

77. Rauch 2018, 2021; Sloman & Fernbach 2017.

78. Trust in science steady: American Academy of Arts and Sciences 2018. Trust
in academia sinking: Jones 2018.

79. Flaherty 2020. For other examples, see Kors & Silverglate 1998; Lukianoff
2012; Lukianoff & Haidt 2018; and the Heterodox Academy (https://hetero
doxacademy.org/), the Foundation for Individual Rights in Education (https://
www.thefire.org/), and *Quillette* magazine (https://quillette.com/).

80. Haidt 2016.

81. American Academy of Arts and Sciences 2018.

82. Nyhan 2013; Nyhan & Reifler 2012.

83. Willingham 2007.

84. Bond 2009; Hoffrage, Lindsey, et al. 2000; Lilienfeld, Ammirati, & Landfield 2009; Mellers, Ungar, et al. 2014; Morewedge, Yoon, et al. 2015; Willingham 2007.

85. Kahan, Wittlin, et al. 2011; Stanovich 2021.

86. Ellickson 1991; Ridley 1997.

87. Rauch 2021; Sloman & Fernbach 2017.

88. Eisenstein 2012.

89. Kräenbring, Monzon Penza, et al. 2014.

90. See "Wikipedia: List of policies and guidelines," https://en.wikipedia.org
/wiki/Wikipedia:List_of_policies_and_guidelines, and "Wikipedia: Five pillars," https://en.wikipedia.org/wiki/Wikipedia:Five_pillars.

91. Social media reform: Fox 2020; Lyttleton 2020. Some early analyses: Pennycook, Cannon, & Rand 2018; Pennycook & Rand 2020a.

92. Joyner 2011; Tetlock 2015.

93. Pinker 2018, pp. 380–81.

94. Elster 1998; Fishkin 2011.
95. Mercier & Sperber 2011.

CHAPTER 11: WHY RATIONALITY MATTERS

1. Singer 1981/2011, p. 88.
2. For a trenchant analysis of "conflict versus mistake" as drivers of human progress, see Alexander 2018.
3. These examples are discussed in chapters 4–9; see also Stanovich 2018; Stanovich, West, & Toplak 2016.
4. Stanovich 2018.
5. http://whatstheharm.net/index.html. Many of his examples are backed by scientific reports, listed in http://whatstheharm.net/scientificstudies.html. Farley stopped maintaining the site around 2009, but sporadically reports examples in his Twitter feed @WhatsTheHarm, https://twitter.com/whatstheharm.
6. Bruine de Bruin, Parker, & Fischhoff 2007.
7. Ritchie 2015.
8. Bruine de Bruin, Parker, & Fischhoff 2007. See also Parker, Bruine de Bruin, et al. 2018 for an eleven-year follow-up, and Toplak, West, & Stanovich 2017 for similar results. In 2020, the economist Mattie Toma and I replicated the result in a survey of 157 Harvard students taking my Rationality course (Toma 2020).
9. Pinker 2011; Pinker 2018. Related conclusions: Kenny 2011; Norberg 2016; Ridley 2010; and the websites *Our World in Data* (https://ourworldindata.org/) and *Human Progress* (https://www.humanprogress.org/).
10. Roser, Ortiz-Ospina, & Ritchie 2013, accessed Dec. 8, 2020; Pinker 2018, chaps. 5, 6.
11. Pinker 2018, chap. 7.
12. Roser 2016, accessed Dec. 8, 2020; Pinker 2018, chap. 8.
13. Pinker 2011, chaps. 5, 6; Pinker 2018, chap. 11. Related conclusions: J. Goldstein 2011; Mueller 2021; Payne 2004.
14. Road map to solving the climate crisis: Goldstein-Rose 2020.
15. Pinker 2011, chaps. 4, 7; Pinker 2018, chap. 15. Related conclusions: Appiah 2010; Grayling 2007; Hunt 2007; Payne 2004; Shermer 2015; Singer 1981/2011.
16. Alexander 2018.
17. Pinker 2011, chap. 4; see also Appiah 2010; Grayling 2007; Hunt 2007; Payne 2004.
18. Welzel 2013, p. 122; see Pinker 2018, p. 228 and note 45, and pp. 233–35 and note 8.
19. *Concerning Heretics, Whether They Are to Be Persecuted*, quoted in Grayling 2007, pp. 53–54.
20. Mueller 2021.
21. Erasmus 1517/2017.
22. Beccaria 1764/2010; my blend of two translations.
23. Pinker 2018, pp. 211–13.
24. Bentham & Crompton 1785/1978.

25. Bentham 1823, chap. 19.
26. Singer 1981/2011.
27. Davis 1984.
28. Locke 1689/2015, 2nd treatise, chap. VI, sect. 61.
29. Locke 1689/2015, 2nd treatise, chap. IV, sect 22.
30. Astell 1730/2010.
31. Wollstonecraft 1792/1995.
32. Douglass 1852/1999.

REFERENCES

Abelson, R. P. 1986. Beliefs are like possessions. *Journal for the Theory of Social Behaviour, 16*, 223–50. https://doi.org/10.1111/j.1468-5914.1986.tb00078.x.

Abito, J. M., & Salant, Y. 2018. The effect of product misperception on economic outcomes: Evidence from the extended warranty market. *Review of Economic Studies, 86*, 2285–318. https://doi.org/10.1093/restud/rdy045.

Acerbi, A. 2019. Cognitive attraction and online misinformation. *Palgrave Communications, 5*, 1–7. https://doi.org/10.1057/s41599-019-0224-y.

Aggarwal, C. C. 2018. *Neural networks and deep learning.* New York: Springer.

Ainslie, G. 2001. *Breakdown of will.* New York: Cambridge University Press.

Alexander, S. 2018. Conflict vs. mistake. *Slate Star Codex.* https://slatestarcodex.com/2018/01/24/conflict-vs-mistake/.

Ali, R. 2011. *Dear colleague letter* (policy guidance from the assistant secretary for civil rights). US Department of Education. https://www2.ed.gov/about/offices/list/ocr/letters/colleague-201104.html.

Allais, M. 1953. Le comportement de l'homme rationnel devant le risque: Critique des postulats et axiomes de l'école Americaine. *Econometrica, 21*, 503–46. https://doi.org/10.2307/1907921.

American Academy of Arts and Sciences. 2018. *Perceptions of science in America.* Cambridge, MA: American Academy of Arts and Sciences. https://www.amacad.org/publication/perceptions-science-america.

Appiah, K. A. 2010. *The honor code: How moral revolutions happen.* New York: W. W. Norton.

Arbital. 2020. Bayes' rule. https://arbital.com/p/bayes_rule/?l=1zq.

Arkes, H. R., Gigerenzer, G., & Hertwig, R. 2016. How bad is incoherence? *Decision, 3*, 20–39. https://doi.org/10.1037/dec0000043.

Arkes, H. R., & Mellers, B. A. 2002. Do juries meet our expectations? *Law and Human Behavior, 26,* 625–39. https://doi.org/10.1023/A:1020929517312.

Armstrong, S. L., Gleitman, L. R., & Gleitman, H. 1983. What some concepts might not be. *Cognition, 13,* 263–308. https://doi.org/10.1016/0010-0277(83)90012-4.

Ashby, F. G., Alfonso-Reese, L. A., Turken, A. U., & Waldron, E. M. 1998. A neuropsychological theory of multiple systems in category learning. *Psychological Review, 105,* 442–81. https://doi.org/10.1037/0033-295X.105.3.442.

Astell, M. 1730/2010. *Some reflections upon marriage. To which is added a preface, in answer to some objections.* Farmington Hills, MI: Gale ECCO.

Bacon, F. 1620/2017. *Novum organum.* Seattle, WA: CreateSpace.

Bankoff, C. 2014. Dick Cheney simply does not care that the CIA tortured innocent people. *New York Magazine,* Dec. 14. https://nymag.com/intelligencer/2014/12/cheney-alright-with-torture-of-innocent-people.html.

Bar-Hillel, M. 1980. The base-rate fallacy in probability judgments. *Acta Psychologica, 44,* 211–33. https://doi.org/10.1016/0001-6918(80)90046-3.

Baron, J. 2012. Applying evidence to social programs. *New York Times,* Nov. 29. https://economix.blogs.nytimes.com/2012/11/29/applying-evidence-to-social-programs/.

Baron, J. 2019. Actively open-minded thinking in politics. *Cognition, 188,* 8–18. https://doi.org/10.1016/j.cognition.2018.10.004.

Baron, J., & Jost, J. T. 2019. False equivalence: Are liberals and conservatives in the United States equally biased? *Perspectives on Psychological Science, 14,* 292–303. https://doi.org/10.1177/1745691618788876.

Basterfield, C., Lilienfeld, S. O., Bowes, S. M., & Costello, T. H. 2020. The Nobel disease: When intelligence fails to protect against irrationality. *Skeptical Inquirer,* May. https://skepticalinquirer.org/2020/05/the-nobel-disease-when-intelligence-fails-to-protect-against-irrationality/.

Batt, J. 2004. *Stolen innocence: A mother's fight for justice—the authorised story of Sally Clark.* London: Ebury Press.

Baumard, N., & Boyer, P. 2013. Religious beliefs as reflective elaborations on intuitions: A modified dual-process model. *Current Directions in Psychological Science, 22,* 295–300. https://doi.org/10.1177/0963721413478610.

Baumeister, R. F., Stillwell, A., & Wotman, S. R. 1990. Victim and perpetrator accounts of interpersonal conflict: Autobiographical narratives about anger. *Journal of Personality and Social Psychology, 59,* 994–1005. https://doi.org/10.1037/0022-3514.59.5.994.

Baumeister, R. F., & Tierney, J. 2012. *Willpower: Rediscovering the greatest human strength.* London: Penguin.

Bazelon, E., & Larimore, R. 2009. How often do women falsely cry rape? *Slate,* Oct. 1. https://slate.com/news-and-politics/2009/10/why-it-s-so-hard-to-quantify-false-rape-charges.html.

BBC News. 2004. Avoid gold teeth, says Turkmen leader. Apr. 7. http://news.bbc.co.uk/2/hi/asia-pacific/3607467.stm.

BBC News. 2020. The Crown: Netflix has "no plans" for a fiction warning. Dec. 6. https://www.bbc.com/news/entertainment-arts-55207871.

Beccaria, C. 1764/2010. *On crimes and punishments and other writings* (R. Davies, trans.; R. Bellamy, ed.). New York: Cambridge University Press.

Bell, E. T. 1947. *The development of mathematics* (2nd ed.). New York: McGraw-Hill.

Bentham, J. 1789. *An introduction to the principles of morals and legislation* (2nd ed.). https://www.econlib.org/library/Bentham/bnthPML.html.

Bentham, J., & Crompton, L. 1785/1978. Offences against one's self: Paederasty (part I). *Journal of Homosexuality*, *3*, 389–405. https://doi.org/10.1300/J082v03n04_07.

Binmore, K. 1991. *Fun and games: A text on game theory*. Boston: Houghton Mifflin.

Binmore, K. 2007. *Game theory: A very short introduction*. New York: Oxford University Press.

Binmore, K. 2008. Do conventions need to be common knowledge? *Topoi*, *27*, 17–27. https://doi-org.ezp-prod1.hul.harvard.edu/10.1007/s11245-008-9033-4.

Blackwell, M. 2020. Black Lives Matter and the mechanics of conformity. *Quillette*, Sept. 17. https://quillette.com/2020/09/17/black-lives-matter-and-the-mechanics-of-conformity/.

Block, W. 1976/2018. *Defending the undefendable*. Auburn, AL: Ludwig von Mises Institute.

Bloom, P. 2003. *Descartes' baby: How the science of child development explains what makes us human*. New York: Basic Books.

Bond, M. 2009. Risk school. *Nature*, *461*, 1189–92, Oct. 28.

Bornstein, D. 2012. The dawn of the evidence-based budget. *New York Times*, May 30. https://opinionator.blogs.nytimes.com/2012/05/30/worthy-of-government-funding-prove-it.

Bornstein, D., & Rosenberg, T. 2016. When reportage turns to cynicism. *New York Times*, Nov. 14. https://www.nytimes.com/2016/11/15/opinion/when-reportage-turns-to-cynicism.html.

Braverman, B. 2018. Why you should steer clear of extended warranties. *Consumer Reports*, Dec. 22. https://www.consumerreports.org/extended-warranties/steer-clear-extended-warranties/.

Breyer, S. 1993. *Breaking the vicious circle: Toward effective risk regulation*. Cambridge, MA: Harvard University Press.

Bruine de Bruin, W., Parker, A. M., & Fischhoff, B. 2007. Individual differences in adult decision-making competence. *Journal of Personality and Social Psychology*, *92*, 938–56. https://doi.org/10.1037/0022-3514.92.5.938.

Brunvand, J. H. 2014. *Too good to be true: The colossal book of urban legends* (rev. ed.). New York: W. W. Norton.

Bump, P. 2020. Trump's effort to steal the election comes down to some utterly ridiculous statistical claims. *Washington Post*, Dec. 9. https://www.washingtonpost.com/politics/2020/12/09/trumps-effort-steal-election-comes-down-some-utterly-ridiculous-statistical-claims/.

Burns, K. 2010. At veterinary colleges, male students are in the minority. *American Veterinary Medical Association*, Feb. 15. https://www.avma.org/javma-news/2010-02-15/veterinary-colleges-male-students-are-minority.

Caldeira, K., Emanuel, K., Hansen, J., & Wigley, T. 2013. Top climate change scientists' letter to policy influencers. CNN, Nov. 3. https://www.cnn.com/2013/11/03/world/nuclear-energy-climate-change-scientists-letter/index.html.

Campbell, B., & Manning, J. 2018. *The rise of victimhood culture: Microaggressions, safe spaces, and the new culture wars*. London: Palgrave Macmillan.

Caplan, B. 2017. What's wrong with the rationality community. *EconLog*, Apr. 4. https://www.econlib.org/archives/2017/04/whats_wrong_wit_22.html.

Carroll, L. 1895. What the tortoise said to Achilles. *Mind, 4*, 178–80.

Carroll, L. 1896/1977. Symbolic logic. In W. W. Bartley, ed., *Lewis Carroll's Symbolic Logic*. New York: Clarkson Potter.

Carroll, S. M. 2016. *The big picture: On the origins of life, meaning, and the universe itself*. New York: Penguin Random House.

Cesario, J., & Johnson, D. J. 2020. Statement on the retraction of "Officer characteristics and racial disparities in fatal officer-involved shootings." https://doi.org/10.31234/osf.io/dj57k.

Chagnon, N. A. 1997. *Yanomamö* (5th ed.). Fort Worth, TX: Harcourt Brace.

Chapman, L. J., & Chapman, J. P. 1967. Genesis of popular but erroneous psychodiagnostic observations. *Journal of Abnormal Psychology, 72*, 193–204. https://doi.org/10.1037/h0024670.

Chapman, L. J., & Chapman, J. P. 1969. Illusory correlation as an obstacle to the use of valid psychodiagnostic signs. *Journal of Abnormal Psychology, 74*, 271–80. https://doi.org/10.1037/h0027592.

Charlesworth, T. E. S., & Banaji, M. R. 2019. Patterns of implicit and explicit attitudes: I. Long-term change and stability from 2007 to 2016. *Psychological Science, 30*, 174–92. https://doi.org/10.1177/0956797618813087.

Cheng, P. W., & Holyoak, K. J. 1985. Pragmatic reasoning schemas. *Cognitive Psychology, 17*, 391–416. https://doi.org/10.1016/0010-0285(85)90014-3.

Chivers, T. 2019. *The AI does not hate you: Superintelligence, rationality and the race to save the world*. London: Weidenfeld & Nicolson.

Chomsky, N. 1972/2006. *Language and mind* (extended ed.). New York: Cambridge University Press.

Chwe, M. S.-Y. 2001. *Rational ritual: Culture, coordination, and common knowledge*. Princeton, NJ: Princeton University Press.

Clegg, L. F. 2012. Protean free will. Unpublished manuscript, California Institute of Technology. https://resolver.caltech.edu/CaltechAUTHORS:20120328-152031480.

Cohen, I. B. 1997. *Science and the Founding Fathers: Science in the political thought of Thomas Jefferson, Benjamin Franklin, John Adams, and James Madison*. New York: W. W. Norton.

Cohn, A., Maréchal, M. A., Tannenbaum, D., & Zünd, C. L. 2019. Civic honesty around the globe. *Science, 365*, 70–73. https://doi.org/10.1126/science.aau8712.

Cohon, R. 2018. Hume's moral philosophy. In E. N. Zalta, ed., *The Stanford Encyclopedia of Philosophy.* https://plato.stanford.edu/entries/hume-moral/.

Cole, M., Gay, J., Glick, J., & Sharp, D. W. 1971. *The cultural context of learning and thinking.* New York: Basic Books.

Combs, B., & Slovic, P. 1979. Newspaper coverage of causes of death. *Journalism Quarterly, 56,* 837–49.

Cosmides, L. 1989. The logic of social exchange: Has natural selection shaped how humans reason? Studies with the Wason selection task. *Cognition, 31,* 187–276. https://doi.org/10.1016/0010-0277(89)90023-1.

Cosmides, L., & Tooby, J. 1996. Are humans good intuitive statisticians after all? Rethinking some conclusions from the literature on judgment under uncertainty. *Cognition, 58,* 1–73. https://doi.org/10.1016/0010-0277(95)00664-8.

Coyne, J. A. 2015. *Faith versus fact: Why science and religion are incompatible.* New York: Penguin.

Crockett, Z. 2015. The time everyone "corrected" the world's smartest woman. *Priceonomics,* Feb. 19. https://priceonomics.com/the-time-everyone-corrected-the-worlds-smartest/.

Curtis, G. N. 2020. The *Fallacy Files* taxonomy of logical fallacies. https://www.fallacyfiles.org/taxonnew.htm.

Dasgupta, P. 2007. The Stern Review's economics of climate change. *National Institute Economic Review, 199,* 4–7. https://doi.org/10.1177/0027950107077111.

Davis, D. B. 1984. *Slavery and human progress.* New York: Oxford University Press.

Dawes, R. M., Faust, D., & Meehl, P. E. 1989. Clinical versus actuarial judgment. *Science, 243,* 1668–74. https://doi.org/10.1126/science.2648573.

Dawkins, R. 1976/2016. *The selfish gene* (40th anniv. ed.). New York: Oxford University Press.

Dawkins, R. 2006. *The God delusion.* New York: Houghton Mifflin.

Dawson, E., Gilovich, T., & Regan, D. T. 2002. Motivated reasoning and performance on the Wason selection task. *Personality and Social Psychology Bulletin, 28,* 1379–87. https://doi.org/10.1177/014616702236869.

De Freitas, J., Thomas, K., DeScioli, P., & Pinker, S. 2019. Common knowledge, coordination, and strategic mentalizing in human social life. *Proceedings of the National Academy of Sciences, 116,* 13751–58. https://doi.org/10.1073/pnas.1905518116.

de Lazari-Radek, K., & Singer, P. 2012. The objectivity of ethics and the unity of practical reason. *Ethics, 123,* 9–31. https://doi.org/10.1086/667837.

De Zutter, A., Horselenberg, R., & van Koppen, P. J. 2017. The prevalence of false allegations of rape in the United States from 2006–2010. *Journal of Forensic Psychology, 2.* https://doi.org/10.4172/2475-319X.1000119.

Deary, I. J. 2001. *Intelligence: A very short introduction.* New York: Oxford University Press.

DellaVigna, S., & Kaplan, E. 2007. The Fox News effect: Media bias and voting. *Quarterly Journal of Economics, 122,* 1187–234. https://doi.org/10.1162/qjec.122.3.1187.

Dennett, D. C. 2006. *Breaking the spell: Religion as a natural phenomenon.* New York: Penguin.

Dennett, D. C. 2013. *Intuition pumps and other tools for thinking.* New York: W. W. Norton.

Ditto, P. H., Clark, C. J., Liu, B. S., Wojcik, S. P., Chen, E. E., et al. 2019. Partisan bias and its discontents. *Perspectives on Psychological Science, 14,* 304–16. https://doi.org/10.1177/1745691618817753.

Ditto, P. H., Liu, B. S., Clark, C. J., Wojcik, S. P., Chen, E. E., et al. 2019. At least bias is bipartisan: A meta-analytic comparison of partisan bias in liberals and conservatives. *Perspectives on Psychological Science, 14,* 273–91. https://doi.org/10.1177/1745691617746796.

Donaldson, H., Doubleday, R., Hefferman, S., Klondar, E., & Tummarello, K. 2011. Are talking heads blowing hot air? An analysis of the accuracy of forecasts in the political media. Hamilton College. https://www.hamilton.edu/documents/Analysis-of-Forcast-Accuracy-in-the-Political-Media.pdf.

Douglass, F. 1852/1999. What to the slave is the Fourth of July? In P. S. Foner, ed., *Frederick Douglass: Selected speeches and writings.* Chicago: Lawrence Hill.

Duffy, B. 2018. *The perils of perception: Why we're wrong about nearly everything.* London: Atlantic Books.

Eagle, A. 2019. Chance versus randomness. In E. N. Zalta, ed., *The Stanford Encyclopedia of Philosophy.* https://plato.stanford.edu/entries/chance-randomness/.

Earman, J. 2002. Bayes, Hume, Price, and miracles. *Proceedings of the British Academy, 113,* 91–109.

Edwards, A. W. F. 1996. Is the Pope an alien? *Nature, 382,* 202. https://doi.org/10.1038/382202b0.

Einstein, A. 1981. *Albert Einstein, the human side: New glimpses from his archives* (H. Dukas & B. Hoffman, eds.). Princeton, NJ: Princeton University Press.

Eisenstein, E. L. 2012. *The printing revolution in early modern Europe* (2nd ed.). New York: Cambridge University Press.

Eliot, G. 1883/2017. *Essays of George Eliot* (T. Pinney, ed.). Philadelphia: Routledge.

Ellickson, R. C. 1991. *Order without law: How neighbors settle disputes.* Cambridge, MA: Harvard University Press.

Elster, J., ed. 1998. *Deliberative democracy.* New York: Cambridge University Press.

Emerson, R. W. 1841/1993. *Self-reliance and other essays.* New York: Dover.

Erasmus, D. 1517/2017. *The complaint of peace: To which is added, Antipolemus; or, the plea of reason, religion, and humanity, against war.* Miami, FL: HardPress.

Erceg, N., Galić, Z., & Bubić, A. 2019. "Dysrationalia" among university students: The role of cognitive abilities, different aspects of rational thought and self-control in explaining epistemically suspect beliefs. *Europe's Journal of Psychology, 15,* 159–75. https://doi.org/10.5964/ejop.v15i1.1696.

Evans, J. St. B. T. 2012. Dual-process theories of deductive reasoning: Facts and fallacies. In K. J. Holyoak & R. G. Morrison, eds., *The Oxford Handbook of Thinking and Reasoning.* Oxford: Oxford University Press.

Fabrikant, G. 2008. Humbler, after a streak of magic. *New York Times*, May 11. https://www.nytimes.com/2008/05/11/business/11bill.html.

Federal Aviation Administration. 2016. *Pilot's handbook of aeronautical knowledge*. Oklahoma City: US Department of Transportation. https://www.faa.gov /regulations_policies/handbooks_manuals/aviation/phak/media/pilot_hand book.pdf.

Federal Bureau of Investigation. 2019. Crime in the United States, expanded homicide data table 1. https://ucr.fbi.gov/crime-in-the-u.s/2019/crime-in-the -u.s.-2019/tables/expanded-homicide-data-table-1.xls.

Feller, W. 1968. *An introduction to probability theory and its applications*. New York: Wiley.

Fiddick, L., Cosmides, L., & Tooby, J. 2000. No interpretation without repre-sentation: The role of domain-specific representations and inferences in the Wason selection task. *Cognition*, 77, 1–79. https://doi.org/10.1016/S0010-0277 (00)00085-8.

Finkel, E. J., Bail, C. A., Cikara, M., Ditto, P. H., Iyengar, S., et al. 2020. Political sectarianism in America. *Science*, 370, 533–36. https://doi.org/10.1126/science .abe1715.

Fishkin, J. S. 2011. *When the people speak: Deliberative democracy and public consulta-tion*. New York: Oxford University Press.

Flaherty, C. 2020. Failure to communicate: Professor suspended for saying a Chinese word that sounds like a racial slur in English. *Inside Higher Ed*. https://www.insidehighered.com/news/2020/09/08/professor-suspended -saying-chinese-word-sounds-english-slur.

Fodor, J. A. 1968. *Psychological explanation: An introduction to the philosophy of psy-chology*. New York: Random House.

Fox, C. 2020. Social media: How might it be regulated? BBC News, Nov. 12. https://www.bbc.com/news/technology-54901083.

Frank, R. H. 1988. *Passions within reason: The strategic role of the emotions*. New York: W. W. Norton.

Frederick, S. 2005. Cognitive reflection and decision making. *Journal of Economic Perspectives*, 19, 25–42. https://doi.org/10.1257/089533005775196732.

French, C. 2012. Precognition studies and the curse of the failed replications. *The Guardian*, Mar. 15. http://www.theguardian.com/science/2012/mar/15 /precognition-studies-curse-failed-replications.

Friedersdorf, C. 2018. Why can't people hear what Jordan Peterson is actually saying? *The Atlantic*, Jan. 22. https://www.theatlantic.com/politics/archive/2018 /01/putting-monsterpaint-onjordan-peterson/550859/.

Friesen, J. P., Campbell, T. H., & Kay, A. C. 2015. The psychological advantage of unfalsifiability: The appeal of untestable religious and political ideologies. *Journal of Personality and Social Psychology*, 108, 515–29. https://doi.org/10.1037 /pspp0000018.

Galton, F. 1886. Regression towards mediocrity in hereditary stature. *Journal of the Anthropological Institute of Great Britain and Ireland*, 15, 246–63.

Gampa, A., Wojcik, S. P., Motyl, M., Nosek, B. A., & Ditto, P. H. 2019. (Ideo) logical reasoning: Ideology impairs sound reasoning. *Social Psychological and Personality Science, 10*, 1075–83. https://doi.org/10.1177/1948550619829059.

Gardner, M. 1959. Problems involving questions of probability and ambiguity. *Scientific American, 201*, 174–82.

Gardner, M. 1972. Why the long arm of coincidence is usually not as long as it seems. *Scientific American, 227*.

Gelman, A., & Loken, E. 2014. The statistical crisis in science. *American Scientist, 102*, 460–65.

Gelman, S. A. 2005. *The essential child: Origins of essentialism in everyday thought.* New York: Oxford University Press.

Gettier, E. L. 1963. Is justified true belief knowledge? *Analysis, 23*, 121–23.

Gigerenzer, G. 1991. How to make cognitive illusions disappear: Beyond "heuristics and biases." *European Review of Social Psychology, 2*, 83–115. https://doi.org/10.1080/14792779143000033.

Gigerenzer, G. 1996. On narrow norms and vague heuristics: A reply to Kahneman and Tversky. *Psychological Review, 103*, 592–96. https://doi.org/10.1037/0033-295X.103.3.592.

Gigerenzer, G. 1998. Ecological intelligence: An adaptation for frequencies. In D. D. Cummins & C. Allen, eds., *The evolution of mind.* New York: Oxford University Press.

Gigerenzer, G. 2004. Gigerenzer's Law of Indispensable Ignorance. *Edge.* https://www.edge.org/response-detail/10224.

Gigerenzer, G. 2006. Out of the frying pan into the fire: Behavioral reactions to terrorist attacks. *Risk Analysis, 26*, 347–51. https://doi.org/10.1111/j.1539-6924.2006.00753.x.

Gigerenzer, G. 2008a. The evolution of statistical thinking. In G. Gigerenzer, ed., *Rationality for mortals: How people cope with uncertainty.* New York: Oxford University Press.

Gigerenzer, G. 2008b. *Rationality for mortals: How people cope with uncertainty.* New York: Oxford University Press.

Gigerenzer, G. 2011. What are natural frequencies? *BMJ, 343*, d6386. https://doi.org/10.1136/bmj.d6386.

Gigerenzer, G. 2014. Breast cancer screening pamphlets mislead women. *BMJ, 348*, g2636. https://doi.org/10.1136/bmj.g2636.

Gigerenzer, G. 2015. On the supposed evidence for libertarian paternalism. *Review of Philosophy and Psychology, 6*, 361–83. https://doi.org/10.1007/s13164-015-0248-1.

Gigerenzer, G. 2018a. The Bias Bias in behavioral economics. *Review of Behavioral Economics, 5*, 303–36. https://doi.org/10.1561/105.00000092.

Gigerenzer, G. 2018b. Statistical rituals: The replication delusion and how we got there. *Advances in Methods and Practices in Psychological Science, 1*, 198–218. https://doi.org/10.1177/2515245918771329.

Gigerenzer, G., & Garcia-Retamero, R. 2017. Cassandra's regret: The psychology of not wanting to know. *Psychological Review, 124*, 179–96.

Gigerenzer, G., Hertwig, R., Van Den Broek, E., Fasolo, B., & Katsikopoulos, K. V. 2005. "A 30% chance of rain tomorrow": How does the public understand probabilistic weather forecasts? *Risk Analysis: An International Journal*, *25*, 623–29. https://doi.org/10.1111/j.1539-6924.2005.00608.x.

Gigerenzer, G., & Kolpatzik, K. 2017. How new fact boxes are explaining medical risk to millions. *BMJ*, *357*, j2460. https://doi.org/10.1136/bmj.j2460.

Gigerenzer, G., Krauss, S., & Vitouch, O. 2004. The null ritual: What you always wanted to know about significance testing but were afraid to ask. In D. Kaplan, ed., *The Sage Handbook of Quantitative Methodology for the Social Sciences*. Thousand Oaks, CA: Sage.

Gigerenzer, G., Swijtink, Z., Porter, T., Daston, L., Beatty, J., et al. 1989. *The empire of chance: How probability changed science and everyday life*. New York: Cambridge University Press.

Gilbert, B. 2019. The 10 most-viewed fake-news stories on Facebook in 2019 were just revealed in a new report. *Business Insider*, Nov. 6. https://www.businessinsider.com/most-viewed-fake-news-stories-shared-on-facebook-2019-2019-11.

Gilovich, T., Vallone, R., & Tversky, A. 1985. The hot hand in basketball: On the misperception of random sequences. *Cognitive Psychology*, *17*, 295–314. https://doi.org/10.1016/0010-0285(85)90010-6.

Glaeser, E. L. 2004. Psychology and the market. *American Economic Review*, *94*, 408–13. http://www.jstor.org/stable/3592919.

Goda, G. S., Levy, M. R., Manchester, C. F., Sojourner, A., & Tasoff, J. 2015. The role of time preferences and exponential-growth bias in retirement savings. *National Bureau of Economic Research Working Paper Series*, no. 21482. https://doi.org/10.3386/w21482.

Goldstein, J. S. 2010. Chicken dilemmas: Crossing the road to cooperation. In I. W. Zartman & S. Touval, eds., *International cooperation: The extents and limits of multilateralism*. New York: Cambridge University Press.

Goldstein, J. S. 2011. *Winning the war on war: The decline of armed conflict worldwide*. New York: Penguin.

Goldstein, J. S., & Qvist, S. A. 2019. *A bright future: How some countries have solved climate change and the rest can follow*. New York: PublicAffairs.

Goldstein, J. S., Qvist, S. A., & Pinker, S. 2019. Nuclear power can save the world. *New York Times*, Apr. 6. https://www.nytimes.com/2019/04/06/opinion/sunday/climate-change-nuclear-power.html.

Goldstein, R. N. 2006. *Betraying Spinoza: The renegade Jew who gave us modernity*. New York: Nextbook/Schocken.

Goldstein, R. N. 2010. *36 arguments for the existence of God: A work of fiction*. New York: Pantheon.

Goldstein, R. N. 2013. *Plato at the Googleplex: Why philosophy won't go away*. New York: Pantheon.

Goldstein-Rose, S. 2020. *The 100% solution: A plan for solving climate change*. New York: Melville House.

Good, I. 1996. When batterer becomes murderer. *Nature*, *381*, 481. https://doi.org/10.1038/381481a0.

Goodfellow, I., Bengio, Y., & Courville, A. 2016. *Deep learning.* Cambridge, MA: MIT Press.

Gould, S. J. 1988. The streak of streaks. *New York Review of Books.* https://www.nybooks.com/articles/1988/08/18/the-streak-of-streaks/.

Gould, S. J. 1999. *Rocks of ages: Science and religion in the fullness of life.* New York: Ballantine.

Gracyk, T. 2020. Hume's aesthetics. In E. N. Zalta, ed., *Stanford Encyclopedia of Philosophy.* https://plato.stanford.edu/archives/sum2020/entries/hume-aesthetics/.

Granberg, D., & Brown, T. A. 1995. The Monty Hall dilemma. *Personality & Social Psychology Bulletin, 21,* 711–23. https://doi.org/10.1177/0146167295217006.

Grayling, A. C. 2007. *Toward the light of liberty: The struggles for freedom and rights that made the modern Western world.* New York: Walker.

Green, D. M., & Swets, J. A. 1966. *Signal detection theory and psychophysics.* New York: Wiley.

Greene, J. 2013. *Moral tribes: Emotion, reason, and the gap between us and them.* New York: Penguin.

Grice, H. P. 1975. Logic and conversation. In P. Cole & J. L. Morgan, eds., *Syntax and semantics,* vol. 3, *Speech acts.* New York: Academic Press.

Haidt, J. 2012. *The righteous mind: Why good people are divided by politics and religion.* New York: Pantheon.

Haidt, J. 2016. Why universities must choose one telos: truth or social justice. *Heterodox Academy,* Oct. 16. https://heterodoxacademy.org/blog/one-telos-truth-or-social-justice-2/.

Hájek, A. 2019. Interpretations of probability. In E. N. Zalta, ed., *The Stanford Encyclopedia of Philosophy.* https://plato.stanford.edu/archives/fall2019/entries/probability-interpret/.

Hallsworth, M., & Kirkman, E. 2020. *Behavioral insights.* Cambridge, MA: MIT Press.

Hamilton, I. A. 2018. Jeff Bezos explains why his best decisions were based off intuition, not analysis. *Inc.,* Sept. 14. https://www.inc.com/business-insider/amazon-ceo-jeff-bezos-says-his-best-decision-were-made-when-he-followed-his-gut.html.

Harris, S. 2005. *The end of faith: Religion, terror, and the future of reason.* New York: W. W. Norton.

Hastie, R., & Dawes, R. M. 2010. *Rational choice in an uncertain world: The psychology of judgment and decision making* (2nd ed.). Los Angeles: Sage.

Henderson, L. 2020. The problem of induction. In E. N. Zalta, ed., *The Stanford Encyclopedia of Philosophy.* https://plato.stanford.edu/archives/spr2020/entries/induction-problem/.

Henrich, J., Heine, S. J., & Norenzayan, A. 2010. The weirdest people in the world? *Behavioral and Brain Sciences, 33,* 61–83. https://doi.org/10.1017/S0140525X0999152X.

Hertwig, R., & Engel, C. 2016. Homo ignorans: Deliberately choosing not to know. *Perspectives on Psychological Science, 11,* 359–72.

Hertwig, R., & Gigerenzer, G. 1999. The "conjunction fallacy" revisited: How intelligent inferences look like reasoning errors. *Journal of Behavioral Decision Making, 12*, 275–305. https://doi.org/10.1002/(SICI)1099-0771(199912)12:4<275::AID-BDM323>3.0.CO;2-M.

Hobbes, T. 1651/1957. *Leviathan.* New York: Oxford University Press.

Hoffrage, U., Lindsey, S., Hertwig, R., & Gigerenzer, G. 2000. Communicating statistical information. *Science, 290*, 2261–62. https://doi.org/10.1126/science.290.5500.2261.

Holland, P. W. 1986. Statistics and causal inference. *Journal of the American Statistical Association, 81*, 945–60. https://doi.org/10.2307/2289064.

Homer. 700 BCE/2018. *The Odyssey* (E. Wilson, trans.). New York: W. W. Norton.

Hood, B. 2009. *Supersense: Why we believe in the unbelievable.* New York: Harper-Collins.

Horowitz, D. L. 2001. *The deadly ethnic riot.* Berkeley: University of California Press.

Hume, D. 1739/2000. *A treatise of human nature.* New York: Oxford University Press.

Hume, D. 1748/1999. *An enquiry concerning human understanding.* New York: Oxford University Press.

Hunt, L. 2007. *Inventing human rights: A history.* New York: W. W. Norton.

Ichikawa, J. J., & Steup, M. 2018. The analysis of knowledge. In E. N. Zalta, ed., *The Stanford Encyclopedia of Philosophy.* https://plato.stanford.edu/entries/knowledge-analysis/.

Ioannidis, J. P. A. 2005. Why most published research findings are false. *PLoS Medicine, 2*, e124. https://doi.org/10.1371/journal.pmed.0020124.

James, W. 1890/1950. *The principles of psychology.* New York: Dover.

Jarvis, S., Deschenes, O., & Jha, A. 2019. *The private and external costs of Germany's nuclear phase-out.* https://haas.berkeley.edu/wp-content/uploads/WP304.pdf.

Jenkins, S. 2020. The Crown's fake history is as corrosive as fake news. *The Guardian,* Nov. 16. http://www.theguardian.com/commentisfree/2020/nov/16/the-crown-fake-history-news-tv-series-royal-family-artistic-licence.

Jeszeck, C. A., Collins, M. J., Glickman, M., Hoffrey, L., & Grover, S. 2015. Retirement security: Most households approaching retirement have low savings. United States Government Accountability Office. https://www.gao.gov/assets/680/670153.pdf.

Johnson, D. J., & Cesario, J. 2020. Reply to Knox and Mummolo and Schimmack and Carlsson: Controlling for crime and population rates. *Proceedings of the National Academy of Sciences, 117*, 1264–65. https://doi.org/10.1073/pnas.1920184117.

Johnson, D. J., Tress, T., Burkel, N., Taylor, C., & Cesario, J. 2019. Officer characteristics and racial disparities in fatal officer-involved shootings. *Proceedings of the National Academy of Sciences, 116*, 15877–82. https://doi.org/10.1073/pnas.1903856116.

Johnson, S. 1963. *The letters of Samuel Johnson with Mrs. Thrale's genuine letters to him* (R. W. Chapman, ed.). New York: Oxford University Press.

Jones, J. M. 2018. Confidence in higher education down since 2015. *Gallup Blog*, Oct. 9. https://news.gallup.com/opinion/gallup/242441/confidence-higher -education-down-2015.aspx.

Joyner, J. 2011. Ranking the pundits: A study shows that most national columnists and talking heads are about as accurate as a coin flip. *Outside the Beltway*, May 3. https://www.outsidethebeltway.com/ranking-the-pundits/.

Kaba, M. 2020. Yes, we mean literally abolish the police. *New York Times*, June 12. https://www.nytimes.com/2020/06/12/opinion/sunday/floyd-abolish-defund -police.html.

Kahan, D. M. 2013. Ideology, motivated reasoning, and cognitive reflection. *Judgment and Decision Making*, *8*, 407–24. http://dx.doi.org/10.2139/ssrn.2182588.

Kahan, D. M. 2015. Climate-science communication and the measurement problem. *Political Psychology*, *36*, 1–43. https://doi.org/10.1111/pops.12244.

Kahan, D. M., Hoffman, D. A., Braman, D., Evans, D., & Rachlinski, J. J. 2012. "They saw a protest": Cognitive illiberalism and the speech-conduct distinction. *Stanford Law Review*, *64*, 851–906.

Kahan, D. M., Peters, E., Dawson, E. C., & Slovic, P. 2017. Motivated numeracy and enlightened self-government. *Behavioural Public Policy*, *1*, 54–86. https:// doi.org/10.1017/bpp.2016.2.

Kahan, D. M., Peters, E., Wittlin, M., Slovic, P., Ouellette, L. L., et al. 2012. The polarizing impact of science literacy and numeracy on perceived climate change risks. *Nature Climate Change*, *2*, 732–35. https://doi.org/10.1038/nclimate1547.

Kahan, D. M., Wittlin, M., Peters, E., Slovic, P., Ouellette, L. L., et al. 2011. The tragedy of the risk-perception commons: Culture conflict, rationality conflict, and climate change. *Yale Law & Economics Research Paper*, *435*. http:// dx.doi.org/10.2139/ssrn.1871503.

Kahneman, D. 2002. Daniel Kahneman—facts. *The Nobel Prize*. https://www .nobelprize.org/prizes/economic-sciences/2002/kahneman/facts/.

Kahneman, D. 2011. *Thinking, fast and slow*. New York: Farrar, Straus and Giroux.

Kahneman, D., Slovic, P., & Tversky, A. 1982. *Judgment under uncertainty: Heuristics and biases*. New York: Cambridge University Press.

Kahneman, D., & Tversky, A. 1972. Subjective probability: A judgment of representativeness. *Cognitive Psychology*, *3*, 430–54. https://doi.org/10.1016/0010 -0285(72)90016-3.

Kahneman, D., & Tversky, A. 1979. Prospect theory: An analysis of decisions under risk. *Econometrica*, *47*, 263–91. https://doi.org/10.1142/9789814417358_0006.

Kahneman, D., & Tversky, A. 1996. On the reality of cognitive illusions. *Psychological Review*, *103*, 582–91. https://doi.org/10.1037/0033-295X.103.3.582.

Kaplan, R. D. 1994. The coming anarchy. *The Atlantic*. https://www.theatlantic .com/magazine/archive/1994/02/the-coming-anarchy/304670/.

Kelemen, D., & Rosset, E. 2009. The human function compunction: Teleological explanation in adults. *Cognition*, *111*, 138–43. https://doi.org/10.1016/j.cognition .2009.01.001.

Kendler, K. S., Kessler, R. C., Walters, E. E., MacLean, C., Neale, M. C., et al. 2010. Stressful life events, genetic liability, and onset of an episode of major depression in women. *Focus, 8*, 459–70. https://doi.org/10.1176/foc.8.3.foc459.

Kenny, C. 2011. *Getting better: Why global development is succeeding—and how we can improve the world even more.* New York: Basic Books.

Kessler, G., Rizzo, S., & Kelly, M. 2020. Trump is averaging more than 50 false or misleading claims a day. *Washington Post*, Oct. 22. https://www.washing tonpost.com/politics/2020/10/22/president-trump-is-averaging-more -than-50-false-or-misleading-claims-day/.

King, G., Keohane, R. O., & Verba, S. 1994. *Designing social inquiry: Scientific inference in qualitative research.* Princeton, NJ: Princeton University Press.

Kingdon, J. 1993. *Self-made man: Human evolution from Eden to extinction?* New York: Wiley.

Kissinger, H. 2018. How the Enlightenment ends. *The Atlantic*, June. https:// www.theatlantic.com/magazine/archive/2018/06/henry-kissinger-ai-could -mean-the-end-of-human-history/559124/.

Knox, D., & Mummolo, J. 2020. Making inferences about racial disparities in police violence. *Proceedings of the National Academy of Sciences, 117*, 1261–62. https://doi.org/10.1073/pnas.1919418117.

Kors, A. C., & Silverglate, H. A. 1998. *The shadow university: The betrayal of liberty on America's campuses.* New York: Free Press.

Kräenbring, J., Monzon Penza, T., Gutmann, J., Muehlich, S., Zolk, O., et al. 2014. Accuracy and completeness of drug information in Wikipedia: A com- parison with standard textbooks of pharmacology. *PLoS ONE, 9*, e106930. https://doi.org/10.1371/journal.pone.0106930.

Krämer, W., & Gigerenzer, G. 2005. How to confuse with statistics, or: The use and misuse of conditional probabilities. *Statistical Science, 20*, 223–30. https:// doi.org/10.1214/088342305000000029.

Kunda, Z. 1990. The case for motivated reasoning. *Psychological Bulletin, 108*, 480–98. https://doi.org/10.1037/0033-2909.108.3.480.

Laibson, D. 1997. Golden eggs and hyperbolic discounting. *Quarterly Journal of Economics, 112*, 443–77. https://doi.org/10.1162/003355397555253.

Lake, B. M., Ullman, T. D., Tenenbaum, J. B., & Gershman, S. J. 2017. Building machines that learn and think like people. *Behavioral and Brain Sciences, 39*, 1–101. https://doi.org/10.1017/S0140525X16001837.

Lane, R. 2021. A truth reckoning: Why we're holding those who lied for Trump accountable. *Forbes*, Jan. 7. https://www.forbes.com/sites/randalllane/2021/01 /07/a-truth-reckoning-why-were-holding-those-who-lied-for-trump-ac countable/?sh=5fedd2605710.

Lankford, A., & Madfis, E. 2018. Don't name them, don't show them, but report everything else: A pragmatic proposal for denying mass killers the attention they seek and deterring future offenders. *American Behavioral Scientist, 62*, 260–79. https://doi.org/10.1177/0002764217730854.

Lee, R. B., & Daly, R., eds. 1999. *The Cambridge Encyclopedia of Hunters and Gath- erers.* Cambridge, UK: Cambridge University Press.

Lehrer, J. 2010. The truth wears off. *New Yorker*, Dec. 5. https://www.newyorker
.com/magazine/2010/12/13/the-truth-wears-off.

Leibniz, G. W. 1679/1989. On universal synthesis and analysis, or the art of discovery and judgment. In L. E. Loemker, ed., *Philosophical papers and letters*.
New York: Springer.

Levitt, S. D., & Dubner, S. J. 2009. *Freakonomics: A rogue economist explores the
hidden side of everything*. New York: William Morrow.

Lewis, D. K. 1969. *Convention: A philosophical study*. Cambridge, MA: Harvard
University Press.

Lewis, M. 2016. *The undoing project: A friendship that changed our minds*. New York:
W. W. Norton.

Lewis, M. K. 2016. *Too dumb to fail: How the GOP betrayed the Reagan revolution to
win elections (and how it can reclaim its conservative roots)*. New York: Hachette.

Lewis-Kraus, G. 2016. The great A.I. awakening. *New York Times Magazine*, Dec.
14, p. 12. https://www.nytimes.com/2016/12/14/magazine/the-great-ai-awak
ening.html.

Liberman, M. Y. 2004. If P, so why not Q? *Language Log*, Aug. 5. http://itre.cis
.upenn.edu/~myl/languagelog/archives/001314.html.

Lichtenstein, S., & Slovic, P. 1971. Reversals of preference between bids and
choices in gambling decisions. *Journal of Experimental Psychology, 89*, 46–55.
https://doi.org/10.1037/h0031207.

Liebenberg, L. 1990. *The art of tracking: The origin of science*. Cape Town: David
Philip.

Liebenberg, L. 2013/2021. *The origin of science: The evolutionary roots of scientific
reasoning and its implications for tracking science* (2nd ed.). Cape Town: Cyber-
Tracker. https://cybertracker.org/downloads/tracking/Liebenberg-2013-The
-Origin-of-Science.pdf.

Liebenberg, L. 2020. Notes on tracking and trapping: Examples of hunter-
gatherer ingenuity. Unpublished manuscript. https://stevenpinker.com/files
/pinker/files/liebenberg.pdf.

Liebenberg, L., //Ao, /A., Lombard, M., Shermer, M., Xhukwe, /U., et al. 2021.
Tracking science: An alternative for those excluded by citizen science. *Citizen
Science: Theory and Practice, 6(1)*, 6. https://doi.org/10.5334/cstp.284.

Lilienfeld, S. O., Ammirati, R., & Landfield, K. 2009. Giving debiasing away:
Can psychological research on correcting cognitive errors promote human
welfare? *Perspectives on Psychological Science, 4*, 390–98. https://doi.org/10.1111
/j.1745-6924.2009.01144.x.

Locke, J. 1689/2015. *The second treatise of civil government*. Peterborough, Ont.:
Broadview Press.

Lockwood, A. H., Welker-Hood, K., Rauch, M., & Gottlieb, B. 2009. *Coal's assault on human health: A report from Physicians for Social Responsibility*. https://
www.psr.org/blog/resource/coals-assault-on-human-health/.

Loftus, E. F., Doyle, J. M., Dysart, J. E., & Newirth, K. A. 2019. *Eyewitness testimony: Civil and criminal* (6th ed.). Dayton, OH: LexisNexis.

Lord, C. G., Ross, L., & Lepper, M. R. 1979. Biased assimilation and attitude polarization: The effects of prior theories on subsequently considered evidence. *Journal of Personality and Social Psychology, 37,* 2098–109. https://doi.org/10.1037/0022-3514.37.11.2098.

Luce, R. D., & Raiffa, H. 1957. *Games and decisions: Introduction and critical survey.* New York: Dover.

Lukianoff, G. 2012. *Unlearning liberty: Campus censorship and the end of American debate.* New York: Encounter Books.

Lukianoff, G., & Haidt, J. 2018. *The coddling of the American mind: How good intentions and bad ideas are setting up a generation for failure.* New York: Penguin.

Lynn, S. K., Wormwood, J. B., Barrett, L. F., & Quigley, K. S. 2015. Decision making from economic and signal detection perspectives: Development of an integrated framework. *Frontiers in Psychology, 6.* https://doi.org/10.3389/fpsyg.2015.00952.

Lyttleton, J. 2020. Social media is determined to slow the spread of conspiracy theories like QAnon. Can they? *Millennial Source,* Oct. 28. https://themilsource.com/2020/10/28/social-media-determined-to-slow-spread-conspiracy-theories-like-qanon-can-they/.

MacAskill, W. 2015. *Doing good better: Effective altruism and how you can make a difference.* New York: Penguin.

Maines, R. 2007. Why are women crowding into schools of veterinary medicine but are not lining up to become engineers? *Cornell Chronicle,* June 12. https://news.cornell.edu/stories/2007/06/why-women-become-veterinarians-not-engineers.

Mann, T. E., & Ornstein, N. J. 2012/2016. *It's even worse than it looks: How the American constitutional system collided with the new politics of extremism* (new ed.). New York: Basic Books.

Marcus, G. F. 2000. Two kinds of representation. In E. Dietrich & A. B. Markman, eds., *Cognitive dynamics: Conceptual and representational change in humans and machines.* Mahwah, NJ: Erlbaum.

Marcus, G. F. 2018. The deepest problem with deep learning. *Medium,* Dec. 1. https://medium.com/@GaryMarcus/the-deepest-problem-with-deep-learning-91c5991f5695.

Marcus, G. F., & Davis, E. 2019. *Rebooting AI: Building artificial intelligence we can trust.* New York: Pantheon.

Marlowe, F. 2010. *The Hadza: Hunter-gatherers of Tanzania.* Berkeley: University of California Press.

Martin, G. J., & Yurukoglu, A. 2017. Bias in cable news: Persuasion and polarization. *American Economic Review, 107,* 2565–99. https://doi.org/10.1257/aer.20160812.

Maymin, P. Z., & Langer, E. J. 2021. Cognitive biases and mindfulness. *Humanities and Social Sciences Communications, 8,* 40. https://doi.org/10.1057/s41599-021-00712-1.

Maynard Smith, J. 1982. *Evolution and the theory of games.* New York: Cambridge University Press.

McCarthy, J. 2015. More Americans say crime is rising in U.S. Gallup, Oct. 22. https://news.gallup.com/poll/186308/americans-say-crime-rising.aspx.

McCarthy, J. 2019. Americans still greatly overestimate U.S. gay population. *Gallup.* https://news.gallup.com/poll/259571/americans-greatly-overestimate -gay-population.aspx.

McCawley, J. D. 1993. *Everything that linguists have always wanted to know about logic—but were ashamed to ask* (2nd ed.). Chicago: University of Chicago Press.

McClure, S. M., Laibson, D., Loewenstein, G., & Cohen, J. D. 2004. Separate neural systems value immediate and delayed monetary rewards. *Science, 306,* 503–7. https://doi.org/10.1126/science.1100907.

McGinn, C. 2012. *Truth by analysis: Games, names, and philosophy.* New York: Oxford University Press.

McNeil, B. J., Pauker, S. G., Sox, H. C., Jr., & Tversky, A. 1982. On the elicitation of preferences for alternative therapies. *New England Journal of Medicine, 306,* 1259–62. https://doi.org/10.1056/NEJM198205273062103.

Meehl, P. E. 1954/2013. *Clinical versus statistical prediction: A theoretical analysis and a review of the evidence.* Brattleboro, VT: Echo Point Books.

Mellers, B. A., Hertwig, R., & Kahneman, D. 2001. Do frequency representations eliminate conjunction effects? An exercise in adversarial collaboration. *Psychological Science, 12,* 269–75. https://doi.org/10.1111/1467-9280.00350.

Mellers, B. A., Ungar, L., Baron, J., Ramos, J., Gurcay, B., et al. 2014. Psychological strategies for winning a geopolitical forecasting tournament. *Psychological Science, 25,* 1106–15. https://doi.org/10.1177/0956797614524255.

Mercier, H. 2020. *Not born yesterday: The science of who we trust and what we believe.* Princeton, NJ: Princeton University Press.

Mercier, H., & Sperber, D. 2011. Why do humans reason? Arguments for an argumentative theory. *Behavioral and Brain Sciences, 34,* 57–111. https://doi.org /10.1017/S0140525X10000968.

Mercier, H., & Sperber, D. 2017. *The enigma of reason.* Cambridge, MA: Harvard University Press.

Mercier, H., Trouche, E., Yama, H., Heintz, C., & Girotto, V. 2015. Experts and laymen grossly underestimate the benefits of argumentation for reasoning. *Thinking & Reasoning, 21,* 341–55. https://doi.org/10.1080/13546783.2014 .981582.

Michel, J.-B., Shen, Y. K., Aiden, A. P., Veres, A., Gray, M. K., The Google Books Team, Pickett, J. P., Hoiberg, D., Clancy, D., Norvig, P., Orwant, J., Pinker, S., Nowak, M., & Lieberman-Aiden, E. 2011. Quantitative analysis of culture using millions of digitized books. *Science, 331,* 176–82.

Millenson, J. R. 1965. An inexpensive Geiger gate for controlling probabilities of events. *Journal of the Experimental Analysis of Behavior, 8,* 345–46.

Miller, J. B., & Sanjurjo, A. 2018. Surprised by the hot hand fallacy? A truth in the law of small numbers. *Econometrica, 86,* 2019–47. https://doi.org/10.3982 /ECTA14943.

Miller, J. B., & Sanjurjo, A. 2019. A bridge from Monty Hall to the hot hand: The principle of restricted choice. *Journal of Economic Perspectives, 33*, 144–62. https://doi.org/10.1257/jep.33.3.144.

Mischel, W., & Baker, N. 1975. Cognitive appraisals and transformations in delay behavior. *Journal of Personality and Social Psychology, 31*, 254–61. https://doi.org/10.1037/h0076272.

Mlodinow, L. 2009. *The drunkard's walk: How randomness rules our lives.* New York: Vintage.

Moore, D. W. 2005. Three in four Americans believe in paranormal. Gallup, June 16. https://news.gallup.com/poll/16915/three-four-americans-believe-paranormal.aspx.

Morewedge, C. K., Yoon, H., Scopelliti, I., Symborski, C. W., Korris, J. H., et al. 2015. Debiasing decisions: Improved decision making with a single training intervention. *Policy Insights from the Behavioral and Brain Sciences, 2*, 129–40. https://doi.org/10.1177/2372732215600886.

Mueller, J. 2006. *Overblown: How politicians and the terrorism industry inflate national security threats, and why we believe them.* New York: Free Press.

Mueller, J. 2021. *The stupidity of war: American foreign policy and the case for complacency.* New York: Cambridge University Press.

Myers, D. G. 2008. *A friendly letter to skeptics and atheists.* New York: Wiley.

Nagel, T. 1970. *The possibility of altruism.* Princeton, NJ: Princeton University Press.

Nagel, T. 1997. *The last word.* New York: Oxford University Press.

National Research Council. 2003. *The polygraph and lie detection.* Washington, DC: National Academies Press.

National Research Council. 2009. *Strengthening forensic science in the United States: A path forward.* Washington, DC: National Academies Press.

National Science Board. 2014. *Science and Engineering Indicators 2014.* Alexandria, VA: National Science Foundation. https://www.nsf.gov/statistics/seind14/index.cfm/home.

National Science Board. 2020. *The State of U.S. Science and Engineering 2020.* Alexandria, VA: National Science Foundation. https://ncses.nsf.gov/pubs/nsb20201/.

Nature editors. 2020a. A four-year timeline of Trump's impact on science. *Nature,* Oct. 5. https://doi.org/10.1038/d41586-020-02814-3.

Nature editors. 2020b. In praise of replication studies and null results. *Nature, 578*, 489–90. https://doi.org/10.1038/d41586-020-00530-6.

Nickerson, R. S. 1996. Hempel's paradox and Wason's selection task: Logical and psychological puzzles of confirmation. *Thinking & Reasoning, 2*, 1–31. https://doi.org/10.1080/135467896394546.

Nickerson, R. S. 1998. Confirmation bias: A ubiquitous phenomenon in many guises. *Review of General Psychology, 2*, 175–220. https://doi.org/10.1037/1089-2680.2.2.175.

Nolan, D., Bremer, M., Tupper, S., Malakhoff, L., & Medeiros, C. 2019. *Barnstable County high crash locations: Cape Cod Commission.* https://www.capecodcommission.org/resource-library/file/?url=/dept/commission/team/tr/Reference/Safety-General/Top50CrashLocs_2018Final.pdf.

Norberg, J. 2016. *Progress: Ten reasons to look forward to the future*. London: One-world.

Nordhaus, W. 2007. Critical assumptions in the Stern Review on climate change. *Science, 317*, 201–2. https://doi.org/10.1126/science.1137316.

Norenzayan, A., Smith, E. E., Kim, B., & Nisbett, R. E. 2002. Cultural prefer-ences for formal versus intuitive reasoning. *Cognitive Science, 26*, 653–84.

Norman, A. 2016. Why we reason: Intention-alignment and the genesis of human rationality. *Biology and Philosophy, 31*, 685–704. https://doi.org/10.1007/s10539-016-9532-4.

Norman, A. 2021. *Mental immunity: Infectious ideas, mind parasites, and the search for a better way to think*. New York: HarperCollins.

Nyhan, B. 2013. Building a better correction: Three lessons from new research on how to counter misinformation. *Columbia Journalism Review*. http://archives.cjr.org/united_states_project/building_a_better_correction_nyhan_new_misperception_research.php.

Nyhan, B. 2018. Fake news and bots may be worrisome, but their political power is overblown. *New York Times*, Feb. 13. https://www.nytimes.com/2018/02/13/upshot/fake-news-and-bots-may-be-worrisome-but-their-political-power-is-overblown.html.

Nyhan, B., & Reifler, J. 2012. *Misinformation and fact-checking: Research findings from social science*. Washington, DC: New America Foundation.

Nyhan, B., & Reifler, J. 2019. The roles of information deficits and identity threat in the prevalence of misperceptions. *Journal of Elections, Public Opinion and Parties, 29*, 222–44. https://doi.org/10.1080/17457289.2018.1465061.

O'Keefe, S. M. 2020. One in three Americans would not get COVID-19 vaccine. Gallup, Aug. 7. https://news.gallup.com/poll/317018/one-three-americans-not-covid-vaccine.aspx.

Open Science Collaboration. 2015. Estimating the reproducibility of psycho-logical science. *Science, 349*. https://doi.org/10.1126/science.aac4716.

Paresky, P., Haidt, J., Strossen, N., & Pinker, S. 2020. The New York Times sur-rendered to an outrage mob. Journalism will suffer for it. *Politico*, May 14. https://www.politico.com/news/magazine/2020/05/14/bret-stephens-new-york-times-outrage-backlash-256494.

Parker, A. M., Bruine de Bruin, W., Fischhoff, B., & Weller, J. 2018. Robustness of decision-making competence: Evidence from two measures and an 11-year longitudinal study. *Journal of Behavioral Decision Making, 31*, 380–91. https://doi.org/10.1002/bdm.2059.

Pashler, H., & Wagenmakers, E. J. 2012. Editors' introduction to the special sec-tion on replicability in psychological science: A crisis of confidence? *Per-spectives on Psychological Science, 7*, 528–30. https://doi.org/10.1177/1745691612465253.

Paulos, J. A. 1988. *Innumeracy: Mathematical illiteracy and its consequences*. New York: Macmillan.

Payne, J. L. 2004. *A history of force: Exploring the worldwide movement against habits of coercion, bloodshed, and mayhem*. Sandpoint, ID: Lytton.

Pearl, J. 2000. *Causality: Models, reasoning, and inference.* New York: Cambridge University Press.

Pearl, J., & Mackenzie, D. 2018. *The book of why: The new science of cause and effect.* New York: Basic Books.

Pennycook, G., Cannon, T. D., & Rand, D. G. 2018. Prior exposure increases perceived accuracy of fake news. *Journal of Experimental Psychology: General, 147,* 1865–80. https://doi.org/10.1037/xge0000465.

Pennycook, G., Cheyne, J. A., Koehler, D. J., & Fugelsang, J. A. 2020. On the belief that beliefs should change according to evidence: Implications for conspiratorial, moral, paranormal, political, religious, and science beliefs. *Judgment and Decision Making, 15,* 476–98. https://doi.org/10.31234/osf.io/a7k96.

Pennycook, G., Cheyne, J. A., Seli, P., Koehler, D. J., & Fugelsang, J. A. 2012. Analytic cognitive style predicts religious and paranormal belief. *Cognition, 123,* 335–46. https://doi.org/10.1016/j.cognition.2012.03.003.

Pennycook, G., & Rand, D. G. 2020a. The cognitive science of fake news. https://psyarxiv.com/ar96c.

Pennycook, G., & Rand, D. G. 2020b. Who falls for fake news? The roles of bullshit receptivity, overclaiming, familiarity, and analytic thinking. *Journal of Personality, 88,* 185–200. https://doi.org/10.1111/jopy.12476.

Pew Forum on Religion and Public Life. 2009. *Many Americans mix multiple faiths.* Washington, DC: Pew Research Center. https://www.pewforum.org/2009/12/09/many-americans-mix-multiple-faiths/.

Pinker, S. 1994/2007. *The language instinct.* New York: HarperCollins.

Pinker, S. 1997/2009. *How the mind works.* New York: W. W. Norton.

Pinker, S. 1999/2011. *Words and rules: The ingredients of language.* New York: HarperCollins.

Pinker, S. 2002/2016. *The blank slate: The modern denial of human nature.* New York: Penguin.

Pinker, S. 2007. *The stuff of thought: Language as a window into human nature.* New York: Viking.

Pinker, S. 2010. The cognitive niche: Coevolution of intelligence, sociality, and language. *Proceedings of the National Academy of Sciences, 107,* 8993–99. https://doi.org/10.1073/pnas.0914630107.

Pinker, S. 2011. *The better angels of our nature: Why violence has declined.* New York: Viking.

Pinker, S. 2012. Why are states so red and blue? *New York Times,* Oct. 24. http://opinionator.blogs.nytimes.com/2012/10/24/why-are-states-so-red-and-blue/?_r=0.

Pinker, S. 2015. Rock star psychologist Steven Pinker explains why #thedress looked white, not blue. *Forbes,* Feb. 28. https://www.forbes.com/sites/matthewherper/2015/02/28/psychologist-and-author-stephen-pinker-explains-thedress/.

Pinker, S. 2018. *Enlightenment now: The case for reason, science, humanism, and progress.* New York: Viking.

Pinker, S., & Mehler, J., eds. 1988. *Connections and symbols.* Cambridge, MA: MIT Press.

Pinker, S., & Prince, A. 2013. The nature of human concepts: Evidence from an unusual source. In S. Pinker, ed., *Language, cognition, and human nature: Selected articles.* New York: Oxford University Press.

Plato. 399–390 BCE/2002. Euthyphro (G. M. A. Grube, trans.). In J. M. Cooper, ed., *Plato: Five dialogues—Euthyphro, Apology, Crito, Meno, Phaedo* (2nd ed.). Indianapolis: Hackett.

Polderman, T. J. C., Benyamin, B., de Leeuw, C. A., Sullivan, P. F., van Bochoven, A., et al. 2015. Meta-analysis of the heritability of human traits based on fifty years of twin studies. *Nature Genetics, 47,* 702–9. https://doi.org/10.1038/ng.3285.

Popper, K. R. 1983. *Realism and the aim of science.* London: Routledge.

Poundstone, W. 1992. *Prisoner's dilemma: John von Neumann, game theory, and the puzzle of the bomb.* New York: Anchor.

President's Council of Advisors on Science and Technology. 2016. *Report to the President: Forensic science in criminal courts: ensuring scientific validity of feature-comparison methods.* https://obamawhitehouse.archives.gov/sites/default/files/microsites/ostp/PCAST/pcast_forensic_science_report_final.pdf.

Priest, G. 2017. *Logic: A very short introduction* (2nd ed.). New York: Oxford University Press.

Proctor, R. N. 2000. *The Nazi war on cancer.* Princeton, NJ: Princeton University Press.

Pronin, E., Lin, D. Y., & Ross, L. 2002. The bias blind spot: Perceptions of bias in self versus others. *Personality and Social Psychology Bulletin, 28,* 369–81. https://doi.org/10.1177/0146167202286008.

Purves, D., & Lotto, R. B. 2003. *Why we see what we do: An empirical theory of vision.* Sunderland, MA: Sinauer.

Rachels, J., & Rachels, S. 2010. *The elements of moral philosophy* (6th ed.). Columbus, OH: McGraw-Hill.

Raemon. 2017. What exactly is the "Rationality Community?" *LessWrong,* Apr. 9. https://www.lesswrong.com/posts/s8yvtCbbZW2S4WnhE/what-exactly-is-the-rationality-community.

Railton, P. 1986. Moral realism. *Philosophical Review, 95,* 163–207. https://doi.org/10.2307/2185589.

Rauch, J. 2018. The constitution of knowledge. *National Affairs,* Fall 2018. https://www.nationalaffairs.com/publications/detail/the-constitution-of-knowledge.

Rauch, J. 2021. *The constitution of knowledge: A defense of truth.* Washington, DC: Brookings Institution Press.

Richardson, J., Smith, A., Meaden, S., & Flip Creative. 2020. Thou shalt not commit logical fallacies. https://yourlogicalfallacyis.com/.

Richardson, L. F. 1960. *Statistics of deadly quarrels.* Pittsburgh: Boxwood Press.

Ridley, M. 1997. *The origins of virtue: Human instincts and the evolution of cooperation.* New York: Viking.

Ridley, M. 2010. *The rational optimist: How prosperity evolves.* New York: Harper-Collins.

Ritchie, H. 2018. Causes of death. *Our World in Data.* https://ourworldindata.org/causes-of-death.

Ritchie, S. 2015. *Intelligence: All that matters*. London: Hodder & Stoughton.

Ropeik, D. 2010. *How risky is it, really? Why our fears don't always match the facts.* New York: McGraw-Hill.

Rosch, E. 1978. Principles of categorization. In E. Rosch & B. B. Lloyd, eds., *Cognition and categorization*. Hillsdale, NJ: Erlbaum.

Rosen, J. 1996. The bloods and the crits. *New Republic*, Dec. 9. https://newrepublic.com/article/74070/the-bloods-and-the-crits.

Rosenthal, E. C. 2011. *The complete idiot's guide to game theory*. New York: Penguin.

Roser, M. 2016. Economic growth. *Our World in Data*. https://ourworldindata.org/economic-growth.

Roser, M., Ortiz-Ospina, E., & Ritchie, H. 2013. Life expectancy. *Our World in Data*. https://ourworldindata.org/life-expectancy.

Roser, M., Ritchie, H., Ortiz-Ospina, E., & Hasell, J. 2020. Coronavirus pandemic (COVID-19). *Our World in Data*. https://ourworldindata.org/coronavirus.

Rosling, H. 2019. *Factfulness: Ten reasons we're wrong about the world—and why things are better than you think*. New York: Flatiron.

Roth, G. A., Abate, D., Abate, K. H., Abay, S. M., Abbafati, C., et al. 2018. Global, regional, and national age-sex-specific mortality for 282 causes of death in 195 countries and territories, 1980–2017: A systematic analysis for the Global Burden of Disease Study 2017. *The Lancet, 392*, 1736–88. https://doi.org/10.1016/S0140-6736(18)32203-7.

Rumelhart, D. E., Hinton, G. E., & Williams, R. J. 1986. Learning representations by back-propagating errors. *Nature, 323*, 533–36. https://doi.org/10.1038/323533a0.

Rumelhart, D. E., McClelland, J. L., & PDP Research Group. 1986. *Parallel distributed processing: Explorations in the microstructure of cognition*, vol. 1, *Foundations*. Cambridge, MA: MIT Press.

Rumney, P. N. S. 2006. False allegations of rape. *Cambridge Law Journal, 65*, 128–58. https://doi.org/10.1017/S0008197306007069.

Russell, B. 1950/2009. *Unpopular essays*. Philadelphia: Routledge.

Russell, B. 1969. Letter to Mr. Major. In B. Feinberg & R. Kasrils, eds., *Dear Bertrand Russell: A selection of his correspondence with the general public, 1950–1968*. London: Allen & Unwin.

Russett, B., & Oneal, J. R. 2001. *Triangulating peace: Democracy, interdependence, and international organizations*. New York: W. W. Norton.

Sá, W. C., West, R. F., & Stanovich, K. E. 1999. The domain specificity and generality of belief bias: Searching for a generalizable critical thinking skill. *Journal of Educational Psychology, 91*, 497–510. https://doi.org/10.1037/0022-0663.91.3.497.

Saenen, L., Heyvaert, M., Van Dooren, W., Schaeken, W., & Onghena, P. 2018. Why humans fail in solving the Monty Hall dilemma: A systematic review. *Psychologica Belgica, 58*, 128–58. https://doi.org/10.5334/pb.274.

Sagan, S. D., & Suri, J. 2003. The madman nuclear alert: Secrecy, signaling, and safety in October 1969. *International Security, 27*, 150–83.

Saldin, R. P., & Teles, S. M. 2020. *Never Trump: The revolt of the conservative elites.* New York: Oxford University Press.

Salganik, M. J., Lundberg, I., Kindel, A. T., Ahearn, C. E., Al-Ghoneim, K., et al. 2020. Measuring the predictability of life outcomes with a scientific mass collaboration. *Proceedings of the National Academy of Sciences, 117*, 8398–403. https://doi.org/10.1073/pnas.1915006117.

Satel, S. 2008. *When altruism isn't enough: The case for compensating kidney donors.* Washington, DC: AEI Press.

Savage, I. 2013. Comparing the fatality risks in United States transportation across modes and over time. *Research in Transportation Economics, 43*, 9–22. https://doi.org/10.1016/j.retrec.2012.12.011.

Savage, L. J. 1954. *The foundations of statistics.* New York: Wiley.

Schelling, T. C. 1960. *The strategy of conflict.* Cambridge, MA: Harvard University Press.

Schelling, T. C. 1984. The intimate contest for self-command. In T. C. Schelling, ed., *Choice and consequence: Perspectives of an errant economist.* Cambridge, MA: Harvard University Press.

Schneps, L., & Colmez, C. 2013. *Math on trial: How numbers get used and abused in the courtroom.* New York: Basic Books.

Scott-Phillips, T. C., Dickins, T. E., & West, S. A. 2011. Evolutionary theory and the ultimate–proximate distinction in the human behavioral sciences. *Perspectives on Psychological Science, 6*, 38–47. https://doi.org/10.1177/1745691610393528.

Scribner, S., & Cole, M. 1973. Cognitive consequences of formal and informal education. *Science, 182*, 553–59. https://doi.org/10.1126/science.182.4112.553.

Seebach, L. 1994. The fixation with the last 10 percent of risk. *Baltimore Sun,* Apr. 13. https://www.baltimoresun.com/news/bs-xpm-1994-04-13-1994103157-story.html.

Selvin, S. 1975. A problem in probability. *American Statistician, 29*, 67. https://www.jstor.org/stable/2683689.

Serwer, A. 2006. The greatest money manager of our time. *CNN Money,* Nov. 15. https://money.cnn.com/magazines/fortune/fortune_archive/2006/11/27/8394343/index.htm.

Shackel, N. 2014. Motte and bailey doctrines. https://blog.practicalethics.ox.ac.uk/2014/09/motte-and-bailey-doctrines/.

Sherman, C. 2019. The shark attack that changed Cape Cod forever. *Boston Magazine,* May 14. https://www.bostonmagazine.com/news/2019/05/14/cape-cod-sharks/.

Shermer, M. 1997. *Why people believe weird things.* New York: Freeman.

Shermer, M. 2008. The doping dilemma: Game theory helps to explain the pervasive abuse of drugs in cycling, baseball, and other sports. *Scientific American, 298*, 82–89. https://www.jstor.org/stable/26000562?seq=1.

Shermer, M. 2011. *The believing brain: From ghosts and gods to politics and conspiracies.* New York: St. Martin's Press.

Shermer, M. 2015. *The moral arc: How science and reason lead humanity toward truth, justice, and freedom.* New York: Henry Holt.

Shermer, M. 2020a. COVID-19 conspiracists and their discontents. *Quillette*, May 7. https://quillette.com/2020/05/07/covid-19-conspiracists-and-their-discontents/.

Shermer, M. 2020b. The top ten weirdest things countdown. *Skeptic*. https://www.skeptic.com/reading_room/the-top-10-weirdest-things/.

Shermer, M. 2020c. Why people believe conspiracy theories. *Skeptic, 25,* 12–17.

Shtulman, A. 2017. *Scienceblind: Why our intuitive theories about the world are so often wrong.* New York: Basic Books.

Shubik, M. 1971. The dollar auction game: A paradox in noncooperative behavior and escalation. *Journal of Conflict Resolution, 15,* 109–11. https://doi.org/10.1177/002200277101500111.

Simanek, D. 1999. Horse's teeth. https://www.lockhaven.edu/~dsimanek/horse.htm.

Simmons, J. P., Nelson, L. D., & Simonsohn, U. 2011. False-positive psychology: Undisclosed flexibility in data collection and analysis allows presenting anything as significant. *Psychological Science, 22,* 1359–66. https://doi.org/10.1177/0956797611417632.

Simon, H. A. 1956. Rational choice and the structure of the environment. *Psychological Review, 63,* 129–38. https://doi.org/10.1037/h0042769.

Singer, P. 1981/2011. *The expanding circle: Ethics and sociobiology.* Princeton, NJ: Princeton University Press.

Sloman, S. A. 1996. The empirical case for two systems of reasoning. *Psychological Bulletin, 119,* 3–22. https://doi.org/10.1037/0033-2909.119.1.3.

Sloman, S. A., & Fernbach, P. 2017. *The knowledge illusion: Why we never think alone.* New York: Penguin.

Slovic, P. 1987. Perception of risk. *Science, 236,* 280–85. https://doi.org/10.1126/science.3563507.

Slovic, P. 2007. "If I look at the mass I will never act": Psychic numbing and genocide. *Judgment and Decision Making, 2,* 79–95. https://doi.org/10.1007/978-90-481-8647-1_3.

Slovic, P., & Tversky, A. 1974. Who accepts Savage's axiom? *Behavioral Science, 19,* 368–73. https://doi.org/10.1002/bs.3830190603.

Soave, R. 2014. Ezra Klein "completely supports" "terrible" Yes Means Yes law. *Reason,* Oct. 13. https://reason.com/2014/10/13/ezra-klein-completely-supports-terrible/.

Social Progress Imperative. 2020. 2020 Social Progress Index. https://www.socialprogress.org/.

Sowell, T. 1987. *A conflict of visions: Ideological origins of political struggles.* New York: Quill.

Sowell, T. 1995. *The vision of the anointed: Self-congratulation as a basis for social policy.* New York: Basic Books.

Sperber, D. 1997. Intuitive and reflective beliefs. *Mind & Language, 12,* 67–83. https://doi.org/10.1111/j.1468-0017.1997.tb00062.x.

Sperber, D., Cara, F., & Girotto, V. 1995. Relevance theory explains the selection task. *Cognition, 57,* 31–95. https://doi.org/10.1016/0010-0277(95)00666-M.

Spinoza, B. 1677/2000. *Ethics* (G. H. R. Parkinson, trans.). New York: Oxford University Press.

Stango, V., & Zinman, J. 2009. Exponential growth bias and household finance. *Journal of Finance, 64*, 2807–49. https://doi.org/10.1111/j.1540-6261.2009.01518.x.

Stanovich, K. E. 2012. On the distinction between rationality and intelligence: Implications for understanding individual differences in reasoning. In K. J. Holyoak & R. G. Morrison, eds., *The Oxford Handbook of Thinking and Reasoning*. New York: Oxford University Press.

Stanovich, K. E. 2018. How to think rationally about world problems. *Journal of Intelligence, 6(2)*. https://doi.org/10.3390/jintelligence6020025.

Stanovich, K. E. 2020. The bias that divides us. *Quillette*, Sept. 26. https://quillette.com/2020/09/26/the-bias-that-divides-us/.

Stanovich, K. E. 2021. *The bias that divides us: The science and politics of myside thinking*. Cambridge, MA: MIT Press.

Stanovich, K. E., & Toplak, M. E. 2019. The need for intellectual diversity in psychological science: Our own studies of actively open-minded thinking as a case study. *Cognition, 187*, 156–66. https://doi.org/10.1016/j.cognition.2019.03.006.

Stanovich, K. E., & West, R. F. 1998. Cognitive ability and variation in selection task performance. *Thinking and Reasoning, 4*, 193–230.

Stanovich, K. E., West, R. F., & Toplak, M. E. 2016. *The rationality quotient: Toward a test of rational thinking*. Cambridge, MA: MIT Press.

Statista Research Department. 2019. Beliefs and conspiracy theories in the U.S.—Statistics & Facts. Aug. 13. https://www.statista.com/topics/5103/beliefs-and-superstition-in-the-us/#dossierSummary__chapter5.

Stenger, V. J. 1990. *Physics and psychics: The search for a world beyond the senses*. Buffalo, NY: Prometheus.

Stevenson, B., & Wolfers, J. 2008. Economic growth and subjective well-being: Reassessing the Easterlin Paradox. *Brookings Papers on Economic Activity*, 1–87. https://doi.org/10.3386/w14282.

Stoppard, T. 1972. *Jumpers: A play*. New York: Grove Press.

Stuart, E. A. 2010. Matching methods for causal inference: A review and a look forward. *Statistical Science, 25*, 1–21. https://doi.org/10.1214/09-STS313.

Suits, B. 1978/2014. *The grasshopper: Games, life, and utopia* (3rd ed.). Peterborough, Ont.: Broadview Press.

Sunstein, C. R., & Vermeule, A. 2008. Conspiracy theories. *John M. Olin Program in Law and Economics Working Papers, 387*. https://dx.doi.org/10.2139/ssrn.1084585.

Swets, J. A., Dawes, R. M., & Monahan, J. 2000. Better decisions through science. *Scientific American, 283*, 82–87.

Sydnor, J. 2010. (Over)insuring modest risks. *American Economic Journal: Applied Economics, 2*, 177–99. https://doi.org/10.1257/app.2.4.177.

Sykes, C. J. 2017. *How the right lost its mind*. New York: St. Martin's Press.

Taber, C. S., & Lodge, M. 2006. Motivated skepticism in the evaluation of political beliefs. *American Journal of Political Science, 50*, 755–69. https://doi.org/10.1111/j.1540-5907.2006.00214.x.

Talwalkar, P. 2013. The taxi-cab problem. *Mind Your Decisions*, Sept. 5. https://mindyourdecisions.com/blog/2013/09/05/the-taxi-cab-problem/.

Tate, J., Jenkins, J., Rich, S., Muyskens, J., Fox, J., et al. 2020. Fatal force. https://www.washingtonpost.com/graphics/investigations/police-shootings-database/, retrieved Oct. 14, 2020.

Temple, N. 2015. The possible importance of income and education as co-variates in cohort studies that investigate the relationship between diet and disease. *F1000Research*, *4*, 690. https://doi.org/10.12688/f1000research.6929.2.

Terry, Q. C. 2008. *Golden Rules and Silver Rules of humanity: Universal wisdom of civilization*. Berkeley, CA: AuthorHouse.

Tetlock, P. E. 1994. Political psychology or politicized psychology: Is the road to scientific hell paved with good moral intentions? *Political Psychology*, *15*, 509–29. https://doi.org/10.2307/3791569.

Tetlock, P. E. 2002. Social functionalist frameworks for judgment and choice: Intuitive politicians, theologians, and prosecutors. *Psychological Review*, *109*, 451–71. https://doi.org/10.1037/0033-295X.109.3.451.

Tetlock, P. E. 2003. Thinking the unthinkable: Sacred values and taboo cognitions. *Trends in Cognitive Sciences*, *7*, 320–24. https://doi.org/10.1016/S1364-6613(03)00135-9.

Tetlock, P. E. 2009. *Expert political judgment: How good is it? How can we know?* Princeton, NJ: Princeton University Press.

Tetlock, P. E. 2015. All it takes to improve forecasting is keep score. Paper presented at the Seminars about Long-Term Thinking, San Francisco, Nov. 23.

Tetlock, P. E., & Gardner, D. 2015. *Superforecasting: The art and science of prediction*. New York: Crown.

Tetlock, P. E., Kristel, O. V., Elson, S. B., Green, M. C., & Lerner, J. S. 2000. The psychology of the unthinkable: Taboo trade-offs, forbidden base rates, and heretical counterfactuals. *Journal of Personality and Social Psychology*, *78*, 853–70. https://doi.org/10.1037/0022-3514.78.5.853.

Thaler, R. H., & Sunstein, C. R. 2008. *Nudge: Improving decisions about health, wealth, and happiness*. New Haven, CT: Yale University Press.

Thomas, K. A., De Freitas, J., DeScioli, P., & Pinker, S. 2016. Recursive mentalizing and common knowledge in the bystander effect. *Journal of Experimental Psychology: General*, *145*, 621–29. https://doi.org/10.1037/xge0000153.

Thomas, K. A., DeScioli, P., Haque, O. S., & Pinker, S. 2014. The psychology of coordination and common knowledge. *Journal of Personality and Social Psychology*, *107*, 657–76. https://doi.org/10.1037/a0037037.

Thompson, C. 2020. QAnon is like a game—a most dangerous game. *WIRED Magazine*, Sept. 22. https://www.wired.com/story/qanon-most-dangerous-multiplatform-game/.

Thompson, D. A., & Adams, S. L. 1996. The full moon and ED patient volumes: Unearthing a myth. *American Journal of Emergency Medicine*, *14*, 161–64. https://doi.org/10.1016/S0735-6757(96)90124-2.

Tierney, J. 1991. Behind Monty Hall's doors: Puzzle, debate, and answer. *New York Times*, July 21. https://www.nytimes.com/1991/07/21/us/behind-monty -hall-s-doors-puzzle-debate-and-answer.html.

Tierney, J., & Baumeister, R. F. 2019. *The power of bad: How the negativity effect rules us and how we can rule it.* New York: Penguin.

Todd, B. 2017. Introducing longtermism. https://80000hours.org/articles/future -generations/.

Tollefson, J. 2020. How Trump damaged science—and why it could take decades to recover. *Nature, 586*, 190–94, Oct. 5. https://www.nature.com/articles /d41586-020-02800-9.

Toma, M. 2020. Gen Ed 1066 decision-making competence survey. Harvard University.

Tooby, J., & Cosmides, L. 1993. Ecological rationality and the multimodular mind: Grounding normative theories in adaptive problems. In K. I. Mank-telow & D. E. Over, eds., *Rationality: Psychological and philosophical perspectives.* London: Routledge.

Tooby, J., Cosmides, L., & Price, M. E. 2006. Cognitive adaptations for *n*-person exchange: The evolutionary roots of organizational behavior. *Managerial and Decision Economics, 27*, 103–29. https://doi.org/10.1002/mde.1287.

Tooby, J., & DeVore, I. 1987. The reconstruction of hominid behavioral evolu-tion through strategic modeling. In W. G. Kinzey, ed., *The evolution of human behavior: Primate models.* Albany: SUNY Press.

Toplak, M. E., West, R. F., & Stanovich, K. E. 2017. Real-world correlates of performance on heuristics and biases tasks in a community sample. *Journal of Behavioral Decision Making, 30*, 541–54. https://doi.org/10.1002/bdm.1973.

Trivers, R. L. 1971. The evolution of reciprocal altruism. *Quarterly Review of Biology, 46*, 35–57. https://doi.org/10.1086/406755.

Tversky, A. 1969. Intransitivity of preferences. *Psychological Review, 76*, 31–48. https://doi.org/10.1037/h0026750.

Tversky, A. 1972. Elimination by aspects: A theory of choice. *Psychological Review, 79*, 281–99. https://doi.org/10.1037/h0032955.

Tversky, A., & Kahneman, D. 1971. Belief in the law of small numbers. *Psycho-logical Bulletin, 76*, 105–10. https://doi.org/10.1037/h0031322.

Tversky, A., & Kahneman, D. 1973. Availability: A heuristic for judging fre-quency and probability. *Cognitive Psychology, 5*, 207–32. https://doi.org/10.1016 /0010-0285(73)90033-9.

Tversky, A., & Kahneman, D. 1974. Judgment under uncertainty: Heuristics and biases. *Science, 185*, 1124–31. https://doi.org/10.1126/science.185.4157.1124.

Tversky, A., & Kahneman, D. 1981. The framing of decisions and the psychology of choice. *Science, 211*, 453–58. https://doi.org/10.1126/science.7455683.

Tversky, A., & Kahneman, D. 1982. Evidential impact of base rates. In D. Kah-neman, P. Slovic, & A. Tversky, eds., *Judgment under uncertainty: Heuristics and biases.* New York: Cambridge University Press.

Tversky, A., & Kahneman, D. 1983. Extensions versus intuitive reasoning: The conjunction fallacy in probability judgment. *Psychological Review, 90*, 293–315.

Twain, M. 1897/1989. *Following the equator.* New York: Dover.

Uscinski, J. E., & Parent, J. M. 2014. *American conspiracy theories.* New York: Oxford University Press.

Vaci, N., Edelsbrunner, P., Stern, E., Neubauer, A., Bilalić, M., et al. 2019. The joint influence of intelligence and practice on skill development throughout the life span. *Proceedings of the National Academy of Sciences, 116,* 18363–69. https://doi.org/10.1073/pnas.1819086116.

van Benthem, J. 2008. Logic and reasoning: Do the facts matter? *Studia Logica, 88,* 67–84. https://doi.org/10.1007/s11225-008-9101-1.

van Prooijen, J.-W., & van Vugt, M. 2018. Conspiracy theories: Evolved functions and psychological mechanisms. *Perspectives on Psychological Science, 13,* 770–88. https://doi.org/10.1177/1745691618774270.

VanderWeele, T. J. 2014. Commentary: Resolutions of the birthweight paradox: competing explanations and analytical insights. *International Journal of Epidemiology, 43,* 1368–73. https://doi.org/10.1093/ije/dyu162.

Varian, H. 2006. Recalculating the costs of global climate change. *New York Times,* Dec. 14. https://www.nytimes.com/2006/12/14/business/14scene.html.

Vazsonyi, A. 1999. Which door has the Cadillac? *Decision Line,* 17–19. https://web.archive.org/web/20140413131827/http://www.decisionsciences.org/DecisionLine/Vol30/30_1/vazs30_1.pdf.

Venkataraman, B. 2019. *The optimist's telescope: Thinking ahead in a reckless age.* New York: Riverhead Books.

von Neumann, J., & Morgenstern, O. 1953/2007. *Theory of games and economic behavior* (60th anniversary commemorative ed.). Princeton, NJ: Princeton University Press.

vos Savant, M. 1990. Game show problem. *Parade,* Sept. 9. https://web.archive.org/web/20130121183432/http://marilynvossavant.com/game-show-problem/.

Vosoughi, S., Roy, D., & Aral, S. 2018. The spread of true and false news online. *Science, 359,* 1146–51. https://doi.org/10.1126/science.aap9559.

Wagenaar, W. A., & Sagaria, S. D. 1975. Misperception of exponential growth. *Perception & Psychophysics, 18,* 416–22. https://doi.org/10.3758/BF03204114.

Wagenaar, W. A., & Timmers, H. 1979. The pond-and-duckweed problem: Three experiments on the misperception of exponential growth. *Acta Psychologica, 43,* 239–51. https://doi.org/10.1016/0001-6918(79)90028-3.

Walker, C., Petulla, S., Fowler, K., Mier, A., Lou, M., et al. 2019. 10 years. 180 school shootings. 356 victims. CNN, July 24. https://www.cnn.com/interactive/2019/07/us/ten-years-of-school-shootings-trnd/.

Wan, W., & Shammas, B. 2020. Why Americans are numb to the staggering coronavirus death toll. *Washington Post,* Dec. 21. https://www.washingtonpost.com/health/2020/12/21/covid-why-we-ignore-deaths/.

Warburton, N. 2007. *Thinking from A to Z* (3rd ed.). New York: Routledge.

Wason, P. C. 1966. Reasoning. In B. M. Foss, ed., *New horizons in psychology.* London: Penguin.

Weber, M. 1922/2019. *Economy and society: A new translation* (K. Tribe, trans.). Cambridge, MA: Harvard University Press.

Weissman, M. B. 2020. Do GRE scores help predict getting a physics Ph.D.? A comment on a paper by Miller et al. *Science Advances, 6,* eaax3787. https://doi .org/10.1126/sciadv.aax3787.

Welzel, C. 2013. *Freedom rising: Human empowerment and the quest for emancipation.* New York: Cambridge University Press.

Wilkinson, W. 2019. *The density divide: Urbanization, polarization, and populist backlash.* Washington, DC: Niskanen Center. https://www.niskanencenter .org/the-density-divide-urbanization-polarization-and-populist-backlash/.

Williams, D. 2020. Motivated ignorance, rationality, and democratic politics. *Synthese,* 1–21.

Willingham, D. T. 2007. Critical thinking: Why is it so hard to teach? *American Educator, 31,* 8–19. https://doi.org/10.3200/AEPR.109.4.21-32.

Wittgenstein, L. 1953. *Philosophical investigations.* New York: Macmillan.

Wolfe, D., & Dale, D. 2020. "It's going to disappear": A timeline of Trump's claims that Covid-19 will vanish. Oct. 31. https://www.cnn.com/interactive /2020/10/politics/covid-disappearing-trump-comment-tracker/.

Wolfe, J. M., Kluender, K. R., Levi, D. M., Bartoshuk, L. M., Herz, R. S., et al. 2020. *Sensation & perception* (6th ed.). Sunderland, MA: Sinauer.

Wollstonecraft, M. 1792/1995. *A Vindication of the rights of woman: With strictures on political and moral subjects.* New York: Cambridge University Press.

Wood, T., & Porter, E. 2019. The elusive backfire effect: Mass attitudes' steadfast factual adherence. *Political Behavior, 41,* 135–63. https://doi.org/10.1007 /s11109-018-9443-y.

Yang, A. 2020. The official website for the Yang 2020 campaign. www.yang 2020.com.

Yglesias, M. 2020a. Defund police is a bad idea, not a bad slogan. *Slow Boring,* Dec. 7. https://www.slowboring.com/p/defund-police-is-a-bad-idea-not-a.

Yglesias, M. 2020b. The End of Policing left me convinced we still need policing. *Vox,* June 18. https://www.vox.com/2020/6/18/21293784/alex-vitale-end-of -policing-review.

Young, C. 2014a. The argument against affirmative consent laws gets Voxjacked. *Reason,* Oct. 15. https://reason.com/2014/10/15/the-argument-against-affirma tive-consent/.

Young, C. 2014b. Crying rape. *Slate,* Sept. 18. https://slate.com/human-interest /2014/09/false-rape-accusations-why-must-we-pretend-they-never-happen .html.

Zelizer, V. A. 2005. *The purchase of intimacy.* Princeton, NJ: Princeton University Press.

Ziman, J. M. 1978. *Reliable knowledge: An exploration of the grounds for belief in science.* New York: Cambridge University Press.

INDEX OF BIASES
AND FALLACIES

active open-mindedness, lack of, 310–11, 356–57n67

ad hominem fallacy, 19–20, 90–91, 92–93, 291

affective fallacy, 92–93, 291

affirming the consequent, 83, 85, 139

Allais paradox, 188–90

all-or-none causation fallacy, 260, 269

appeal to emotion, 92

a priori–a posteriori confusion. *See* post hoc probability

argument from authority, 3–4, 90, 291

availability bias and heuristic, 11, 119–23, 125–27, 320, 347n22

bandwagon fallacy, 90, 93, 291

base rate neglect, 154–57, 349–50nn6,27

begging the question, 89

bias bias, 291

biased assimilation, 290–91

biased evaluation, 291, 294

bounded rationality, 184–88

broken-leg problem, overestimation of, 279–80

burden of proof fallacy, 89

clinical vs. actuarial judgment, 278–80

close-mindedness. *See* active open-mindedness, lack of

cluster illusion, 146–48

cognitive reflection, lack of, 8–10, 311

collider fallacy, *261*, 262–63

confirmation bias, 13–14, 142–43, 216, 290, 342n26

conjunction fallacy (Linda problem), 26–29, 115, 116, 156

correlation implies causation, 245–47, 251–52, 312, 321, 323–24, 329–30

data snooping, 145–46, 160

denying the antecedent, 83, 294

dieter's fallacy, 101

discounting the future too steeply, 320

dread risk, 122

exponential growth bias, 10–12, 320–21

expressive rationality, 297–98

false dichotomy, 100

forbidden base rates, 62, 163–66

framing effects, 117–18, 168–70, 178, 188–92, 192–96, 321, 323, 349–50n27

garden of forking paths, 145, 185, 348n60

genetic fallacy, 91, 92–93, 291

guilt by association, 91

heretical counterfactuals, 64–65
hot hand fallacy, 131
hot hand fallacy fallacy, 131–32
hyperbolic discounting. *See* myopic
    temporal discounting

illusory correlation, 245–46, 251–52, 321
imaginability. *See* availability heuristic
intransitivity, 176, 185–88
irrelevant alternatives, sensitivity to,
    177–78, 188–92, 350n8

loss aversion, 192–94

mañana fallacy, 101
Meadow's fallacy (multiplying
    probabilities of interdependent
    events), 129–30, 131
money pump, 176, 180, 185, 187–88
monocausal fallacy, 260, 272–73
motte-and-bailey tactic (moving the
    goalposts), 88
moving the goalposts (motte-and-
    bailey), 88
myopic temporal discounting, 52–56, 54
myside bias, 294–96, 297, 312–13, 316, 317,
    357n73
mythology mindset, 301–9

no true Scotsman fallacy, 88

openness to evidence, lack of, 310–11,
    356–57n67
outrage, communal, 123–27
overconfidence, 20, 29–30, 33, 115, 216,
    255, 323

paradoxical tactics, 58–62
paradox of the heap, 101
post hoc probability, 141–48, 160, 321
preference reversal, 52–53, 55
probability neglect, 11, 28, 321

propensity confused with probability,
    21–22, 118, 139–40, 198, 216
prosecutor's fallacy, 140–41

questionable research practices,
    145–46, 160

regression to the mean, unawareness of,
    254–56, 320, 353n13
regret avoidance, 17, 190
representativeness heuristic, 27, 155–56
resistance to evidence. *See* openness to
    evidence, lack of

selective exposure, 290–91
sexism, 19–20
slippery slope fallacy, 100–101
so-what-you're-saying-is, 88
special pleading, 88
stereotyping, 99–100, 108–09. *See also*
    representativeness
straw man, 88, 291
sunk cost fallacy, 237–38, 320, 323
System 1 thinking. *See* cognitive
    reflection, lack of

taboos, 62, 124, 166. *See also* forbidden
    base rates; heretical
    counterfactuals; taboo tradeoffs
taboo tradeoffs, 62–64, 184, 350n15
Texas sharpshooter fallacy, 142–46,
    160, 321
Tragedy of the Rationality Commons,
    298, 315–17
tu quoque (what-aboutery), 89

unreflective thinking, 8–10, 311

virtus dormitiva, 11–12, 53, 89

what-aboutery (*tu quoque*), 89
Winner's Curse, 256

INDEX

Abelson, Robert, 299
abortion, 79, 100–101, 295, 311
academia
    academic freedom in, 41
    ad hominem fallacy and, 91
    argument from authority and, 90
    confidence in, sinking, 313–14
    critical race theory, 123
    informal fallacies and intellectual life
        of, 92–93
    left-wing monoculture, 297, 313–14
    suppression of opinions in, 43, 313–14
    viewpoint diversity, lack of, 313–14
    See also education; universities
accidental deaths, 120–22, 139–40, 199
Achilles, 38
Active Open-Mindedness, 310–11, 324
    See also openness to evidence
ad hominem fallacy, 19–20, 90–91,
        92–93, 291
adversarial collaboration, 29
Aeneid, 302
affective fallacy, 92–93, 291
affirming the antecedent, 80
affirming the consequent, 83, 85, 139
Afghanistan, US invasion of, 122, 124
African Americans
    Black Lives Matter, 26,
        124–25

as percentage of population, 120
    police killings of, 123, 124–25, 141
    See also racism; slavery
AI. See artificial intelligence
algorithm, as term, 346n34
Allais, Maurice, 188
Allais paradox, 188–90
all-or-none causation fallacy, 260, 269
alternative medicine, 258
analytic vs. synthetic propositions, 94
anarchism, 244
AND, as logical connector, 75–76
    See also conjunction
animals, cruelty to, 334–35
anti-vax movement, 284, 287, 304, 304–5,
        310, 311, 321
appeal to emotion, 92
a priori–a posteriori confusion. See post hoc
        probability
Arab Spring, 124
argument from authority, 3–4, 90, 291
Aristotle, 81, 99, 101
Arkes, Hal, 219–20
artificial intelligence (AI)
    great awakening, 106–7, 346n34
    unpredictability of human behavior and
        accuracy of, 281
    See also deep learning networks
Astell, Mary, 336

astrology, 14, 90, 142, 158, 285–86, 305, 311, 322, 354–55n8
atheism, 302
auctions, 236–37, 238
auditory perception, 146, 211
autism–vaccine claims, 287, 311
autocracy
    dissidents in, 299
    imposition of beliefs, 43, 245–46
    peace and, 266, 269, 270
    rationality and, 245–46
availability heuristic
    COVID-19 and, 11
    dangers of, 126–27
    definition, 119
    ignorance of progress and, 125–26, 324
    journalism and, 120, 125–27
    and killings, social response to, 122–23
    and memory, 119–20
    risk assessment by, 120–23, 320
    world events and, 120–22, 347n22

Bacon, Francis, 14, 94–95, 142, 346n21
bandwagon fallacy, 90, 93, 291
Bannon, Steve, 313
bargaining, 59–60, 235, 236
Barrie, J. M., 264
base rates
    change in, 167
    conditional probabilities and, 134, 138–39, 141
    conjunction probabilities and, 130
    forbidden, 62, 163–66
    neglect of, 154–57, 349–50nn6,27
    as priors in Bayesian reasoning, 167–69
    random sampling, 168
    as reference class, 167
    relevance of, 167–68
Baumeister, Roy, 124
Bayesian reasoning
    base-rate neglect, 154–57, 349–50nn6,27
    base-rates, forbidden, 62, 163–66
    causal Bayesian networks, 260–63, 261, 263
    changes in base rates, 167
    credence in a hypothesis, 151
    definition, 149–50, 151
    equation for, 151–53

"extraordinary claims require extraordinary evidence," 159, 161
    forecasting and, 162–63
    medicine and, 150–51, 152, 153–54, 167, 169–70, 321
    and miracles, argument against, 158–59
    morality of, 166–69
    myside bias and, 297
    numerical calculation of, 154
    odds, 159
    political commentary and, 162–63
    posterior probability, 151, 153
    random sampling of examples, 168
    reference class, 167–68
    reframing of single-event probabilities into frequencies, 168–70, 349–50n27
    replicability crisis and, 159–61
    San people and, 4, 166
    Signal Detection Theory and, 202, 205, 213, 214, 351n6
    subjectivist probability and, 115, 151
    updating, 157–58
    visualization of, 170–71
—"LIKELIHOOD" term in equation, 152–54
    bell curves as plot of, 205, 205, 351
    definition, 152, 351n6, 352n6, 352n21
    statistical significance as, 224–25, 352n21
—"MARGINAL" term in equation, 153–54
—"PRIOR" term in equation, 152–53
    base rates used as, problems with, 167–69
    low priors, and decreased credence, 157–63
    low priors, and signal detection, 213, 214
    myside bias and, 297
Bayes, Reverend Thomas, 151, 158
Beccaria, Cesare, 332–33
begging the question, 89
beliefs
    distal vs. testable, 298–99
    factually unsupported, 300, 301–3, 304
    imposed by force, 43–44, 245–46, 290
    justified true, 36–37, 344n1
    realist vs. mythological, 298–303
    reflective vs. intuitive, 298–99
bell curve, 204–5
    Central Limit Theorem and, 205, 351n5
    fat-tailed, 204–5

regression to the mean and, 253
signal noise detection and, 206–11, 219
Belling the Cat, 231, 232
Bem, Daryl, 159–60
Bentham, Jeremy, 333–35
The Better Angels of Our Nature (Pinker),
   183–84
Bezos, Jeff, 251
bias bias, 291
biased assimilation, 290–91
biased evaluation, 291, 294
Biden, Joe, 6, 130
Black Lives Matter, 26, 124–25
Blackstone, William, 217–18
Bodin, Jean, 335
bounded rationality, 184–88
Boy or Girl paradox, 137
brain, vs. deep learning networks, 107–9
Breyer, Stephen, 191
broken-leg problem, 279–80
Bruine de Bruin, Wändi, 323–24
burden of proof fallacy, 89
Bush, George W., 11, 218
Butler, Judith, 90
butterfly effect, 114

Calvin, John, 330–31
Carlin, George, 283, 299
Carroll, Lewis, 38, 75, 96, 225
Carroll, Sean, 160
Castellio, Sebastian, 330–31
Categorical Imperative (Kant), 69
causation
  all-or-none fallacy, 269
  conditions and, 259, 260
  confounding (epiphenomena), 246, 257,
    260, 263, 265
  confounding, ruling out, 267–68,
    270–71
  counterfactuals, 64, 257, 259, 264
  definition, 256–57
  fundamental problem of, 258
  mechanisms, 258–59
  overdetermination and, 259, 260
  paradoxes of, 259–61, 260
  preemption and, 259, 260
  probabilistic, 259–60
  reverse causation, 246, 263, 267, 269–70
  San people and awareness of, 3, 4–5
  temporal stability, 258

unit homogeneity, 258
See also randomized controlled trial
—CAUSAL NETWORKS
  overview, 260
  Bayesian networks, 260–63, *261*, 263
  causal chain, 261–62, *261*
  causal collider fallacy, *261*, 262–63
  causal fork, *261*, 262
  endogeneity, 263
  Matthew Effect, 263–64, 354n21
  multicollinearity, 263
  unpredictability of human behavior
    and, 280–81
—MULTIPLE CAUSES
  overview, 272–73
  interacting causes, 273–76, *273–74*
  interaction, 273, 275, 277–78
  main effect, 273, 274, 275, 278
  nature and nurture, 273, 275–77, *276*
  regression equations and, 278
  talent and practice, 272–73, 277–78, *278*
Chagnon, Napoleon, 307
chaos, 114
Chapman, Loren and Jean, 251
cheater detection, 15–16
chemtrails, 299
chess, 277–78, *278*
Chicken game, 59–61, 235–36, 237,
    344n29
children
  adoption of, 63
  heights of, 252–54
  IQs of, 252–53
  low birth weights, 262–63
  probability of boy vs. girl at birth,
    128–29, 132–33, 136–38
  ultimate vs. proximate goals and, 46–47
China, 24
choice architecture, 56
Chomsky, Noam, 71, 90
CIA, 91
Circe, 53, 55
circular explanations, 89
Clark, Sally, 129–30
Clegg, Liam, 231
climate change
  availability bias and perception of
    energy sources, 121–22, 347n22
  avoidance of sectarian symbolism, 312
  discounting the future and, 51–52

climate change (*cont.*)
  game theory and, 227–28, 242–44
  Public Goods games and, 242–43, 244
—DENIAL OF
  argument from authority and, 90, 91
  credibility of universities and, 314
  openness to evidence vs., 311
  politicization of, 295, 310, 312
  Trump and, 284
clinical vs. actuarial judgment, 278–80
Clinton, Bill, 25, 101
Clinton, Hillary, 82, 117, 285, 299, 306
close-mindedness. *See* active
      open-mindedness
cluster illusion, 146–48
coal, deaths due to, 121–22, 191, 347n22
cognitive illusions
  bias against recognition of, 291
  definition, 29–30
  interpretation of questions and, 32–33
  professionals are vulnerable to, 321
  recognition of, 312
  visual illusions and, 30
cognitive reflection, 8–10, 311
Cognitive Reflection Test, 8–10, 50
coincidence, prevalence of, 143–44, 287
coin flips, 22, 114, 146, 192
Cole, Michael, 96–97
collider fallacy, *261*, 262–63
common knowledge, 124, 234
communal outrage, 124–27
competition, 173
complement of an event, 128, 133–34
computers, 38–39, 41, 106–7
conditional probability, 128
  affirming the consequent and, 139–41
  base-rate differences, 134, 138–39, 141
  definition, 134
  and enumeration of possibilities,
      137–38, 169–70
  independent events defined via, 136–37
  language and ambiguities in, 140–41
  notation for, 134
  San people and, 4
  statistical significance and, *225*
  Venn diagrams illustrating, 134–36, *136*
  See also Bayesian reasoning
conditionals (IF-THEN)
  antecedent (IF), 77
  confirmation bias and, 13–14, 342n26

consequent (THEN), 77
  definition, 77
  everyday conversation vs., 78–80
  false antecedents, 78–79
  knowledge and, 14–16
  syllogism, 12
  truth table for, 77–78
  Wason selection task, 13–15, 77, 83,
      290, 291–92
confirmation bias
  definition, 13
  forensics and, 216
  post hoc probability illusion and, 142–43
  superstitions and, 14, 142
  the Wason selection task and, 13–14,
      290, 342n26
conflicts of interest, 93
confounding (epiphenomena), 246, 257,
      260, 263, 265, 267–68, 270–71
conjunction fallacy
  definition, 25
  in forecasting, 23–26, 343n46
  Linda problem, 26–29, 115, 116, 156
  relative frequency and, 28–29
conjunction of events, probability of,
      128–32, 137
connectionist nets. *See* deep learning
conservation, the San people and, 5
conservatives vs. liberals. *See* left and right
      (political)
consistency, 82
conspiracy theories
  adherents of, equivocal, 298–99
  beliefs in, not based on truth of, 302
  COVID-19, 283–84
  as entertainment, 303, 308
  and evolution of ideas, 308–9
  openness to evidence vs., 311
  police, reporting to, 299, 308
  popularity of, 286
  as predating social media, 287
  real conspiracies and, 307–8
  reflective vs. intuitive, 299
  rumor and, 308
  signal detection and, 307–8
consumers
  extended warranties, 197–98
  as money pumps, 176, 180, 185, 187–88
contradiction, anything follows
      from, 81–82

conventions and standards, 234–35
conversation, rules of, 10, 21, 28, 30, 78–80, 87–88, 308, 343n43
cooperation
  in the Prisoner's Dilemma, 239–42
  in Public Goods games, 242–44
coordination games, 233–35
correlation
  causation not implied by, 245–47, 251–52, 312, 321, 323–24, 329–30
  coefficient (r), 250–51
  cross-lagged panel correlation, 269–70
  definition, 247
  definition of "prediction," 247
  illusory, 245–46, 251–52, 321
  San people and, 4
  scatterplots, 247–52, 270–71
  See also causation; regression
Cosmides, L., 169
counterfactuals, 64, 257, 259, 264
  heretical, 64–65
COVID-19, 2, 193–94, 242, 283
  exponential growth bias and, 11–12
  media fear mongering, 126–27
  misinformation, 245, 283–84, 296, 316
Coyne, Jerry, 302
creationism, 173, 295, 305, 311
credit card debt, 11, 320–21
crib death, 129–30
Crick, Francis, 158
crime
  availability bias and perceptions of, 126
  confirmation bias and, 13–14
  Great American Crime Decline, 126
  gun control and rates of, 292–93
  and punishment, 332–33
  rational ignorance and, 58
  regression to the mean and, 255–56
  signal detection and, 202, 216–21, 352n17
  statistical independence and, 129
  See also homicide; judicial system
critical race theory, 123
critical theory, 35–36
critical thinking, 34, 36, 40, 87, 287, 314, 320
  definition, 74
  San people and, 3–4
  stereotypes and failures of, 19–20, 27
  teaching, 82, 87, 314–15

The Crown (TV series), 303
CSI (TV show), 216
Cuban Missile Crisis, 236

d′, 214–16, 218–21, 352n17
Darwin, Charles, 173
data, vs. anecdotes, xiv, 119–22, 125, 167, 300, 312, 314
data snooping, 145–46, 160
Dawes, Robyn, 175
Dawkins, Richard, 302, 308
Dean, James. *See Rebel Without a Cause*
death, 196, 197, 304
death penalty, 221, 294, 311, 333
deductive logic, 73–84, 95–100, 102, 108–9
deep learning networks
  biases perpetuated by, 107, 165
  the brain compared to, 107–9
  definition, 102
  error back-propagation and, 105–6
  hidden layers of neurons in, 105–7
  intuition as demystified by, 107–8
  logical inference distinguished from, 107
  terms for, 102
  two-layer networks, 103–5
De Freitas, Julian, 343n46
demagogues, 125, 126
democracy
  checks and balances in, 41, 316, 317
  corrosion of truth as undermining, 309
  data as a public good and, 119
  education and information access predicts, 330
  and peace, 88, 264, 266, 269–72, 327
  presumption of innocence in, 218
  and risk literacy, importance of, 171
  and science, trust in, 145
  Trump and threats to, 126, 130–31, 284, 313
Democratic Party and Democrats
  COVID-19 conspiracy theories involving, 283
  expressive rationality and, 298
  politically motivated numeracy and, 292–94
  See also left and right (political); politics
Dennett, Daniel, 231, 302
denying the antecedent, 83, 294
denying the consequent, 80–81
deontic logic, 84

dependence among events
  conjunctions and, 128–31, 137
  defined via conditional
    probability, 137
  falsely assuming, 131
  the "hot hand" in basketball and,
    131–32
  the judicial system and, 129–30
  selection of events and, 132
  voter fraud claims and, 130–31
depression, 276–77, *276*, 280
Derrida, Jacques, 90
Descartes, René, 40
deterministic systems, 114
Dick, Philip K., 298
dieter's fallacy, 101
digital media
  ideals of, 316
  truth-serving measures needed by, 314,
    316–17
  Wikipedia, 316
  See also media; social media
Dilbert cartoons, *91*, 112–13, *112*, *117*
DiMaggio, Joe, 147–48
discounting the future, 47–56, 320
discrimination, forbidden base rates
    and, 163–66
disenchantment of the world
    (Weber), 303
disjunction of events, probability of,
    128, 132–34
disjunctions (OR), definition, 77
disjunctive addition, 81
disjunctive syllogism, 81
distributions, statistical, 203–5
  bell curve (normal or Gaussian), 204–5
  bimodal, 204
  fat-tailed, 204–5
Ditto, Peter, 293–94, 297
DNA as forensic technique, 216
domestic violence, 138–39
Dostoevsky, Fyodor, 289
Douglass, Frederic, 338–39
dread risk, 122
dreams, 13, 304
Dr. Strangelove (film), 61
drug war, 84
dualism, mind-body, 304
Dubner, Stephen, 266
Duck Soup (film), 294–95

ecological rationality, 16, 95–98, 153, 214,
    246, 313–14
economics, 62–64, 173–74, 265, 322
education
  benefits of college education, 264
  college admissions, 262, 263, 266–67, 294
  curricula teaching rationality, 314–15
  democratic values as predicted by, 330
  mission of, 43, 93
  randomized controlled trials and, 265
  standardized testing, 262
  statistical competence as priority in, 171
  See also academia; science—education;
    universities
Einstein, Albert, 86, 90, 111, 113
Eliot, George, 73
Elizabeth II, 303
Emerson, Ralph Waldo, 82
emotions
  appeal to (informal fallacy), 92
  availability bias and, 120
  emotional threats, 60
  Escalation games and, 238
  expected utility and, 179, 181–83, 190–92
  goals of reasoning and, 45, 46–47,
    344n11
  paradoxical tactics and, 44–45, 344n11
  Tit for Tat strategy, 241–42, 243–44
  See also goals—time-frame conflicts
empathy, 230, 338
empirical propositions, 94–95
endogeneity, 263
Enlightenment
  abolitionist movement of, 335–36
  AI seen as threat to, 207
  argument against criminalization of
    homosexuality, 333–34
  argument against cruel punishment,
    332–33
  argument against cruelty to animals,
    334–35
  digital media and ideal of, 316
  feminism, 336–37
  and mythology zone vs. reality, 303
  universal realism as creed of, 301
entertainment
  conspiracy theories as, 303, 308
  gambling as, 183
  the mythology zone as, 306–7, 309
  paradoxical tactics depicted in, 61–62

the paranormal as, 305
political discourse as, 303
enthymemes, 85
environment, 63, 191, 242–43, 244
   See also climate change
epidemiology, 257, 259–60, 270
epiphenomena. *See* confounding
epistemological crisis, 284
equality
   as expanding circle of sympathy, 340
   policies signaling commitment to, 165
   propositional reasoning and, 108–9
   See also impartiality; moral progress
Erasmus, Desiderius, 331–32
Erdös, Paul, 18, 20
Erdös number, 18, 342n37
Escalation games, 236–38
essences, 302, 304–5, 321
ethnicity, and forbidden base rates, 163–66
Euclidean geometry, 95–96, 289
eugenics, 91
European Union, 327
evidence
   evidential interpretation of probability,
      115–16
   inadmissible, 57–58
   judicial, 57–58, 217–21
   openness to, 310–11, 356–57n67
   probability and truth and, 27–28
   social justice and, 42
   strength of. *See* Bayesian reasoning
evolution, 69, 173, 288, 308–9
excludability, 265
exclusive or (xor), 104–5, 133
expected utility
   Beccaria's argument against cruel
      punishment and, 332–33
   calculation of, 179–80
   and certainty, 182–83, 188–90, 192
   definition, 179
   diminishing marginal utility and,
      181–84, *181*
   emotions and, 179, 181–83, 190–92
   Erasmus's argument against war
      and, 331–32
   intuitive calculation of, 180
   lives and, 183–84
   loss-aversion and, 192–94
   maximization of, as rational choice,
      174, 179

money and, 181–83
possibility, certainty and, 190–92, 195–96
risk-aversion and, 182–83, 192
as term, 179
   See also rational choice
exponential future discounting, 50–52, 53, *54*
exponential growth bias, 10–12, 320–21
expressive rationality, 297–98
extraterrestrials, 286

Facebook, 127
fact-checking, 41, 300–301, 314, 316, 317–18
Fail Safe (film), 61
fairness, 164, 165, 217
fake news
   ease of spreading, 6
   as entertainment, 306–7
   fictionalized history as, 303
   historical, 287–88
   the irrationality crisis and, 284–85
   openness to evidence vs., 311
   religious miracles as, 287
   rumor and, 308
false dichotomy, 100
family resemblance categories
   classical logic vs., 98–101, 346n28
   definition, 99
   informal fallacies arising
      from, 100–101
   the representativeness heuristic and, 155
   true/false values and, 100
   See also deep learning
famine, reduction of, 326
Farley, Tim, 321–23
Fauci, Anthony, 283–84
feminism, Enlightenment and, 336–37
finances, 237–38, 320–21
financial industry, 142–43, 165–66, 188, 327
Fischer, Bobby, 144
Fischhoff, Baruch, 323–24
Floyd, George, 123, 124–25
focal points, 124, 234–35
Forbes magazine, 313
forbidden base rates, 62, 163–66
forecasting
   accuracy of regression equations vs.
      human experts, 278–80
   the conjunction rule and, 22–29
   importance of, 22–23
   superforecasters, 162–63

forecasting (*cont.*)
  unpredictability of human behavior,
    280–81
  weather, 114, 127, 133, 220
Fore people, 153
formal fallacies
  affirming the consequent, 83, 85, 139
  definition, 74, 83
  denying the antecedent, 83, 294
  detecting, 84
  formal reconstruction exposes, 85–87
fortuitous randomization, 267
Foucault, Michel, 90
Fox News, 267–68
framing effects, 178, 188–96, 321, 323
Frank and Ernest (cartoon), *255*
Franklin, Benjamin, 196
Freakonomics (Levitt and Dubner), 266
Frederick, Shane, 9–10, 50, 342nn17,19
freedom of speech, 41, 313–14
French Revolution, 337
frequency
  as factor in memory recall, 119
  probability reframed as, 28–29, 117–18,
    168–70, 349–50n27
frequentist interpretation of probability,
    116, 117, 118, 128
Freud, Sigmund, 80, 90
full moon, and hospital ERs, 251–52
future, discounting of, 47–56, 320
  See also goals—time-frame conflicts
fuzzy categories. *See* deep learning; family
    resemblance categories

Galileo, 144
Galton, Francis, 252, 253
gambler's fallacy, 20–21, 131, 132
gambling, 6, 20–21, 115, 231
  choice of games, 179–80
  as entertainment, 183
  lucky streaks, 148
  rational choice theory and, 176–80,
    182–83, 188, 188–92, 321, 350n23
games, as family resemblance
    category, 98–99
game theory
  Chicken game, 59–62
  common knowledge and, 234
  communal outrage and, 123–25
  coordination games, 233–35

definition, 228
empathy and, 230
Escalation games, 59–61, 235–38,
    344n29
focal points, 124, 234–35
and homicides, reactions to, 123
myside bias and, 297–98
Nash equilibrium, 230, 232, 236, 240
outguessing standoffs, 230–32, 234
Prisoner's Dilemma, 238–42, 244
promises/statements of intent and, 234
Public Goods games, 242–44
randomness as strategy, 228, 230–31
win–win outcomes, 232
zero-sum games, 229–32
garden of forking paths, 145, 185, 348n60
Gardner, Martin, 143–44, 342n33
Gates, Bill, 283
GDP per capita, 247–50, 263–64, 270–72,
    271, 354n21
Gelman, Andrew, 348n60
gender
  bias, 19–20, 40, 164, 165, 166,
    263, 336–37
  of children, probability, 128–29,
    132–33, 136–38
  crying and, 273
  forbidden base rates and, 163–66
general linear model, 262
genes, and proximate vs. ultimate
    motives, 46–47
genetically modified foods (GMOs),
    122, 304–5
genetic fallacy, 91, 92–93, 291
genetic testing, 57
genius, 84, 275–76, 277–78, *278*
Gigerenzer, Gerd, 28, 118, 166,
    169–70, 197
Gilovich, Tom, 131
goals
  overview, 45–46
  actuarial prediction, 163–64, 165
  and definition of rationality, 36–37,
    70, 164
  fairness as goal, 164
  incompatible, 46–47
  as matter of taste, 45, 344n11
  —TIME-FRAME CONFLICTS
  exponential future discounting, 50–52,
    53, *54*

hyperbolic future discounting, 52–56, *54*
   living for the present, 47–50
   the marshmallow dilemma, 47–48, 50
   myopic future discounting, 52–53, 55
   Odyssean self-control and, 53, 55–56
   self-control and, 48
   social discounting rate, 51–52
God
   argument for belief in (Pascal), 175
   doesn't play dice (Einstein), 111
   morality and (Plato), 67
   outside testable reality, 302
   See also religion
Golden Rule (and variants), 68–69
Goldstein, Rebecca Newberger, 289
Google, 108
Gore, Al, 312
Gould, Stephen Jay, 27, 67, 147–48
government
   checks and balances in, 41, 316
   elections and, 130–31, 317
   libertarian paternalism, 56
   profiling and, 163–66
   rule of law, 241, 243, 244
   the social contract, 244
   taxes, 242, 244
   and wealth, achievement of, 327
   See also democracy; policy
gravity, 113, 258–59
Grimshaw, Jane, 225
Guatemala, 91
guilt by association, 91
gun control, 292–93, 295, 298
Guthrie, Arlo, 57

happiness–income correlation, 247–50,
      263–64, 354n21
Harris, Sam, 302
Hastie, Reid, xvi, 175
health, 321
   changing views of best practices for, 48
   epiphenomena and, 257
   myopic future discounting of, 52–53, 55
   taboo tradeoffs and, 63–64
   ultimate vs. proximate goals and, 47
   See also medicine; mental health
health care system, 64
Heaven's Gate cult, 304
Henry V, 303
Heraclitus, 258

heretical counterfactuals, 64–65
Hertwing, Ralph, 28, 29, 349–50n27
Hillel, Rabbi, 68
histograms, 203–4
history, fictionalized, 302, 303
Hitchens, Christopher, 302
HIV/AIDS, 25, 26
Hobbes, Thomas, 244
Holiday, Billie, 263
Holocaust, 67, 184, 286
homicides
   availability bias and response to, 122–23
   conditional probability and, 138–39
   game theory and reactions to, 123
   police killings of African Americans,
      123, 124–25, 141
   rampage shooters, 121–22, 126, 156
   rate of, 122–23
   See also crime; judicial system; terrorism
Homo sapiens, 1–2
hot hand fallacy, 131
hot hand fallacy fallacy, 131–32
human-made threats, 122
Hume, David, 45, 66, 67, 94, 158, 226, 227,
      242, 256–57
hunger, reduction of, 326
hunter-gatherers
   Bayesian reasoning and, 4, 166
   "cognitive niche," 6
   ecological rationality and, 96–97
   mythological mindset and, 301
   real conspiracies and, 307
   San peoples and rationality, 2–5, 96,
      166, 341n4
hyperbolic future discounting, 52–56, *54*
hypochondria, 84, 139, 153–54, 155
hypothesis
   alternative, 223
   as conditional on the evidence, 151
   falsification of, 14
   null, 222–23
   See also Bayesian reasoning; statistical
      decision theory

Ibsen, Henrik, 346n17
ideas
   as memes evolving to reproduce, 308–9
   strength of, vs. the virtue of the source
      for, 91–93, 339–40
identity, beliefs and, 93

IF-THEN statements. *See* conditional
    probability; conditionals
Iliad, 302
illusory correlation, 245–46, 251–52, 321
imaginability, 25
    See also availability heuristic
immigration, misconception of rate, 120
impartiality
    as the core of morality, 68–69, 317, 340
    as the core of rationality, 317
    fairness and, 164
impeachment of a U.S. president, 75
inconsistency, and arguments for moral
    progress, 329
independent events
    causation and, 131–32
    conditional probabilities and, 136–37
    conjunction probabilities and, 128–31
    definition, 128
    false assumptions about, 131
    gambling and, 20–21
inductive logic, 74, 77, 226
    See also Bayesian reasoning
Industrial Revolution, 326–27
inequitable dangers, 122
informal fallacies
    ad hominem, 19–20, 90–91, 92–93, 291
    affective, 92–93
    appeal to emotion, 92
    argument from authority, 3–4, 90, 291
    bandwagon fallacy, 90, 93, 291
    begging the question, 89
    burden of proof, 89
    circular explanations, 89
    and the context of a statement, 93
    definition, 74, 87
    desire to win arguments and, 87–88,
        290, 291
    dieter's, 101
    false dichotomy, 100
    genetic, 91, 92–93, 291
    guilt by association, 91
    mañana, 101
    moving the goalposts
        (motte-and-bailey), 88
    no true Scotsman, 88
    paradox of the heap, 101
    slippery slope, 100–101
    so-what-you're-saying-is, 88
    special pleading, 88

straw man, 88, 291
    tendentious presuppositions, 89
    tu quoque (what-aboutery), 89
innumeracy, 155, 295
    of the press, 125–27, 314
instrumental variable regression, 267–68
insurance, 64, 165, 182, 191, 198, 321
intelligence
    cognitive fallacies and, 321
    openness to evidence and, 311
    rationality of groups and, 291–92
    vs. reasoning competence, 323–24
interaction, statistical, 273, 275, 277–78
interest and interest rates, 48, 50
international relations, 23–24, 60–61, 236,
        240, 269–71, 343n46
intransitivity, 176, 185–88
intuition
    deep learning as demystifying, 107
    expected utility calculations and, 180
intuitive dualism, 304
intuitive essentialism, 304–5, 321
intuitive teleology, 305
intuitive vs. reflective beliefs, 299
invalid inferences. *See* formal fallacies
Ioannidis, John, 162
Iran, 23–24, 91
Iraq, US invasion of, 122, 124, 286
irrationality
    as commonplace accusation, 6
    rational irrationality, 60–62
    See also paradoxical tactics
—CRISIS OF
    overview, 283–86
    comfort as explanation, 288
    explanations of, unsatisfactory, 286–88
    functionality in the real world, 288
    and impartiality, 304, 317
    motivated reasoning and, 289–92, 297,
        298, 310
    self and other and, 304, 317
    social media blamed for, 287–88
    as threat, 309
    zones of belief. *See* mythology mindset;
        reality mindset
    See also cognitive illusions; informal
        fallacies; Index of Biases and
        Fallacies
irrelevant alternatives, sensitivity to,
        177–78, 188–92, 350n8

James, William, 37
Jefferson, Thomas, 336, 339
Jenkins, Simon, 303
Jewish humor, 83, 89, 141–42, 192, 246, 262, 265, 325
   See also Morgenbesser, Sidney
Jewish people
   conspiracy theories about, 287
   the Golden Rule in religion, 68
   the Holocaust, 67, 184, 286
   as percentage of population, 120
JFK (film), 303
Jindal, Bobby, 357n73
Johnson, Samuel, 111
Johnson, Vinnie "The Microwave," 131
journalism
   cognitive biases and, 125
   data and context, provision of, 127
   editing and fact-checking in, 41, 300–301, 314, 316
   innumeracy of, 125–27, 314
   recommendations for, 127, 314, 316, 317
   and the replicability crisis, 161–62
   "yellow journalism," 125
   See also media; pundits
judicial system
   overview of classic illusions of, 321
   accountability for lying and, 313
   adversarial system of, 41, 316
   correlation implying causation and, 260
   death penalty, 221, 294, 311, 333
   eyewitness testimony, 216, 219
   fairness and, 217
   false convictions, 216–21
   forensic methods in, 216, 219–20
   guilt beyond a reasonable doubt, 217
   inadmissible evidence, 57–58
   juries, 57–58, 63, 202, 217–21
   lie detectors in, 219
   preponderance of the evidence, 217, 218
   presumption of innocence, 217
   Prisoner's Dilemma, 238–42, 244
   probability illusions and, 117–18, 129–30, 131, 138–39
   prosecutor's fallacy, 140–41
   signal detection and, 202, 216–21, 352n17
   See also crime; homicide
Jung, Carl, 144

Kahan, Dan, 293, 295, 297–98
Kahneman, Daniel, 7, 9–10, 11, 25–29, 29, 119, 146, 154–55, 156, 190–95, 254, 342n15, 349–50nn6,27
Kaine, Tim, 82
Kant, Immanuel, 69, 327
Kaplan, Robert, 25
Kardashians, 99, 102
Kennedy, John F., 144, 259, 286, 303
Kennedy, John F., Jr., 33
Keynes, John Maynard, 310
Khomeini, Ayatollah, 65
King, Martin Luther, Jr., 328, 339
Kissinger, Henry, 107
knowledge
   Bayesian reasoning and priors, 157–58
   defined as justified true belief, 36, 344n1
   logic and requirement to ignore, 95–98
   rumors conveying, 308
   and trust in institutions, 313–14
   used in service of goals, 36–37
Kpelle people, 96–97

Langer, Ellen, 342n17
language
   ambiguity of conditionals, 140–41
   conversational habits, 10, 21, 28, 30, 78–80, 87–88, 308, 343n43
   defenses against lying and, 313
   as recursive, 71
Laplace, Pierre-Simon, 113–14
Lardner, Ring, 43
La Rochefoucauld, François de, 173
Law and Order (TV show), 238–39
Lebowitz, Fran, 325
left and right (political)
   Bayesian reasoning and, 297
   expressive rationality, 297–98
   intellectual roots of, 296
   mask-wearing during pandemic and, 296
   moral and ideological alignments, 296
   moral superiority and, 296
   motivated numeracy and, 292–94
   openness to evidence and, 311
   political bias as asymmetrical, 312–13, 357n73
   political bias as bipartisan, 295–96, 297, 312

left and right (political) (*cont.*)
  as religious sects, 296
  rise of, factors in, 296–97
  and science, sympathy vs. hostility, 284, 295, 297, 312
  as tribes, 296
  views of protests, 294–95
  See also Democratic Party and Democrats; myside bias; politics; Republican Party and Republicans
Lehrer, Jonah, 353n13
Leibniz, Gottfried Wilhelm, 74, 93–94, 98, 101
Let's Make a Deal. See Monty Hall dilemma
Levitt, Steven, 266
LGBTQ people, 101, 120, 166, 204, 311, 333–34
liberals vs. conservatives. *See* left and right (political)
libertarianism, 56, 63, 244, 296
Liebenberg, Louis, 2, 4
life, dollar value on, 63–64
life expectancy, 325
life outcomes and rationality
  base-rate awareness, 157
  failures of rationality, 321–23
  improvements via rationality, 320–24
  rational choice and, 196, 196–99
  unpredictable, 280–81
lightning deaths, 134, *135*
likelihood (technical term), 152–54, 349n4, 351n6, 352n21
  and statistical significance, 224–25
  See also under Bayesian reasoning
Lilienfeld, Scott, 90
Lincoln, Abraham, 144
Linda problem, 26–29, 115, 116, 156
Linnaean categorization, 102, 108
Locke, John, 335–36
Loftus, Elizabeth, 216
logic
  connector words, 75–76
  vs. content, 14–16, 75, 95–98
  vs. conversation, 10, 21, 28, 30, 78–80, 87–88, 308, 343n43
  deductive, 73–74
  vs. deep learning, 107
  deontic, 84
  ecological rationality as enhanced by, 16

empirical propositions vs., 94–95
  enthymemes, 85
  equality and, 108–9
  form vs. content, 16, 75, 95–98, 294
  justification of, by logic, 39
  vs. knowledge, 95–98
  Leibniz's dream, 74, 93–94
  modal, 84
  political bias and, 294
  predicate, 84
  propositional, 75, 84, 108–9
  as rationality, 85
  in recursive cognition, 108
  San people and, 3
  soundness vs. validity, 82–83
  temporal, 84
  truth tables, 76–78
  See also conditionals; disjunctions; formal fallacies; inductive logic; informal fallacies
—RULES OF VALID INFERENCE
  absurd conclusions yielded by, 82
  affirming the antecedent, 80
  definition, 80, 82
  denying the consequent, 80–81
  disjunctive addition, 81
  disjunctive syllogism (process of elimination), 81
  ecological rationality and, 97
  invalid inferences. *See* formal fallacies
  Principle of Explosion (from contradiction, all follows), 81–82
  sound arguments vs., 82–83
  validity vs. soundness, 83
loss aversion, 192–94
lotteries, 176–78, 182–83, 188–92
Lotto, Beau, 30
Love Story (film), 76–78, 263
lucky streaks, 148

Madison, James, 41
Madman Theory, 60–61, 236
Maduro, Nicolás, 23, 24
main effect, 273, 274, 275, 278
Maine, USS, 124, 125
The Maltese Falcon (film), 61–62
mañana fallacy, 101
Mao Zedong, 245
margins of error, 196
marshmallow test, 47–48, 50

Marx, Chico, 294–95
Marx, Karl, 90
Masons, 36, 40
mathematicians
    caution about mindless use of statistical
        formulas, 166–67
    mansplaining, 18
    paragon of rationality, 74
    simulation vs. proof, 20
    stereotype of, 20
Matthew Effect, 263–64, 354n21
Maymin, Philip, 342n17
Meadow's fallacy (multiplying
        probabilities of interdependent
        events), 129–30, 131
media
    accountability for lying/disinformation,
        313, 314, 316, 317
    availability bias driven by,
        120, 125–27
    consumer awareness of biases in, 127
    correlation confused with causation
        and, 256, 260, 353n13
    cynicism bred by, 126–27
    innumeracy of, 125–27, 314
    negativity bias, 125–26
    politically partisan, 296
    rational choice portrayal by, 173–74
    reforms for rationality, 127, 314, 316, 317
    tabloids, 287–88, 306
    and truth-seeking, 316
    the Winner's Curse and, 256, 353n13
    See also digital media; entertainment;
        journalism; social media
medical quackery, 6, 90, 284, 304, 321, 322
    celebrity doctors, 305
    COVID-19 and, 284
    harms caused by, 322
    hidden mechanisms posited by, 258
    intuitive essentialism and, 304–5, 321
    science laureates and, 90
medicine
    base-rate neglect in diagnosis, 155
    Bayesian reasoning in, 150–51, 152,
        153–54, 167, 169–70, 321
    correlation and causation, 251–52
    COVID-19, 2, 283
    disease control, 325
    drug trials, 58, 264
    evidence-based, 317

expected utility of treatments,
        192–94, 198–99
    false positives, 169–70, 198
    frequencies and, 169–70
    randomized controlled trials and, 264
    rational ignorance and, 57
    rationality and progress in, 325–26
    sensitivity of tests, 150, 154, 169, 352n10
    signal detection and, 202, 211, 213, 220
    specificity of tests, 352n10
    taboo tradeoffs in, 63–64
    See also health; mental health
Meehl, Paul, 279, 280
Mellers, Barbara, 29, 219–20
memes, 144, 308–9
mental health, 251, 256, 276–77, 276, 280
Mercier, Hugo, 87, 291, 298–99, 308, 313
meteorology, 114, 127, 133, 220
The Mikado (Gilbert and Sullivan), 27
military, 63, 220, 231, 295
Miller, Bill, 143
Miller, Joshua, 132
mind-body dualism, 304
miracles, Bayesian argument vs., 158–59
Mischel, Walter, 47
Mlodinow, Len, 143
modal logic, 84
modernity
    vs. ecological rationality, 96–98
    vs. mythology mindset, 303–4
    See also progress
modus ponens, 80
modus tollens, 80–81
Molière, *virtus dormitiva*, 11–12, 89
money
    diminishing marginal utility of,
        181–83, 247
    dollar value on human life, 63–64
    See also finances; GDP per capita
money pump, 176, 180, 185, 187–88
monocausal fallacy, 260, 272–73
Montesquieu, 333, 335
Monty Hall dilemma, 16–22, 115, 342n33
morality
    of Bayesian base rates, 62, 163–66
    discounting the future and, 51–52
    God and, 67
    Golden Rule (and variants), 68–69
    heretical counterfactuals, 64–65
    impartiality as the core, 68–69, 317, 340

morality (*cont.*)
    marginal utility of lives and, 183–84
    the mythology mindset and, 300, 307
    relativism and, 42, 66–67
    self-interest and sociality, 69
    taboo tradeoffs and, 64
    See also moral progress
moral progress
    overview, 328–29
    analogizing oppressed groups, 335
    animals, cruelty to, 334–35
    democracy, 335–36, 339
    feminism, 336–37
    homosexuality, persecution of, 333–34
    ideas vs. proponents, 339–40
    rationality as driver of, 329–30, 340
    redistribution of wealth and, 182
    religious persecution, 301–2, 330
    sadistic punishment, 332–33
    slavery, 334, 335, 336, 337, 338, 339
    war, 331
Morgenbesser, Sidney, 81, 82, 178
Morgenstern, Oskar, 175, 197, 228, 350n1
motivated reasoning, 289–92, 297, 298,
        310. *See also* irrationality––crisis of;
        myside bias
motte-and-bailey tactic, 88
moving the goalposts, 88
multicollinearity, 263
Muslims, 65
Myers, David, 40
myopic temporal discounting,
        52–53, 55
myside bias
    Bayesian reasoning as justification, 297
    bipartisan, 295–96, 297, 312, 357n73
    countermeasures, 316, 317
    definition, 294
    game theory and, 297–98
    moral superiority and, 307
    perception of events, 294–95
    pervasiveness of, 295–96
    politically congenial logic, 294
    politically motivated numeracy, 292–94
    Wikipedia marginalizing, 316
    See also left and right (political)
mythology mindset, 298–303. *See also*
        conspiracy theories; fake news;
        irrationality––crisis of;
        paranormal; superstitions

Nagel, Thomas, 40
Nash equilibrium, 230, 232, 236, 240
Nash, John, 230
national myth, 302, 307
natural selection. *See* evolution
nature and nurture, 273, 275–76,
        275–77, 276
Nazi Germany, 91, 124
negativity bias, 196, 351n36
Neuman, Alfred E., 6
neural networks. *See* deep learning
"New Atheists," 302
Newton, Sir Isaac, 113, 144
New Yorker cartoons, 49, 92, 186, 266
New Yorker muddle, 256
Neyman, Jerzy, 221–22
9/11, 122, 124, 218
9/11 "Truthers," 286, 299
Nixon, Richard, 60
Niyazov, Saparmurat, 245–47, 251
Nobel Prize, 197, 327
noise. *See* Signal Detection Theory
normative models of rationality, 7–8.
        See also Bayesian reasoning;
        causation; correlation; game
        theory; logic; probability; rational
        choice; Signal Detection Theory
NOT, as logical connector, 75
    See also complement of an event
no true Scotsman fallacy, 88
novel threats, 122
nuclear power, 121–22, 191, 347n22
nuisance variables. *See* confounding
numeracy, politically motivated, 292–94
numerology, 143–44

odds, 159, 188
Odyssean self-control, 53, 55–56, 238
Odysseus, 53, 55, 56–57
Ono, Yoko, 285
openness to evidence, 310–11,
        356–57n67
    See also Active Open-Mindedness
organ donation, 63
OR statements. *See* disjunction of events;
        disjunctions
outguessing standoffs, 230–32, 234
out-of-body experiences, 304
out-of-control threats, 122
outrage, communal, 124–27

overconfidence, 20, 29–30, 33, 115, 216, 255, 323
overdetermination, 259, *260*

Pakistan, 88
Palin, Sarah, 79
pandemics
    exponential growth bias and, 11–12
    marginal utility of lives and, 183
    Public Goods games and, 242
    rational choice theory and, 193–94
    See also COVID-19
paradoxical tactics, 58–62
paradox of the heap, 101
parallel distributed processing. *See* deep learning
paranormal
    Bayesian argument against, 158–60
    as entertainment, 305
    mind-body dualism and, 304
    openness to evidence and denial of, 311
    prevalence of beliefs in, 6, 285–86, 354–55n8
    prevalence of rejection of, 309–10
    See also pseudoscience
Parent, Joseph, 287
Parker, Andrew, 323–24
Parker, Theodore, 328
Pascal, Blaise, 174–75
passions
    proximate vs. ultimate motives, 46–47
    reason and, 45–46, 344n11
    See also emotions; goals; morality
pattern associators. *See* deep learning
patterning, random vs. nonrandom, 112–13
Paxton, Ken, 130–31
peace, democracy and, 264, 266, 269–72, 327
Peanuts (cartoon), *286*, 298
Pearl, Judea, 261
Pearson, Egon, 221–22
peer group, expressive rationality and, 297–98
peer review, 41, 58, 160, 300–301, 316
Pence, Mike, 82–83
Pennycook, Gordon, 310–11
perceptrons. *See* deep learning
p-hacking, 145
Pizzagate conspiracy theory, 299, 302, 304
plane crashes as risk, 33, 120, *121*, 122

Plato, *Euthyphro*, 67
poker, 231
police
    and correlation–causation confusion, 260
    evidence-based evaluation of, 317
    killing African Americans, 123, 124–25, 141
    reporting concerns to, 299, 308
policy
    avoiding sectarian symbolism in, 312
    behavioral insights from cognitive science, 56
    deliberative democracy and, 317
    discounting the future, 51–52
    evidence-based, 316, 317
    libertarian paternalism, 56
    randomized controlled trials to test, 265
    rational choice axioms and, 191, 193–94
    signaling equality and fairness, 165
    taboo tradeoffs and, 63–64
    See also government
political commentary. *See* pundits
politics
    affirming the consequent and, 83–84
    cliques, coalitions, sects, and tribes, 296
    fake news impact on, 288
    gerrymandering, 296
    mythology mindset, 303
    sound arguments, 82–83
    taboo tradeoffs and, 64
    voter fraud claims, 130–31
    See also Democratic Party and Democrats; left and right (political); Republican Party and Republicans
polling
    availability and media coverage, 126
    propensity/probability confusion, 118
polygraph tests, 219
Pope, as space alien, 128, 141
Popper, Karl, 14
posterior probability, 151, 153
post hoc probability, 141–48, 160, 321
postmodernism, 35–36
post-truth era, 39–40, 284, 295–96, 303, 313, 314
poverty
    extreme poverty, improvements in, 126
    reasoning competence and, 324
practical reason, 37
predicate calculus, 84

prediction
  in correlation and regression, 247
  human vs. regression, 278–80
  See also forecasting
preemption, 259, *260*
preference reversal, *52–53*, 55
Prince (musician), 35
priors. *See under* Bayesian reasoning
Prisoner's Dilemma, 238–42, 244
  See also Public Goods games
probabilistic causation, 259–60
probability
  of a complement, 128, 133–34
  of a conjunction, 128–32, 137
  of a disjunction, 128, 132–34
  classical definition, 115
  confusion about meaning of, 114–15
  evidential interpretation, 115–16
  ignorance, quantifying, 22
  Monty Hall dilemma, 16–22, 115,
    342n33
  posterior, 151, 153
  the present evidence and, 27–28
  propensity and, 115
  risk estimation, 118–27
  subjectivist interpretation, 115, 116,
    151, 194–96
  See also availability heuristic;
    conditional probability; frequentist
    interpretation; single-event
    probabilities
probability neglect, 11, 28, 321
processes, random, 112
profiling
  forbidding of, 62, 163–66
  rare traits and false positives, 156–57
progress
  evidence for, 325–28
  explaining, 320, 325–28
  See also moral progress; rationality—
    effects of
promises, 61, 234
Pronin, Emily, 291
propensity vs. probability, 21–22, 115
  accidents at home and, 139–40
  definition, 21–22
  DNA forensics and, 216
  medical choices and, 198
  the Monty Hall dilemma and, 21–22
  polling and, 118

  rocket failures and, 118
propositional calculus, 75, 84, 108–9
prosecutor's fallacy, 140–41
Prospect theory, 194–95, *194–95*
prostitution, 64
pseudoscience
  distinguished from science, 14
  mind-body dualism and, 304
  science education and, 305–6
  See also irrationality—crisis of;
    mythology mindset
psychotherapy, 251, 256
public goods
  data as, 119
  definition, 242
  government investment in, 327
  Public Goods games, 242–44
  rationality as, 315
  risk literacy as, 171
public health
  guilt by association fallacy and, 91
  rationality and progress in, 325
  See also COVID-19; pandemics
pundits, 25–26, 162–63, 317
Putin, Vladimir, 23
Pyrrhic victory, 237, 238

QAnon, 284, 287, 299, 306
quantum mechanics, randomness and, 113
questionable research practices, 145–46,
    160, 353n13

racism, 124–25, 141
  forbidden base rates and, 62, 163–66
rampage shooters, 123, 126, 156
randomized controlled trial (RCT),
    264–68
randomness, 112–14
  the cluster illusion and, 146–47
  coin flips and, 114, 146
  See also probability
random sampling, and base-rates, 168
random variables, 203
rape, false accusations of, 218, 260, 352n15
rational choice theory
  overview, 174–75, 350n1
  context should not matter in, 178
  death and, 196
  Erasmus's argument against war, 331–32
  mathematical form of, 174

as maximizing expected utility, 174, 179
odds and, 188
Prospect theory as alternative,
    194–95, *194–95*
risk and, 177
self-interest confused with, 174, 181
in Signal Detection Theory, 202, 211–14
uncertainty vs. risk, 177, 185
as unloved theory, 173–74
and values, consistency of, 175, 180–81,
    197–99
and values, discovery of, 180–81
—AXIOMS
overview, 175
alternative terms for, 350nn6–8,10
Closure, 176–77
Commensurability, 176, 184, 350n6
Consistency, 178
Consolidation, 177, 185, 350n7
Independence (from Irrelevant
    Alternatives), 177–78,
    188–92, 350n8
Interchangeability, 178–79, 196, 350n10
Transitivity, 176, 185–88
—VIOLATION OF THE AXIOMS
bounded rationality, 184–88
framing, 188–92, 321, 323
framing of rewards, 192–96, 321, 323
process of elimination, 185, 186–87
risks vs. rewards, 187–88
satisficing vs. optimizing, 185
savored by psychologists, 196–97
small differences, ignoring, 186
taboo tradeoff, 184, 350n15
rationality
    Active Open-Mindedness/Openness to
        Evidence and, 310–11, 324,
        356–57n67
    as a kit of cognitive tools, 6–7
    arguments against, self-refuting, 39–40
    definition, 36–37, 70, 164
    and emotions, 44–45, 344n11
    enhanced by relevance, concreteness, 8
    epistemic humility and, 40
    expressive, 297–98
    and goals, making choices consistent
        with, 175
    groups foster, 291–92, 317
    paradoxical tactics, 58–62
    rational choice theory as benchmark, 174

rational ignorance, 56–58
rational incapacity, 53–56, 58–60
rational irrationality, 60–62. *See also*
    Madman Theory
reality as motivating, 41–42, 288, 298,
    309, 320
reasoning about reasoning, 37–41,
    70–71
rejection of the paranormal and, 309–10
as uncool, 35–36
winning arguments as adaptive function
    of, 87–88, 291
See also ecological rationality; goals;
    irrationality—crisis of; normative
    models; taboos
—EFFECTS OF
overview, 319–20
life outcomes, 320–24
material progress, 324–28
moral progress. *See* moral progress
—RECOMMENDATIONS TO STRENGTHEN
accountability for lying and
    disinformation, 313, 314, 316–17
avoidance of sectarian symbolism, 312
educational curricula, 314–15
evidence-based evaluation, 312, 317
incentive structures, 315–17
in journalism, 314, 316, 317
norms valorizing, 311–13, 315
in punditry, 317
"Republican party of stupid,"
    312–13, 357n73
scientists in legislatures, 312
in social media, 313, 316–17
viewpoint diversity in higher ed,
    313–14
Rationality Community, 149–50, 312
Rationality Quotient, 311
Rawls, John, 69
realism, universal, 300–301
reality, motivating rationality, 41–42, 288,
    298, 309, 320
reality mindset
    Active Open-Mindedness/Openness to
        Evidence and, 310–11, 324,
        356–57n67
    definition, 299–300
    mythology mindset, border with, 303
    as unnatural, 300–301
reason, in definition of "rationality," 36

Rebel Without a Cause, 59, 236,
        344n29
reciprocity, norms of, 5
recursion, 71, 108
reflectiveness
    Cognitive Reflection Test, 8–11, 50
    definition, 311
    intelligence correlating with, 311
    openness to evidence correlating
        with, 311
    and reasoning competence, 323, 324
    resistance to cognitive illusions and, 311
    unreflective thinking, 8–10, 311
    of weird beliefs, 299
regression
    correlation coefficient (*r*), 250–51
    definition, 248, 252
    equation for, 272, 278–81
    general linear model, 272
    human vs., accuracy of, 278–80
    instrumental variable regression,
        267–68
    multiple regression, 270–72, *271*
    regression line, 248–49, 253–54, 270
    residuals, 249–50, 270–71
regression discontinuity, 266–67
regression line, 248–49, 253–54, 270
regression to the mean
    and bell curve distribution, 253
    definition, 252–53
    imperfect correlation as producing, 254
    regression line, 253–54
    scatterplots and, 253–54
    as statistical phenomenon, 253
    unawareness of, 254–56, 320, 353n13
regret avoidance, 17, 190
relativism and relativists
    argument against rationality, 39–40
    as hypocritical, 42
    morality and, 42, 66–67
religion
    argument from authority, 90
    the cluster illusion and, 147
    forbidden base rates and, 163–66
    the Golden Rule in, 68–69
    heretical counterfactuals, 64–65
    monotheism, 40
    the mythology mindset and, 301–2, 307
    persecution of, progress against, 330–31
    See also God

Rendezvous game, 233–34
replicability crisis in science
    Bayesian reasoning failures and, 159–61
    preregistration as remedy, 145–46
    questionable research practices and,
        145–46, 160, 353n13
    science journalism and, 161–62
    statistical significance and, 225
    Texas sharpshooter fallacy, 144–46, 160
    Winner's Curse, 256
representativeness heuristic, 27, 155–56
Republican Party and Republicans
    calling Democrats socialists, 83–84
    expressive rationality and, 298
    Fox News and, 267, 268, 296
    pizza classified as vegetable by, 101
    politically motivated numeracy, 292–94
    rehabilitating, 312–13, 357n73
    See also left and right (political); politics
reputation, 47, 237, 242, 308, 313
resistance to evidence. *See* openness to
        evidence
retirement savings, 6, 11, 48, 50–51, 55–56
reverse causation, 246, 263
    ruling out, 267, 269–70
right–left. *See* left and right politics
risk, distinguished from uncertainty, 177
risk perception
    availability bias distortions,
        120–23, 320
    dread risks, 122
    estimation by tallying, 119
    human-made threats, 122
    inequitable, 122
    literacy in, 171
    novel threats and, 122
Romeo and Juliet, 37, 41, 175
Roser, Max, 126
Rosling, Hans and Ola, 126
Rosling-Rönnlund, Anna, 126
rule of law, 241, 243, 244
rumors, accuracy of, 308
Rumsfeld, Donald, 177
Rushdie, Salman, *The Satanic Verses*, 65
Russell, Bertrand, 1, 89, 115, 300, 310
Russia, 23, 296, 343n46

safety, 101, 123, 327–28
    taboo tradeoffs and, 63–64
Sagan, Carl, 149, 159

Sanjurjo, Adam, 132
San peoples, 2–5, 96, 166, 341n4
Savage, Leonard, 175, 185–86, 350n1
Scarry, Elaine, 139
scatterplots, 247–52, 253–54, 270–71
Schelling, Thomas, 58, 123–24, 235
school shootings, 121–22, 156
science
    argument from authority and, 90
    confidence in, 313
    left vs. right embrace of, 284, 295,
        297, 312
    legislators who are scientists, 312
    peer review, 41, 58, 160,
        300–301, 316
    politicized issues, 310
    preregistration of study details, 145–46
    randomized controlled trials, 264–68
    rational ignorance and, 58
    testability and, 299, 316
    Trump and rejection of norms of, 284
    See also replicability crisis
—EDUCATION
    not undermining pseudoscience, 305–6
    vs. sacred mythological
        beliefs, 305
    shallow understanding in educated
        people, 295, 305
Scientific Revolution, 94–95
Scissors-Paper-Rock game, 229–31
selection task. *See* Wason
        selection task
selective exposure, 290–91
self-control, 53, 55–56, 238
self-fulfilling prophecies, 164–65
self-interest, and rationality, 68–69
Selvin, Steven, 342n33
Seuss, Dr., 185, 258
sexism, 19–20
sexual orientation, 166, 204
    See also LGBTQ people
Shakespeare, William, ix, 297, 303
shark bite fatalities, 121
Shaw, George Bernard, 68
Shermer, Michael, 286
Signal Detection Theory
    actionable decisions reached via, 202–3
    Bayesian reasoning and, 202, 205, 213,
        214, 351n6
    conspiracy theories and, 307–8

    definition, 202
    expected utility and, 202, 211–14
    ideal observer, 213, 218
    judicial system, 202, 216–21, 352n17
    See also statistical significance
—NOISE VS. SIGNAL
    overview, 202
    bell curve distributions, 203–5, 351n5
    correct rejections, 208
    cutoffs (criterion or response bias),
        206–7, 211–14, 217–21
    fallacy of confusing, 211
    false alarms, 208
    hits, 207
    misses, 208
    noise as always present, 205–6
    proportions of possibilities and
        relaxation vs. tightening of the
        cutoff, 209–11, 220–21
    proportions of results add up to 100
        percent, 209
    response bias vs. sensitivity, 211
    sensitivity (d'), improving, 214–16,
        218–21, 352n17
Simon, Herbert, 184–85
Simpson, Homer, 6, 51, 52
Simpson, O. J., 138–39
Sinclair, Upton, 289–90
Singer, Peter, 319, 335
single-event probabilities
    reframed as frequencies, 117–18,
        168–70, 349–50n27
    subjectivist probability and, 116
Siri, 108
slavery, 67, 89
    abolition and rationality, 334, 335, 336,
        337, 338, 339
slippery slope fallacy, 100–101
Slovic, Paul, 122
SMBC cartoon, 120, *121*
smoke alarms, 261
smoking, 257, 259–61, 262–63
social animals, cooperation in, 241–42
social contract, 244
sociality, and morality, 68–69, 241–42
social justice
    overview, 42
    mission of education and, 93
    persuasion vs. force and, 43–44
    rationality as requirement of, 42

social media
  accountability for lying, 313, 316–17
  blamed for irrationality crisis, 287–88
  blamed for left–right split, 296
  news consumers and, 127
  "What's the Harm?" (Farley), 321–23
  See also fake news
socioeconomic status, 324, 330
"Sophie's choice," 184
sound arguments, 82–83
Sowell, Thomas, 245
so-what-you're-saying-is, 88
special pleading, 88
Sperber, Dan, 87, 291, 299
Spinoza, Baruch, 1, 68
Spock, Mr., 6, 35, 39, 73
spoiler alerts, 57
sports
  family resemblance categories and, 100
  hot hand in basketball, 131–32
  lucky streaks, 147–48
  myside bias and, 295
  outguessing standoff in, 230–31
  Prisoner's Dilemma and Public Goods,
    240, 244
  Sports Illustrated jinx, 255
Stalin, Josef, 183
standards and conventions, 234–35
Stanovich, Keith, 295, 297, 311, 356–57n67
star constellations, 146–47
statistical decision theory, 221–26
statistical distribution. *See* distributions,
    statistical
statistical significance
  overview, 221–22
  alternative hypothesis, 223
  as Bayesian likelihood, 224–25, 352n21
  critical value, placement of, 223
  definition, 224
  null hypothesis, 222–23
  scientists' misunderstanding of, 224–26
  statistical power and, 223
  Type I & II errors, 223–24, 225
stereotypes
  base-rate neglect and, 155–56
  in the conjunction fallacy (Linda
    problem), 156
  failures of critical thinking, 19–20, 27
  family resemblance categories and,
    99–100, 105, 155

  illusory correlation and, 251–52
  vs. propositional reasoning, 108–9
  random sampling vs., 168
  representativeness heuristic, 27, 155–56
Stone, Oliver, *JFK*, 303
Stoppard, Tom, *Jumpers*, 44–45, 66
straw man, 88, 291
Styron, William, *Sophie's Choice*, 184
subjective reality, claims for, 39
subjectivist interpretation of probability,
    115, 116, 151, 194–96
sucker's payoff, 239, 242, 244
suicide, 156
Suits, Bernard, 346n28
sunk cost fallacy, 237–38, 320, 323
Sunstein, Cass, 56
Superfund sites, 191
superstitions
  the cluster illusion and, 147
  and coincidences, prevalence of,
    143–44, 287
  confirmation bias and, 14, 142
  openness to evidence and, 311
  prevalence of, 285–86, 354–55n8
syllogisms, 12, 81
synchronicity, 144, 305
System 1 & 2
  defined, 10
  equality and System 2, 108–9
  the Monty Hall dilemma and, 20
  rational choice and, 187
  reflective and unreflective thinking,
    8–10, 311
  visual illusions and, 30

taboos
  and communal outrages, 123–25
  definition, 62
  forbidden base rates, 62, 163–66
  heretical counterfactuals, 64–65
  taboo on discussing taboo, 166
  taboo tradeoffs, 62–64, 184, 350n15
  victim narratives, 124
taboo tradeoffs, 62–64, 184, 350n15
talent and practice, 272–73, 277–78, *278*
Talking Heads, 35
Talleyrand, Charles-Maurice de, 337
tautologies, 80
  See also begging the question; circular
    explanations

taxicab problem, 155, 168, 170, *171*
television, 216, 238–39, 267–68, 303, 305
temporal discounting, 47–56, 320
temporal logic, 84
temporal stability, 258
tendentious presuppositions, 89
terrorism
    availability bias and, 122
    Bayesian reasoning and prediction of,
        162–63
    man carrying own bomb joke,
        127– 28, 138
    media coverage and, 126
    paradoxical tactics and, 60
    profiling and, 156–57
    torture of terrorists, 218
Tetlock, Philip, 62–65, 162–66
Texas sharpshooter fallacy, 142–46,
    160, 321
Thaler, Richard, 56
theocracies, 43
theoretical reason, 37
#thedress, 32
threats, and paradoxical tactics, 58, 60
Three Cards in a Hat, 138
The Threepenny Opera (Brecht), 121
time, San people and, 3
    See also goals—time-frame conflicts
Tit for Tat strategy, 241–42, 243–44
Tooby, J., 169
Toplak, M. F., 356–57n67
trade and investment, international, 327
Tragedy of the Carbon Commons,
    242–44, 328
Tragedy of the Commons, 242, 243–44, 315
Tragedy of the Rationality Commons,
    298, 315–17
Trivers, Robert, 241
trolley problem, 97
Trump, Donald, 6, 60, 82–83, 88, 92, 126,
    130–31, 145, 245, 283–84, 284, 285,
    288, 303, 306, 310, 312–13, 313
truth tables, 76–78
tu quoque (what-aboutery), 89
Turkmenistan, 245–47, 251
Tversky, Amos, 7, 25–29, 119, 131, 146,
    154–55, 156, 186–87, 190–95, 196,
    254, 342n15, 349–50nn6,27
Twain, Mark, 201
Twitter, 313, 316, 321–23

uncertainty, distinguished from risk, 177
United Nations, 327
unit homogeneity, 258
universal basic income (UBI), 85–87
universal realism, 300–301
universities
    academic freedom in, 41
    benefits of college education, 264
    college admissions, 262, 263,
        266–67, 294
    sexual misconduct policies, 218
    suppression of opinions in, 43, 313–14
    viewpoint diversity, lack of, 313–14
    See also academia; education
unreflective thinking, 8–10, 311
    See also System 1 & 2
urban legends, 287, 306, 308
Uscinski, Joseph, 287
US Constitution, 75, 333
US Department of Education, 218
USSR, 60, 89, 122

vaccines, 284, 325. *See also* anti–vax
    movement
values
    rational choice theory and, 175,
        180–81, 197–99
    signal detection theory and, 218–21
    See also goals
variables, random, 203
vegetables, category, 100, 101, 102–5
vehicles
    autonomous, 31, 108
    climate change and, 242–43
    driving as risk, 120, 121, 122
    plane crashes as risk, 33, 120,
        121, 122
    sales/bargaining, 59, 235
    SUVs, 101
victim narratives, 124
video games, 315
Vietnam War, 60
Vigen, Tyler, 252
virtus dormitiva, 11–12, 53, 89
visual system, 30–32, 37, 108, 170–71
Voltaire, 333
Volunteer's Dilemma game, 231–32, 234
von Neumann, John, 6, 175, 197,
        228, 350n1
vos Savant, Marilyn, 17–18, 19–21, 22

Wagnerian operas, 302
war
  of attrition, 238
  conspiracy and, 307
  democracy as reducing, 264, 266,
    269–72, 327
  the diminishing marginal
    utility of human lives
    and, 183–84
  famine and, 326
  moral progress and reduction
    of, 331–32
  9/11 and, 122, 124
  probability of, computing from
    complement, 134
  rationality and reduction of, 327
warranties, extended, 197–98
Wason, Peter, 13
Wason selection task, 13–15, 77, 83,
  290, 291–92
wealth
  rationality as explanation of growth,
    326–27
  redistribution of, 182, 243
weather forecasting, 114, 127,
  133, 220
Weber, Max, 303

weird beliefs, 286. *See also* conspiracy
    theories; medical quackery;
    paranormal; superstitions
WEIRD societies, 301
Welch, Edgar, 299
what-aboutery (*tu quoque*), 89
What's My Line? (game show), 168
Wikipedia, 316
Wilde, Oscar, 184
Winner's Curse, 256
wishful thinking, 201–2
Wittgenstein, Ludwig, 98–99, 101
Wollstonecraft, Mary, 336–37
workplace rumors, 308
The World According to Garp, 138
World Wars, 124

XKCD cartoon, 134, *135*, *225*, *281*
xor. *See* exclusive or (xor)

Yang, Andrew, 85–87, 90
Yanomanö people, 307

Zeno's second paradox, 38
zero-sum games, 229–32
Ziman, John, 162
Zorba the Greek, 35